STATISTICAL TESTS OF
NONPARAMETRIC
HYPOTHESES

Asymptotic Theory

STATISTICAL TESTS OF
NONPARAMETRIC HYPOTHESES

Asymptotic Theory

Odile Pons

French National Institute for Agronomical Research, France

NEW JERSEY · LONDON · SINGAPORE · BEIJING · SHANGHAI · HONG KONG · TAIPEI · CHENNAI

Published by

World Scientific Publishing Co. Pte. Ltd.

5 Toh Tuck Link, Singapore 596224

USA office: 27 Warren Street, Suite 401-402, Hackensack, NJ 07601

UK office: 57 Shelton Street, Covent Garden, London WC2H 9HE

Library of Congress Cataloging-in-Publication Data
Pons, Odile, author.
  Statistical tests of nonparametric hypotheses : asymptotic theory / Odile Pons, French National
Institute for Agronomical Research, France.
    pages cm
  Includes bibliographical references and index.
  ISBN 978-9814531740 (hard cover : alk. paper)
  1. Nonparametric statistics--Asymptotic theory. I. Title.
  QA278.8.P66 2014
  519.5'4--dc23
                                        2013027370

British Library Cataloguing-in-Publication Data
A catalogue record for this book is available from the British Library.

In-house Editor: Angeline Fong

Printed in Singapore

Preface

In tests of hypotheses concerning the whole distribution function of samples, the optimal test statistics rely on the empirical processes or on their functionals. The approach is similar for tests about the form of densities and regression curves, Kolmogorov-Smirnov and Cramer-von Mises type statistics can be defined for them and for other functions. The main subjects of the book are tests of hypotheses in regular nonparametric models, they include tests based on empirical processes and smooth estimators of density functions, regression functions and regular functions defining the distribution of point processes and Gaussian diffusion processes. The asymptotic behavior of the statistics and the asymptotic properties of the tests are detailed.

The last part generalizes tests built for samples to sequential tests, especially tests for processes observed continuously in time. There is no unified theory of sequential tests and several approaches are considered.

Many tests have been studied since several decades and special conferences have early been devoted to the theory of the tests, they give an interesting account of the main advances. This book is not exhaustive, its emphasis on the asymptotically optimal nonparametric test originated in the empirical processes theory and in a nonparametric version of Lecam's asymptotic theory of testing hypotheses which are presented and generalized in this book.

There exist many tests of hypotheses in specific models which are not mentioned, a large amount of the statistical litterature has been devoted to linear rank tests, to censored data and other special fields. They are well known and widely used in data analysis though they are not always optimal. The tests for change-points were not so much developed and I have already published an exhaustive study on this subject with change-points in time or in a threshold of covariates in regression models, point processes and time-series. The results are not standard and this theory cannot be generalized to other nonregular models.

The computational aspects of the tests are briefly considered, many statistics do not have a free limiting distribution and the conditions for the use of the limits are not guaranteed, so permutations tests or the bootstrap version of the tests must be performed. They are applied under regularity conditions.

Odile M.-T. Pons
July 2013

Contents

Chapter 1

Introduction

1.1 Definitions

The theory of statistical tests has been developed in several directions. For samples of independent and identically distributed observations of a random variable X, the optimality of the Neyman-Pearson tests relies on the likelihood of the whole sample. A likelihood ratio statistic is the ratio of the densities of the observations under the hypothesis of the test and under its alternatives. A test of hypothesis is characterized by its probabilities of error under the hypothesis and under the alternatives, which determine the critical values of the test statistic and the region of acceptance of the hypothesis. In the control of manufacturing standards, the lifetime of the materials are measured and they must satisfy tests of reliability before their acceptance. For example the lifetime of a lamp must belong to a reliability interval, this constraint is supposed to hold for each lamp. Tests of control are performed periodically during the production and a large amount of measures are produced, their quantiles at a required level of error α are the critical values of the interval. The comparison of their histogram or their empirical distribution to standard curves detects modifications of the curves and the degradation of the products.

The aging effect increases the failure risk of materials and they are controlled in order to avoid the breakdown of a machine with components in parallel. An upper bound leads to anticipate their replacement before a failure. The forecast of the production also relies on the control and on the improvement of the performances of the products.

Time indexed processes are continuous observed by the automatic registration of signals which are instantaneously analyzed by control systems in order to detect abnomalities or failures. The large amount of registrated

1

data requires an on-line treatment to prevent a deviation from to a calibration to initial conditions defining the hypothesis H_0 that the following observations must satisfy. Tests are performed sequentially on updated data and the principles of the tests follow the same rules as in the analysis of a single sample. In geophysics, the prevention of earthquakes relies on the detection of small changes in registrations at frailty points where the signals may increase in frequency or in magnitude. Modifications detected over a long interval or abrupt modifications create an alarm. The probability of false alarms and the probability of an undetected change are the errors of the tests and they must be as small as possible.

The critical values of a test are not always calculated for the exact size n of the sample and their asymptotic values are commonly used for large sample sizes. Some tests statistics are centered and normalized under the null hypothesis and they may be approximated by a normal variable according to the weak law of large number in tests about the mean of a sample, like the statistic $S_n = \hat{\sigma}_n^{-1} \bar{X}_n$ calculated with the empirical mean $\bar{X}_n$ and the empirical variance $\hat{\sigma}_n^2$ of a n-sample of a variable X. This is a test statistic of the hypothesis $EX = 0$.

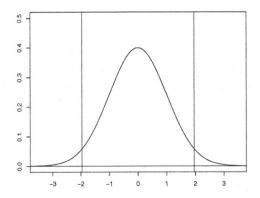

Fig. 1.1   Normal density: the quantiles c_α and $c_{1-\alpha}$ are such that the area of the left and, respectively, right domains under the curve is α.

The distribution function Φ of a normal variable $\mathcal{N}(0,1)$ is symmetric and the normal α-quantiles of a two-sided test are $c_{\frac{\alpha}{2}}$ and $c_{1-\frac{\alpha}{2}} = -c_{\frac{\alpha}{2}}$ such that

$$\frac{1}{2}\alpha = \Phi(c_{\frac{\alpha}{2}}) = 1 - \Phi(1 - c_{\frac{\alpha}{2}}),$$

the error α of a two-sided normal test is represented by the sum of the small areas under the curve in the left and right regions determined by the critical values $\pm c_{\frac{\alpha}{2}}$. In a test with an asymptotically normal statistic under the hypothesis H_0, the hypothesis is rejected if the value of the statistics does not belong to the interval $[-c_{\frac{\alpha}{2}}, c_{\frac{\alpha}{2}}]$. Laplace (1774) presented the calculus of the quantiles of the normal distribution as the fundamental problem of the practical statistics (Pearson, 1920). Most statistical tables have been caculated by finite approximations.

With a sample of Gaussian variables, the distribution of the Student statistic $S_n = \widehat{\sigma}_n^{-1} \bar{X}_n$ is the ratio of two independent variables, the normal variable $\sigma^{-1} \bar{X}_n$ and the square root of the χ_{n-1}^2 variable $(n-1)\widehat{\sigma}_n^2 \sigma^{-2}$. The Student density with parameter ν is defined in $\mathbb{R}$ by the Γ constants as

$$f(t) = \frac{1}{\sqrt{\pi \nu}} \frac{\Gamma\{\frac{1}{2}(\nu+1)\}}{\Gamma(\frac{1}{2}\nu)} \left(1 + \frac{t^2}{\nu}\right)^{-\frac{1}{2}(\nu+1)}.$$

Let $(\Omega, \mathcal{F}, P)$ be a probability space and let X be a real variable defined from $(\Omega, \mathcal{F}, P)$ into a metric space $\mathbb{X}$. A sample of the variable X is a set of independent variables $(X_1, \ldots, X_n)$ having the same distribution as X. A real statistic

$$T_n = T(X_1, \ldots, X_n)$$

is defined as a measurable function in $\mathbb{X}^n$, and in Ω by this map. It is *unbiased* for the estimation of a parameter θ if $ET_n = \theta$ and *consistent* for θ if $\lim_{n \to \infty} ET_n = \theta$. Statistics T_{1n} and T_{2n} are asymptotically equivalent in a class $\mathcal{P}_\Theta$ if there exists θ in Θ such that $T_{1n} - T_{2n}$ tends to zero in probability under P_θ, they are uniformly asymptotically equivalent in $\mathcal{P}_\Theta$ if this convergence holds uniformly in Θ.

A one-sided test of a hypothesis H_0 against an alternative K defined by a statistic T_n has a critical region

$$D_n = D(X_1, \ldots, X_n) = \{T_n > c_n\}$$

where the threshold of the test $c_n = c_n(\alpha)$ depends on the actual distribution F of the variable X under the null hypothesis H_0, or on parameters of its distribution. It is determined by a nominal level as $\alpha = P_{F_0}\{T_n > c_n(\alpha)\}$ for a single hypothesis $H_0 : \{F_0\}$ or by the unknown distribution of X under a composite hypothesis H_0 as

$$\alpha = \sup_{F \in H_0} P_F\{T_n > c_n(\alpha)\}.$$

The hypothesis is rejected in the critical region D_n and it is accepted if $\{T_n \leq c_n(\alpha)\}$. The first kind error α is a fixed probability of false rejection of the hypothesis. The second kind of error of the test is the probability to accept H_0 under the alternative, $1 - \beta_n(\alpha)$ where $\beta_n(\alpha)$ is the power of the test, that is the probability to reject the hypothesis under the alternative. Under the alternative $K = \cup_{F \in H_0} K_F$ such that K_F is the alternative to a distribution F of H_0

$$\beta_n(\alpha) = \sup_{F \in H_0} \inf_{G \in K_F} P_G\{T_n > c_n(\alpha)\}.$$

A *consistent test* has a power larger than its nominal level.

The likelihood ratio of a density f_0 against the alternative of a density f_1 is $L_n = \prod_{i=1}^n f_1(X_i)f_0^{-1}(X_i)$. The conditional means of the likelihood ratio statistics defined for a n-sample by a critical level c_n satisfy the inequalities

$$E_0(L_n | L_n > c_n) = \frac{P_1(L_n > c_n)}{P_0(L_n > c_n)} = \frac{\beta}{\alpha} > c_n,$$

$$E_0(L_n | L_n \leq c_n) = \frac{P_1(L_n \leq c_n)}{P_0(L_n \leq c_n)} = \frac{1 - \beta}{1 - \alpha} \leq c_n$$

and $E_0 L_n = \prod_{i=1}^n E_0\{f_1(X_i)f_0^{-1}(X_i)\} = 1$ for every density f_1.

When α and β are predetermined, they define two critical values $c_n(\alpha)$ and $d_n(\beta)$. The test rejects the hypothesis with the error α if its value is 1 and it accepts H_0 with the error β if its value is 0

$$\phi_n = \begin{cases} 1 & \text{if } T_n > c_n, \\ \kappa & \text{if } d_n < T_n \leq c_n \\ 0 & \text{if } T_n \leq d_n. \end{cases} \tag{1.1}$$

The mean of the tests under the hypothesis and the alternative are

$$E_{H_0} \phi_n = \alpha + \kappa P_0(d_n < T_n \leq c_n),$$

$$E_K(1 - \phi_n) = \beta + (1 - \kappa) P_K(d_n < T_n \leq c_n).$$

The probability to accept H_0 when it is true is $1 - \alpha$. The value κ between 0 and 1 is a probability to accept H_0 in the domain of uncertainty $\{d_n < T_n \leq c_n\}$ of the test, and $1 - \kappa$ is a probability to reject it in the domain of uncertainty.

A two-sided test for an hypothesis H_0 against an alternative K is defined by a statistic $T_n = T_n(F, G)$ for F under H_0 against G under K, its rejection domain at the nominal level α is

$$D_n(\alpha, F, G) = \{T_n(F, G) > c_{1n}(\alpha)\} \cup \{T_n(F, G) < c_{2n}(\alpha)\}$$

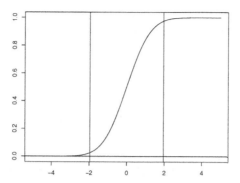

Fig. 1.2   Distribution function Φ of the normal density: $1 - \Phi(c_\alpha) \le 1 - \Phi(-c_\alpha)$.

where $\alpha = P_{F_0}(D_n(\alpha, F_0, G))$ for a simple hypothesis $H_0 : \{F_0\}$ and a simple alternative $K : \{G\}$. With composite hypothesis and alternative, when the distribution F of X under H_0 is unknown, the level is $\alpha = \sup_{F \in H_0} \inf_{G \in K_F} P_F(D_n(\alpha, F, G))$. The power of the test is against a simple alternative $K_F = \{G\}$ is $\beta_n(\alpha, G) = \sup_{F \in H_0} P_G\{D_n(\alpha)\}$ and for a composite alternative K_F, it is

$$\beta_n(\alpha) = \sup_{F \in H_0} \inf_{G \in K_F} P_G\{D_n(\alpha)\}.$$

A test for H_0 against a single alternative G of K_F is *most powerful* at the level α if it power is $\beta(\alpha, H_0, G)$, the probability to reject H_0 under the probability distribution P_G. It is *uniformly most powerful* for H_0 against a composite alternative K_F if it is the most powerful for every alternative distribution of K_F. Acccording to the *Neyman-Pearson lemma*, the uniformly most powerful tests for parametric families of probability density functions is the likelihood ratio test based on the ratio of the density functions $f^{-1}g$, where f satisfies the conditions of the hypothesis H_0 and g the conditions of an alternative K. With composite hypothesis and alternative, the ratio of two functions is replaced by $(\sup_{H_0} f)^{-1} \sup_K g$.

The least favorable probability densities for a test of H_0 against an alternative K is (f_0, f_1) such that f_0 is a density under H_0, f_1 is a density under K and the log-likelihood ratio statistic $S_n = \sum_{i=1}^n \{\log f_1(X_i) - \log f_0(X_i)\}$ satisfies for every n

$$P_F(S_n > c) \le P_0(S_n > c) \le P_1(S_n > c) \le P_G(S_n > c),$$

for all distribution functions F and G with densities satisfying H_0 and, respectively, K and with P_k the probabilities with densities f_k, for $k = 1, 2$.

For example, the least favorable densities for a test of $H_0 : \theta \leq \theta_0$ against the alternative $K : \theta \geq \theta_1$ are $(f_{\theta_0}, f_{\theta_1})$ and the test can be performed with the likelihood ratio of f_{θ_1} over f_{θ_0}. Asymptotically least favorable densities (f_0, f_1) satisfy

$$\lim_n P_F(S_n > c) \leq \lim_n P_0(S_n > c) \leq \lim_n P_1(S_n > c) \leq \lim_n P_G(S_n > c),$$

for all F and G with densities under H_0 and, respectively, K.

A statistic S_n is *sufficient* for the parameter of a probability family $\mathcal{P} = \{P_\theta, \theta \in \Theta\}$ defined in a subspace Θ of $\mathbb{R}^d$ if the distribution of the sample $(X_1, \ldots, X_n)$ depends on θ only through a function of S_n. There exists a measurable real function ϕ in $(\mathbb{X} \times \Theta)$ depending on the observations only through S_n, and a measurable function h in $\mathcal{X}^n$ which does not depend on the parameter, such that

$$\prod_{i=1}^{n} f_\theta(X_i) = \phi(S_n, \theta) h(X_1, \ldots, X_n).$$

The conditional distribution of θ given $S_n = s$, does not depend on the observations. A statistic S_n is *asymptotically sufficient* for the parameter θ if the distribution of the sample has an expansion

$$\prod_{i=1}^{n} f_\theta(X_i) = \phi(S_n, \theta) h(X_1, \ldots, X_n) + R_n(\theta)$$

such that $R_n(\theta)$ converges in probability to zero under P_θ, and the functions ϕ and h satisfy the above properties. According to the families of densities under the hypothesis and the alternative, the existence of a sufficient statistic S_n leads to define uniformly most powerful tests from S_n, a normalized expression of S_n or an equivalent form through a reparametrization. Let us consider a n-sample of Gaussian variables with a mean θ and a variance σ^2, a sufficient statistic for the parameter θ is the empirical mean $\bar{X}_n$, the density of the sample is the product of

$$\phi(\bar{X}_n, \theta) = \exp\left\{ -\frac{n\theta(2\bar{X}_n - \theta^2)}{2\sigma^2} \right\}$$

and of the density h of a sample of Gaussian variables with mean zero and variance σ^2 which do not depend on the parameter θ.

In parametric models, the log-likelihood ratio tests defined by an hypothesis and a fixed alternative have the asymptotic power 1. The log-likelihood ratio tests defined by sequences of local alternatives are generally consistent and their asymptotic power is lower than 1. The classical tests are based on second order expansions of the log-likelihood which provide several asymptotically equivalent statistics, such as the score test and Wald's test. Many results have been published and they have been reviewed in books since several decades (Kendall and Stuart, 1947, Lehmann, 1959, Hájek and Sidák, 1967). Lecam's theory (1956) of tests provides the limiting distribution of the local log-likelihood ratio test statistics under parametric hypothesis and alternatives and their asymptotic equivalence with asymptotically normal test statistics, according to the alternative. The main results are detailed in the next chapter where they are adapted to semi-parametric and nonparametric models. The theory of optimal tests is related to the search of optimal parametric and nonparametric estimators reaching the lowest information bound.

When the hypothesis concerns the whole distribution function of the observed variables, the optimal test statistic is a Kolmogorov-Smirnov statistic based on empirical estimators of the distribution functions, or their functionals, defining the hypothesis H_0. The Cramer-von Mises statistics are empirical L^2 distances of empirical processes and their performances are asymptotically equivalent. Other tests of the same kind are defined for the densities and other functions under regularity conditions.

1.2 Rank tests and empirical distribution functions

The statistical theory has established the optimality of rank tests for specific probability models, with parametric hypotheses and alternatives (Hájek and Sidák, 1967). They rely on Neyman-Pearson's lemma and are related to the quantiles of the empirical distribution functions of the samples, they are briefly presented here.

Let $\mathcal{P}$ be a family of probability measures defined on a measurable space $(\Omega, \mathcal{A})$ and let X be a real variable defined on the probability spaces $(\Omega, \mathcal{F}, \mathcal{P})$, with distribution function F under a probability P_F of $\mathcal{P}$. Nonparametric tests about the distribution of the variable X are based on the empirical distribution function

$$\widehat{F}_n(x) = n^{-1} \sum_{i=1}^{n} 1_{\{X_i \leq x\}}, x \in \mathbb{R},$$

for a n-sample of the variable X. The quantile function of the variable X is defined on $[0, 1]$ by

$$Q(t) = F_X^{-1}(t) = \inf\{x \in \mathbb{R} : F_X(x) \geq t\},$$

it is right-continuous with left-hand limits, like the distribution function. For every uniform variable U, $F_X^{-1}(U)$ has the distribution function F_X and, if F is continuous, then $F(X)$ has an uniform distribution function. The inverse of the distribution function satisfies $F_X^{-1} \circ F_X(x) = x$ for every x in the support of X and $F_X \circ F_X^{-1}(x) = x$ for every continuity point x of F_X. The order statistics of the sample are the variables of the ordered sample $X_{1:n} < X_{2:n} < \cdots < X_{n:n}$ and their ranks are the variables $R_{n,i}$ such that for every $i = 1, \ldots, n$

$$X_i = X_{R_{n,i}:n}.$$

The distribution of the rank statistic $R_{n,k}$ is given by $P(R_{n,k} = j) = n^{-1}$ for every j and $k = 1, \ldots, n$, and the rank vector $R_n = (R_{n,1}, \ldots, R_{n,n})$ has the distribution

$$P(R_n = r) = \frac{1}{n!},$$

for every r in the set of the permutations of $\{1, \ldots, n\}$. The empirical distribution function of the order statistics satisfies

$$\widehat{F}_n(X_{k:n}) = n^{-1}k, \ k = 1, \ldots, n.$$

The empirical quantile function of X is defined on $[0, 1]$ by

$$\widehat{Q}_n(t) = \widehat{F}_n^{-1}(t) = n^{-1} \inf\{k \in \{1, \ldots, n\} : k \geq nt\},$$

so $X_{k:n} = \widehat{Q}_n(n^{-1}k)$, for every $k = 1, \ldots, n$. By the weak convergence of the empirical quantiles $\mathcal{Q}_{Fn}(t) = n^{-\frac{1}{2}}\{\widehat{F}_n^{-1}(t) - F^{-1}(t)\}$ to a Gaussian process, the variable $X_{k:n}$ is a $n^{\frac{1}{2}}$-consistent estimator of $F^{-1}(n^{-1}k)$.

If the variable X has a density f, the density of the order statistic $X_{k:n}$ is determined by

$$f_k(x) = n! \int_x^{+\infty} \cdots \int_{x_{n-1}}^{+\infty} f(x_{k+1}) \cdots f(x_n) \, dx_n \cdots dx_{k+1}$$

$$\times \int_{-\infty}^x \int_{-\infty}^{x_k} \cdots \int_{-\infty}^{x_2} f(x_k) \cdots f(x_2) f(x_1) \, dx_1 \, dx_2 \cdots dx_k$$

$$= n! f(x) \frac{\{1 - F(x)\}^{n-k}}{(n-k)!} \frac{F^{k-1}(x)}{(k-1)!}, \ k = 1, \ldots, n.$$

The minimum X of two independent random variables T and C, with distribution functions F and, respectively, G in $\mathbb{R}_+$, is observed in samples of right-censored times variables such as the follow-up of individuals until a recovery at T or their departure from the study at C for another reason which is considered as a censoring for the medical point of view. The empirical estimator $\widehat{H}_n$ of the distribution function $H = 1 - (1 - F)(1 - G)$ of

X is calculated from a n-sample of $X = T \wedge C$ and its order statistics are $X_{k:n} = \widehat{H}_n^{-1}(n^{-1}k)$, for every $k = 1, \ldots, n$. The indicators $\delta_i = 1_{\{T_i \leq C_i\}}$ attached to the order statistics of the whole sample define the order and the ranks statistics of the uncensored observations T_i such that $\delta_i = 1$ among the X_i, for $i = 1, \ldots, n$. Tests defined from the empirical estimator $\widehat{H}_n$ do not distinguish the variables T and C as it is possible with the ranks indexed by the censoring indicator.

The Kaplan-Meier estimator $\widehat{\overline{F}}_n$ is the optimal estimator of the survival function $\overline{F} = 1 - F$ of the variable X (Wellner, 1982). It jumps at the uncensored observations T_i and the random size of the jumps is modified by the censored observations. The order statistics of this sub-sample cannot be written as the inverse of $\widehat{F}_n$ at the fixed points $n^{-1}k$ for every $k = 1, \ldots, \sum_{i=1}^{n} 1_{\{T_i \leq C_i\}}$. Other test statistics for censored variables are considered in Chapter 7.

1.3 Hypotheses of the tests

The classical hypotheses considered in the theory of the tests are goodness of fit of models, homogeneity of the sample, randomness of the sampling, independence of variables, symmetry of the probability density functions, stochastic dominance. They concern one sample, two samples or k samples of independent or dependent variables. The adequacy of a model for a density or other functions is usually tested by embedding the model of the hypothesis into a larger class of functions where the test is written as a test of null parameters. The relevance of an alternative can be checked by estimating the model under the alternative, the l_n^2-error between the observations and their estimators is used as a criterium for the comparison of embedded models.

Two tests for the same hypothesis differ according to their alternatives. For instance, let $X_1, \ldots, X_{n_1}$ be a sample of a variable X with distribution function F and mean zero, and let $Y_1, \ldots, Y_{n_2}$ be a sample of a variable Y under a change of location, the location parameter θ of the distribution function of Y is estimated by the difference of the means of the samples $\widehat{\theta}_n = \overline{Y}_n - \overline{X}_n$ and the variable $Y - \widehat{\theta}_n$ is asymptotically equivalent to a centered variable.

A test of no change of location of Y as compared to X can be considered as a test of homogeneity of the samples of X and Y or as a tests of equality of their empirical means if their distributions have the same

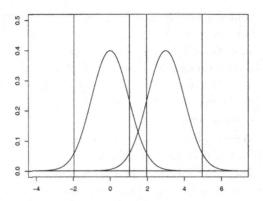

Fig. 1.3   Two Gaussian densities with their critical values.

symmetric form. With non symmetric unimodal distributions, the parameter of interest is the mode, for distributions without constraints, this is the median. In hierarchical models for the comparison of treatments in several groups of items, the comparison of treatment effects is a problem of change of location for non identically distributed observations. Removing the assumption of Gaussian observations, this question is nonparametric (Bickel and Freedman 1981, Beran 1982).

A direct application of the quantiles is the definition of the ROC curve $R = S_1 \circ S_0^{-1}$ of two independent variables X_0 and X_1 with survival functions $S_0 = 1 - F_0$ and, respectively, $S_1 = 1 - F_1$. This is an decreasing function and the primitive of the ROC curve at $S_0^{-1}(t)$ is

$$A(t) = \int_{S_0^{-1}(t)}^{1} S_1 \circ S_0^{-1}(x)\, dx = \int_{t}^{\infty} S_1\, dS_0 = P(X_1 > X_0 > t). \quad (1.2)$$

The comparison of two ROC curves is a comparison of the decreasing estimators $\widehat{S}_{1n} \circ \widehat{S}_{0n}^{-1}$ of two sub-samples of the variables X_0 and X_1. A test of $A(-\infty) = 1$ is a test of stochastic order $X_1 > X_0$. When it is applied to the test of F_0 against G, the functions S_0 and S_1 are respectively the sensitivity of the test $S_0(t) = P_0(T \leq t)$ and its specificity $S_1(t) = P_G(T > t)$.

The tests concerning the form of a density are nonparametric and of a different kind due to the convergence rate of its smooth estimator, they are extended to regression curves, intensities of point processes and to nonparametric diffusions. The tests for samples of independent variables have been modified to take into account censoring or truncation of the sample, and several forms of dependence between the observations of processes. This is the subject of Chapters 6, 7 and 8. Semi-parametric models of point processes with covariates have been extensively studied and applied to data in the literature, they will not be presented.

Let F_0 be a distribution of the observed random variable X and let $(X_1, \ldots, X_n)$ be a vector of independent variables with distribution function $\prod_{i=1}^{n} F(X_i)$. A test for the hypothesis of randomness of a set of random variables $X_1, \ldots, X_n$ is performed by permutations of the variables. For every permution π of $\{1, \ldots, n\}$, $(X_{\pi(1)}, \ldots, X_{\pi(n)})$ has the same distribution as $(X_1, \ldots, X_n)$. A statistic comparing their distributions uniformly over the permutations can be used for a test of randomness.

Let X be a random variable with distribution function F. A goodness of fit test for a single distribution function, $H_0 : F = F_0$, is performed with the empirical process of the sample or with likelihood ratio test statistic if the variable X has a density $f_0 = f_{\theta_0}$ in a parametric class of densities $\mathcal{F} = \{F_\theta, \theta \in \Theta\}$ of $C^2(\Theta)$, with a bounded and open parameter set Θ in $\mathbb{R}^d$ and with a finite and nonsingular information matrix $I = E_0[\{(f_0^{-1}\dot{f}_0)(X)\}^{\otimes 2}]$. Using a reparametrization of the empirical process by the quantiles, the asymptotic distributions of both test statistics are free.

An index for testing the independence of the components X_1 and X_2 of a bivariate random variable X is Kendall's tau. It is defined from the ranks $(R_{1i})_{i \leq n}$ and $(R_{2j})_{j \leq n}$ of the components of a sample $(X_i)_{i \leq n}$ of a two-dimensional variable $X = (X_1, X_2)$

$$\tau_n = \frac{1}{n(n-1)} \sum_{i=1}^{n} \sum_{j \neq i, j=1}^{n} \mathrm{sign}(i-j)\mathrm{sign}(R_{1i} - R_{2j})$$

$$= \frac{1}{n(n-1)} \sum_{i=1}^{n} \left\{ \sum_{j=1}^{i-1} \mathrm{sign}(R_{1i} - R_{2j}) - \sum_{j=i+1}^{n} \mathrm{sign}(R_{1i} - R_{2j}) \right\}.$$

Its mean and variance under the independence are $E\tau_n = 0$ and

$$Var\tau_n = \frac{2(2n+5)}{9n(n-1)}.$$

Let X have a joint distribution function F and marginal distribution functions F_1 and F_2, the dependence function of X is defined by

$$C(s,t) = P\{F_1(X_1) \leq s, F_2(X_2) \leq t\}. \tag{1.3}$$

Under an alternative of positive dependence, it satisfies $C(u,v) \geq uv$ in $[0,1]^2$, under a the negative dependence $C(u,v) \leq uv$ in $[0,1]^2$. Two random variables X and Y are positively dependent if their joint distribution is larger than the product of their marginal distributions

$$P(X \leq x, Y \leq y) \geq P(X \leq x)P(Y \leq y),$$

for all real x and y, this is equivalent to

$$P(X \geq x, Y \geq y) \geq P(X \geq x)P(Y \geq y)$$

and therefore to $C(u, v) \geq uv$ in $[0, 1]^2$.

Oakes (1989) and Genest and Rivest (1993) presented a review of semi-parametric bivariate models for time variables. The Archimedean distributions are defined for a two-dimensional variable $X = (X_1, X_2)$, with survival function $\bar{F}(t) = P(X_1 \geq t_1, X_2 \geq t_2)$, in an additive form by its marginals $\bar{F}_k(t_k) = P(X_k \geq t_k)$, for $k = 1, 2$, and by a monotone dependence function $p : \mathbb{R} \mapsto [0, 1]$ and its inverse $q = p^{-1}$ from $[0, 1]$ to $\mathbb{R}$, as

$$\bar{F}(t) = p \circ \{q \circ \bar{F}_1(t_1) + q \circ \bar{F}_2(t_2)\}. \tag{1.4}$$

Variables X_1 and X_2 with an exponential archimedian distribution are defined by $p_\lambda(y) = e^{\lambda y}$, for λ in $\mathbb{R}_+$ and y in $\mathbb{R}_-$ and the inverse function q is logarithmic in $]0, 1]$. A test of independence is a test of the parameter value $\lambda = 1$. In most semi-parametric models of the joint distribution function, a test of independence is a parametric test for a scalar dependence parameter. Nonparametric tests for independence of variables that do no depend on the marginal distributions of the variables are built on the empirical estimator of their joint distribution function and of their dependence function (1.3).

1.4   Weak convergence of the test statistics

Let F_0 be a distribution of the observed random variable X and let $X_1, \ldots, X_n$ be a sample of X. A test for $H_0 : F = F_0$ against the alternative $K : F \neq F_0$ relies on the empirical process under the null hypothesis

$$\nu_n = n^{\frac{1}{2}}(\widehat{F}_n - F_0).$$

For every integer n, under H_0 the variable $\nu_n(t)$ has the mean zero and the variance $F_0(t)\{1 - F_0(t)\}$, the covariance between $\nu_n(s)$ and $\nu_n(t)$ is

$$K_{F_0}(s, t) = F_0(s \wedge t) - F_0(s)F_0(t)$$

$$= F_0(s) \wedge F_0(t) - F_0(s)F_0(t) = K(F_0(s), F_0(t)), \ (s, t) \in \mathbb{R}^2.$$

The process ν_n converges weakly to the transformed Brownian bridge $B \circ F_0$ as n tends to infinity. Another version of the test can be written with the uniform variable $U = F_0(X)$, its empirical process is

$$W_n(t) = \nu_n \circ F_0^{-1}(t) = n^{\frac{1}{2}}\{\widehat{F}_n \circ F_0^{-1}(t) - t\},$$

for every t in the support of X and its asymptotic distribution is free. Since the supremum of ν_n in $\mathbb{R}$ is identical to the supremum of W_n in $[0,1]$, the statistics ν_n and W_n are asymptotically equivalent.

The Kolmogorov-Smirnov test for H_0 against K has the rejection domain $D_n(\alpha) = \{\sup_{t\in\mathbb{R}} |W_n(t)| > c_\alpha\}$ where $P_{H_0}\{D_n(\alpha)\}$ converges to $\alpha = P\{\sup_{t\in\mathbb{R}} |B| > c_\alpha\}$, as n tends to infinity. Its asymptotic power is the limit of $\beta_n(\alpha) = P_K\{D_n(\alpha)\} = \inf_{F\neq F_0} P_F\{D_n(\alpha)\}$ and it is larger than its level α. The threshold of the tests and its power have been studied in particular by Doob (1949), Donsker (1952).

The distribution functions of real random variables under an alternative of a change of location are $F_\theta(t) = F_0(t - \theta)$, for θ in a subset Θ if $\mathbb{R}$, its first derivative with respect to θ is $\dot{F}_\theta(t) = -f_0(t - \theta)$. The estimator of the parameter under the alternative of a nonzero parameter satisfies $\sum_{i=1}^n f_0(X_i - \widehat{\theta}_n) = 0$, for a sample $(X_i)_{i=1,\ldots,n}$, and F_0 is compared to $F_{\widehat{\theta}_n}$ in the Kolmogorov-Smirnov test. If zero does not belong to the compact closure of Θ, the hypothesis H_0 is not enclosed in the alternative and $\beta_n(\alpha)$ tends to one as n tends to infinity. If zero belongs to compact closure of Θ, $\beta_n(\alpha)$ tends to α as n tends to infinity.

The Cramer-von Mises test statistic is the variable

$$S_n = \int_{\mathbb{R}} W_n^2 \, d\widehat{F}_n = \sum_{i=1}^n \{\widehat{F}_n(X_i) - F_0(X_i)\}^2.$$

Under H_0, it converges weakly to $S = \int_0^1 B^2(t) \, dt$. It is also written in terms of the order statistics $X_{1:n} < X_{2:n} < \cdots < X_{n:n}$

$$S_n = \sum_{k=1}^n \{\widehat{F}_n(X_{k:n}) - F_0(X_{k:n})\}^2$$
$$= \sum_{k=1}^n \left\{\frac{k}{n} - F_0(X_{k:n})\right\}^2.$$

Weighted statistics of the same form as S_n are defined using a positive weight function w defined from $\mathbb{R}$ to $[0,1]$

$$S_n = \int_{\mathbb{R}} W_n^2(x) \, w(x) \, d\widehat{F}_n(x)$$
$$= \sum_{k=1}^n \left\{\frac{k}{n} - F_0(X_{k:n})\right\}^2 w(X_{k:n}).$$

Under H_0, it converges weakly to $S = \int_0^1 B^2(t) \, w \circ F_0^{-1}(t) \, dt$.

The Anderson-Darling test statistic is also defined as a squared L^2-norm, with a normalization of the process W_n by its variance

$$Z_n(t) = \frac{W_n(t)}{\sqrt{F_0(t)\{1 - F_0(t)\}}}, t \in \mathbb{R}.$$

Anderson-Darling (1952) provided analytical forms of the thresholds of this statistic in one-sided and two-sided tests. For every t, under H_0 the variable $A_n(t)$ has the mean zero and the its variance one, however its distribution depends on F_0 via its covariances. The covariance of $A_n(s)$ and $A_n(t)$ is

$$K_{F_0}(s, t) = \frac{F_0(s \wedge t) - F_0(s)F_0(t)}{\sqrt{F_0(t)\{1 - F_0(t)\}}\sqrt{F_0(s)\{1 - F_0(s)\}}}$$

$$= [F_0(s)\{1 - F_0(t)\}]^{\frac{1}{2}}[F_0(t)\{1 - F_0(s)\}]^{-\frac{1}{2}}1_{\{s<t\}}$$

$$+ [F_0(t)\{1 - F_0(s)\}]^{\frac{1}{2}}[F_0(s)\{1 - F_0(t)\}]^{-\frac{1}{2}}1_{\{s\geq t\}},$$

it satisfies $K_{F_0}(s, t) = K_u(F_0(s), F_0(t))$ where K_u is the covariance for the uniform distribution, and $\sup_{(s,t)\in[0,1]} K_u(s, t) = +\infty$.

The Anderson-Darling test statistic is also defined for the comparison of the unknown distributions F anf G of two samples of respective sizes m and n as

$$A_n = \frac{mn}{m + n} \int_{\mathbb{R}} \frac{(\widehat{F}_m - \widehat{G}_n)^2}{\widehat{H}_{m,n}(1 - \widehat{H}_{m,n})}(x)\, d\widehat{H}_{m,n}(x),$$

with the common estimator of F and G under the hypothesis of their equality $\widehat{H}_{m,n} = (m + n)^{-1}(m\widehat{F}_m + n\widehat{G}_n)$. Approximations of these asymptotic statistics and tables of their quantiles, for uniform samples, are given by Shorack and Wellner (1986).

As an example of application, let $X_1, \ldots, X_m$ and $Y_1, \ldots, Y_n$ be independent samples of independent variables X and Y, with respective distribution functions F and G. In a model of change of location and scale of Y with respect to X, they satisfy

$$G(y) = F\left(\frac{y - \mu}{\sigma}\right). \tag{1.5}$$

This is equivalent to

$$G^{-1}(y) = \mu + \sigma F^{-1}(y).$$

Assuming that F and G are continuous and strictly monotone, the parameters are deduced

$$\mu = G^{-1} \circ F(0),$$

$$\sigma = -\frac{\mu}{F^{-1} \circ G(0)}$$

and they are estimated by plugging, with the empirical distribution functions and their quantiles

$$\widehat{\mu}_{nm} = \widehat{G}_n^{-1} \circ \widehat{F}_m(0),$$

$$\widehat{\sigma}_{nm} = -\frac{\widehat{\mu}_{nm}}{\widehat{F}_m^{-1} \circ \widehat{G}_n(0)}.$$

The hypothesis $H_0 : F = G$ in the model (1.5) is the assumption that the parameter (μ, σ) reduces to a single point $(0, 1)$. The estimators $\widehat{\mu}_{nm}$ and $\widehat{\sigma}_{nm}$ are consistent as m and n tend to infinity. Let $P_{\mu,\sigma}$ be a probability measure that yields distribution functions F and G satisfying (1.5). Under the probability $P_{\mu,\sigma}$, $(n^{\frac{1}{2}}(\widehat{\mu}_{nm} - \mu), (m^{-1}n)^{\frac{1}{2}}(\widehat{\sigma}_{nm} - \sigma))$ converges weakly to a Gaussian variable, as a consequence of the weak convergence of the variable $(n^{\frac{1}{2}}\{\widehat{G}_n^{-1} \circ \widehat{F}_m(0) - G^{-1} \circ F(0)\}, m^{\frac{1}{2}}\{\widehat{F}_m^{-1} \circ \widehat{G}_n(0) - F^{-1} \circ G(0)\})$.

A nonparametric test for the hypothesis $H_0 : F = G$ against the general alternative $K : F \neq G$ is based on the Kolmogorov-Smirnov statistic

$$S_{nm} = \sup_{x \in \mathbb{R}}\left(\frac{nm}{n+m}\right)^{\frac{1}{2}}|\widehat{F}_m(x) - \widehat{G}_n(x)|. \tag{1.6}$$

As m and n tend to infinity, S_{nm} converges to the difference of two Brownian bridges under H_0. For every level α of the test, under an alternative $\mu \neq 0$ or $\sigma \neq 1$, S_{nm} tends to $\pm$ infinity and the power of the test tends to 1. From the asymptotic behavior of the estimators of (μ, σ) in (1.5), a test with level α for H_0 is defined by

$$\phi_n = 1 \text{ if } n^{\frac{1}{2}}|\widehat{\mu}_{nm}| > c_{1\alpha} \text{ or } \left(\frac{n}{m}\right)^{\frac{1}{2}}|\widehat{\sigma}_{nm} - 1| > c_{2\alpha},$$

$$\phi_n = 0 \text{ otherwise,}$$

where the critical values are determined by the quantiles of Brownian bridges. Under a fixed alternative, either $n^{\frac{1}{2}}|\widehat{\mu}_{nm}|$ or $(\frac{n}{m})^{\frac{1}{2}}|\widehat{\sigma}_{nm} - 1|$ tend to infinity as n and m tend to infinity. The asymptotic local power of such tests under contiguous alternatives are studied in Chapter 3.

The tests for samples assume independent and identically distributed observations and the asymptotic properties of the statistics are consequences of the weak convergence of the empirical process. They are extended to the general setting of ergodic and weakly dependent processes. For processes with independent increments, the weak convergence of the statistics are obtained like in the Lindenberg central limit theorem for arrays of random variables $(X_{ji})_{i=1,\ldots,n_j, j=1,\ldots,J}$ with respective densities f_j

such that $n^{-1} \sum_{j=1,\ldots,J} f_j$ has a finite limit. The ergodicity is also assumed for the mean function of processes with independent or weakly dependent increments over disjoint intervals. Under an assumption of α-dependence, the covariance of the variables over disjoint intervals tends to zero as their distance tends to infinity and the variance of the sums of variables over disjoint intervals is suposed to converge. A central limit theorem still holds under these conditions (Billingsley 1968).

1.5   Tests for densities and curves

Tests of hypotheses about the form of the distribution function or the density of a random variable are based on their estimation under the constraints of the hypotheses and under their alternatives. Let X be a real random variable with density f defined in a finite or infinite support $\mathcal{I}_X$. A test of monotony or unimodality of a density or another function requires the construction of a monotone or unimodal estimator which can be compared to the usual estimator, without constraint. These questions are of interest in economy, for instance, for a comparison of growing markets, or for a test of a single population against the alternative of a mixture of separate populations.

A unimodal density has a unique maximum at

$$M_f = \inf\{x \text{ in } \mathcal{I}_X : f(x) \geq f(y), \text{ for every } y \in \mathcal{I}_X\}.$$

Its mode M_f is estimated from a sample $X_1, \ldots, X_n$ of X by

$$\widehat{M}_{n,f} = \inf\{x \in \mathcal{I}_X : \widehat{f}_n(x) \geq \widehat{f}_n(y), \text{ for every } y \in \mathcal{I}_X\}, \qquad (1.7)$$

where $\widehat{f}_{n,h}$ is a nonparametric kernel estimator of f defined using a symmetric kernel function $K \geq 0$ with integral one and a bandwidth $h = h_n$ tending to zero as n tends to infinity

$$\widehat{f}_{nh}(x) = n^{-1} \sum_{i=1}^{n} K_h(X_i - x).$$

The next conditions ensure the L^2-convergence of the estimator in the subinterval $\mathcal{I}_{X,h} = \{s \in \mathcal{I}_X; [s - h, s + h] \subset \mathcal{I}_X\}$ of the support.

Condition 1.1.

(1) The kernel function K is a symmetric density such that $\kappa_2 = \int K^2(u)du$ is finite, $|x|^2 K(x) \to 0$ as $|x|$ tends to infinity or K has a compact support with value zero on its frontier.

(2) The density function f defined in $\mathcal{I}_X$ belongs to the class $C^s(I_X)$, with a continuous and bounded derivative of order s, $f^{(s)}$, on $\mathcal{I}_X$.

(3) The kernel function satisfies integrability conditions: the moments $m_{jK} = \int u^j K(u)du$, for $j < s$, m_{sK} and $\int |K'(u)|^\alpha du$, for $\alpha = 1, 2$, are finite.

(4) As $n \to \infty$, h_n tends to zero and $nh_n^{2s+1} = O(1)$ so that $(nh_n)^{\frac{1}{2}}h_n^s$ converges to a finite limit $\gamma_s > 0$.

The optimal global bandwidth minimizing $\|\widehat{f}_{n,h} - f\|_{L^2}$ is a $O(n^{-\frac{1}{2s+1}})$, the condition (4) is therefore the L^2-optimal convergence rate for the estimator of f in $C^s(I_X)$. These conditions are sufficient for the weak converge of the density estimator (Pons, 2011). Similar conditions are written for a regression function defined as the conditional mean function $E(Y|X = x)$, for dependent variables X and Y.

Theorem 1.1. *Under Condition 1.1 for a density f, the process*

$$U_{n,h} = (nh)^{1/2}\{\widehat{f}_{n,h} - f\}I\{\mathcal{I}_{X,h}\}$$

converges weakly to $W_f + \gamma_s b_f$, where W_f is a continuous Gaussian process on $\mathcal{I}_X$ with mean zero and covariance $E\{W_f(x)W_f(x')\} = \delta_{x,x'}\sigma_f^2(x)$, at x and x', $\sigma_f^2(x) = \kappa_2 f(x)$ and $b_f(x) = \frac{1}{s!}m_{sK}f^{(s)}(x) + o(h^s)$.

A monotone estimator $\widetilde{f}_n$ of the density was defined by Grenander (1956) as the derivative of the least concave minorant of the empirical distribution function and the same estimator applies in two steps to unimodal densities. It is piecewise constant, like the histogram.

The histogram is defined as a cumulated empirical distribution on small intervals of equal length h_n, divided by h_n, where the bandwidth h_n and nh_n converge to zero as n tends to infinity. Let $(\Delta_{j,\delta})_{j=1,\ldots,J_{X,\delta}}$ be a partition of $\mathcal{I}_X$ into sub-intervals of length δ and centered at $a_{j,\delta}$. Its bias is $\widetilde{b}_{f,\delta}(x) = \sum_{j \in J_{X,\delta}} 1_{\Delta_{j\delta}}(x)\{f(a_{j\delta}) - f(x)\} + o(\delta) = \delta f^{(1)}(x) + o(\delta)$, it is larger than the bias of kernel estimators. Its variance $\widetilde{v}_f(x)$ is expanded as $\widetilde{v}_{f,\delta}(x) = (n\delta)^{-1}f(x) + o((n\delta)^{-1})$, it has the same order as the variance of the kernel estimator. More results about the histogram are detailed in Appendix A.5. The isotonic histogram of a density function f provides a consistent monotone estimator of a monotone density. Monotone estimators rely on the next isotonization lemma.

Lemma 1.1. *For every real function ϕ in $\mathbb{R}$*

$$\varphi_I(x) = \inf_{v \geq x} \sup_{u \leq x} \frac{1}{v - u}\{\phi(v) - \phi(u)\}$$

is an increasing function and

$$\varphi_D(x) = \sup_{u \leq x} \inf_{v \geq x} \frac{1}{v-u}\{\phi(v) - \phi(u)\}$$

is an decreasing function.

It is easily proved for every real function ϕ. Let $x_1 < x_2$ be real numbers, its proof relies on the comparison of the suprema in the increasing intervals $]-\infty, x_1]$ and $]-\infty, x_2]$ and on the comparison of the minima in the decreasing intervals $[x_2, \infty[\subset [x_1, \infty[$. If ϕ has an increasing first derivative $\phi^{(1)}$ at x, then $\varphi_I(x) = \phi^{(1)}(x)$. If ϕ has a decreasing first derivative $\phi^{(1)}$ at x, then $\varphi_D(x) = \phi^{(1)}(x)$.

As a consequence of Lemma 1.1, if f is increasing, the estimator

$$f^*_{I,n}(x) = \inf_{v \geq x} \sup_{u \leq x} \frac{1}{v-u}\{\widehat{F}_n(v) - \widehat{F}_n(u)\}, \ x \leq M_f,$$

is increasing and it is a piecewise constant estimator of f. If f is decreasing, a decreasing piecewise constant estimator of the density is

$$f^*_{D,n}(x) = \sup_{u \leq x} \inf_{v \geq x} \frac{1}{v-u}\{\widehat{F}_n(v) - \widehat{F}_n(u)\}, \ x > M_f.$$

A unimodal estimator of a density is deduced as

$$f^*_n(x) = \begin{cases} f^*_{I,n}(x) & \text{if } x \leq \widehat{m}_f, \\ f^*_{D,n}(x) & \text{if } x > \widehat{m}_f. \end{cases} \tag{1.8}$$

Another definition of the isotonic estimators of a density is obtained as the right-derivative of the least concave majorant (LCM) of the graph of the points $(X_{i:n}, \widehat{F}_n(X_{i:n}))_{i=1,\ldots,n}$ if the density is decreasing and the left-derivative of the greatest convex minorant (GCM) of the graph of these points if the density is increasing. For monotone densities, it is asymptotically equivalent to the histogram.

The estimators (1.8) and the derivatives of the GCM and LCM graphs are the usual constructions of monotone estimators of a monotone density, the GCM and LCM curves are both increasing for an increasing function or decreasing for a decreasing function. Since the concavity or convexity of a distribution function determines whether its density is increasing or decreasing, these estimators are expected to be identical to (1.8) for increasing or decreasing functions. Smooth estimators of the density can also be considered and the isotonization of an integrated kernel estimator of the density is a smooth estimator of a monotone density. They are defined like (1.8) from the integrated kernel estimator which is a smooth estimator of the distribution function. This smooth version of a monotone estimator is used for testing the monotony and the unimodality of a density in Sections 3.5 and 4.6. The splines estimators are often more variable than kernel estimators.

With right-censored observations, the cumulative intensity Λ of a sample of the time variable T has an estimator of the form $\widehat{\Lambda}_n(t) = \int_0^t Y_n^{-1} \, dN_n$ where N_n is the counting process of the uncensored variables observed until t and Y_n is the process of the variables $T \wedge C$ still unobserved at t. Its derivative λ is estimated by a kernel estimator smoothing $\widehat{\Lambda}_n$ and a smooth isotonic estimator of a monotone or concave function λ is obtained like for a density. They satisfy similar properties described in Section 3.5 for the density and the same tests about their form have the same asymptotic properties.

Smooth isotonic estimators of nonparametric and semi-parametric regression functions and also defined in Sections 3.8 and 4.8. They apply to compare monotone regression functions or the relative monotone curves of k sub-samples, through the estimators $\widehat{m}_{k,n,h} \circ \widehat{m}_{k,n,h}^{-1}$, for $j = 1, \ldots, k-1$.

1.6 Asymptotic levels of tests

Consider the test of a simple hypothesis $\Theta_0 = \{\theta_0\}$ against the alternative $\Theta_1 = \Theta \setminus \{\theta_0\}$, a statistic T_n satisfying the weak law of large numbers under P_{θ_0} provides a test with an asymptotic threshold c_α at the nominal level $\alpha = P_{\theta_0}\{T > c_\alpha\}$. The critical value c_α for the limit of T_n is the limit of critical value $c_{n,\alpha}$ of T_n.

Let $(X_i)_{i=1,\ldots,n}$ be a n-sample of a variable X, a log-likelihood ratio test statistic for the distribution of X is written as a sum $S_n = \sum_{i=1,\ldots,n} \varphi(X_i)$ such that

$$\lim_{n \to \infty} n^{-1} S_n = E_0 \left\{ \log \frac{f_1}{f_0}(X) \right\} = -K(f_0, f_1)$$

where $K(f_0, f_1)$ is the Kullback-Leibler information. The Laplace transform $\psi(t) = E\{f_1(X)f_0^{-1}(X)\}^t$ of $\varphi(X) = \log\{f_1(X)f_0^{-1}(X)\}$ at t equals $\psi_0(t) = \int (f_1^t f_0^{1-t})$ under H_0 and $\psi_1(t) = \int (f_0^{-t} f_1^{1+t})$ under an alternative H_1 with density f_1.

Chernov's theorem yields the limit of the level of the test as n tends to infinity and $n^{-1} c_{n,\alpha}$ converges to c_α

$$\log P_0\{S_n > c_{n,\alpha}\} = \log P_0\{n^{-1}(S_n - E_0 S_n) > n^{-1} c_{n,\alpha} + K(f_0, f_1)\}$$

$$\sim -n I_0(c_\alpha + K(f_0, f_1)),$$

where the bound $I_0(c_\alpha + K(f_0, f_1))$ is the limit of the Cramer transform

$$I_0(n, c_\alpha + K(f_0, f_1)) = \inf_{\lambda > 0}[\lambda\{c_\alpha + K(f_0, f_1)\} - n\log E_0 e^{\lambda\{\varphi(X) - E_0\varphi(X)\}}]$$

$$= \inf_{\lambda > 0}[\lambda c_\alpha - n\log\psi_0(\lambda)] = I_0(n, c_\alpha).$$

As n tends to infinity, the sequence $(c_{\alpha,n})_{n \geq 1}$ satisfies the asymptotic equivalence

$$\alpha \sim \exp\{-nI_0(n^{-1}c_{n,\alpha})\} \sim \exp\{-nI_0(c_\alpha)\}.$$

Under an alternative K, its power $\beta_n(\alpha) = P_K\{S_n > c_{n,\alpha}\}$ has a similar asymptotic behavior

$$\beta_n(\alpha) \sim \exp\{-nI_1(n^{-1}c_{n,\alpha})\}.$$

It is determined under the alternative by the Cramer transform at $c_\alpha - K(f_1, f_0)$, where the Kullback-Leibler information is such that

$$K(f_1, f_0) = E_1\left\{\log\frac{f_1}{f_0}(X)\right\} = \lim_{n \to \infty} n^{-1}S_n,$$

under the alternative, then

$$I_1(n, c_\alpha - K(f_1, f_0)) = \inf_{\lambda > 0}[\lambda\{c_\alpha - K(f_1, f_0))\} - n\log E_1 e^{\lambda\{\varphi(X) - E_1\varphi(X)\}}]$$

$$= \inf_{\lambda > 0}[\lambda c_\alpha - \log\psi_1(\lambda)]) = I_1(n, c_\alpha).$$

Let X be a random variable with a density f_0 under a simple hypothesis H_0 and let $K_n : f_n(x) = f_0(x)\{1 + n^{-\frac{1}{2}}\gamma_n(x)\}$ be a sequence of local alternatives defining the density of X from a sequence of functions $(\gamma_n)_{n \geq 1}$ converging uniformly to a limit γ of $L^2(f_0)$ and such that $\int_{\mathbb{R}} \gamma f_0 = 0$. The Kullback-Leibler information $K(f_n, f_0)$ is approximated as

$$\int_{\mathbb{R}} \log(1 + n^{-\frac{1}{2}}\gamma_n)f_n = \int_{\mathbb{R}} \{n^{-\frac{1}{2}}\gamma_n - (2n)^{-1}\gamma_n^2\}f_n + o(n^{-1})$$

and by the constraint $\int_{\mathbb{R}} \gamma f_0 = 0$

$$K(f_n, f_0) = (2n)^{-1}\int_{\mathbb{R}} \gamma^2 f_0 + o(n^{-1})$$

and $K(f_0, f_n)$ is asymptotically equivalent to $-K(f_n, f_0)$. The limits of $n^{-1}S_n$ under H_0 and the local alternative depend on $\psi_{0,n}(t) = \int_{\mathbb{R}}(1 + n^{-\frac{1}{2}}t\gamma)\, dF_0$ and, respectively, $\psi_{1,n}(t) = \int_{\mathbb{R}}\{1 + n^{-\frac{1}{2}}(1 + t)\gamma\}\, dF_0$.

The likelihood ratio test statistic for a parametric hypothesis H_0 against the alternative H_1 of an upper parametric model for the density of X is

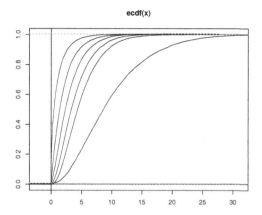

ecdf(x)

Fig. 1.4 Empirical distribution functions of χ_d^2 variables with $d = 1, 2, 3, 4, 5, 10$, simulated with 1000 replicates.

defined by maximizing the density of the sample $L_n(\theta) = \prod_{i=1,\dots,n} f_\theta(X_i)$ under the assumptions H_0 and H_1. It is written as

$$T_n = 2\log\frac{\sup_{\theta\in\Theta_1} L_n(\theta)}{\sup_{\theta\in\Theta_0} L_n(\theta)} = 2\log\frac{L_n(\widehat{\theta}_{1n})}{L_n(\widehat{\theta}_{0n})} = 2\sum_{i=1}^{n}\log\frac{f_{\widehat{\theta}_{1n}}}{f_{\widehat{\theta}_{0n}}}(X_i).$$

The parametric log-likelihood ratio statistic T_n has a χ_d^2 limiting distribution as n tends to infinity, where d is the difference between the dimensions of the parameter spaces under the alternative and the null hypothesis. This is the distribution of the sum of the squares of d independent normal variables $\mathcal{N}(0,1)$, its mean is d and its variance is $2d$. The density of a χ_d^2 variable is the Gamma density

$$\Gamma_\nu(x) = \frac{1}{2^{\frac{\nu}{2}}\Gamma(\frac{\nu}{2})}x^{\frac{\nu-2}{2}}e^{-\frac{x}{2}}, \; x > 0.$$

The asymptotic power of the likelihood ratio test againt sequences of local alternatives is considered in the next chapter.

The accuracy of the asymptotic critical values of tests based on a finite sample size depends on the accuracy of the approximation of the test statistic by a limit under the hypothesis. In the asymptotic theory, the statistic T_n converges weakly to a random variable T or, equivalently, the distribution function of T_n converges uniformly to the distribution function of T as n tends to infinity. By their inversion, the quantiles of T_n which determine the critical values of the test converges to those of T. The asymptotic threshold c of the test satisfies

$$\alpha = \lim_{n\to\infty} \sup_{F\in H_0} P_F\{T_n > c_n(\theta)\} = \sup_{F\in H_0} P_F\{T > c(\theta)\}.$$

The asymptotic power of the test is then

$$\beta(\alpha) = \lim_{n\to\infty} \sup_{F\in H_0} \inf_{G\in K_F} P_\theta\{T_n > c_n\}.$$

A two-sided test defined by a statistic T_n such that the limiting distribution Φ of T_n is symmetric has symmetric asymptotic critical values

$$c_{\frac{\alpha}{2}} = (1-\Phi)^{-1}\left(\frac{\alpha}{2}\right),$$

$$-c_{\frac{\alpha}{2}} = 1 - c_{1-\frac{\alpha}{2}} = \Phi^{-1}\left(\frac{\alpha}{2}\right).$$

By the weak convergence of T_n, for every $\varepsilon > 0$ and for n large enough, $|\Phi_n - \Phi|(c_{n,\frac{\alpha}{2}}) \leq \frac{\varepsilon}{2}$ and $|\Phi_n - \Phi|(-c_{n,\frac{\alpha}{2}}) \leq \frac{\varepsilon}{2}$, which implies

$$|\alpha - 1 + \Phi(c_{n,\frac{\alpha}{2}}) - \Phi(-c_{n,\frac{\alpha}{2}})| \leq \varepsilon$$

and the accuracy of the approximation of $c_{n,\frac{\alpha}{2}}$ by $c_{\frac{\alpha}{2}}$ is determined by the 4th order expansion of $1 - \Phi(x) + \Phi(-x)$

$$|(c_{n,\frac{\alpha}{2}} - c_{\frac{\alpha}{2}})^2 \Phi^{(2)}(c_{\frac{\alpha}{2}}) - \frac{1}{12}(c_{n,\frac{\alpha}{2}} - c_{\frac{\alpha}{2}})^4 \Phi^{(4)}(c_{\frac{\alpha}{2}}) + o((c_{n,\frac{\alpha}{2}} - c_{\frac{\alpha}{2}})^4)| \leq \varepsilon,$$

it depends mainly on the derivatives of Φ and on the accuracy of the approximation of Φ_n by Φ at $c_{\frac{\alpha}{2}}$.

1.7   Permutation and bootstrap tests

Under some hypotheses, the observations are invariant by permutation of their rank in the sample and the critical values of the tests are defined as the mean of statistics under all possible permutations of the observations under H_0. This procedure can be used for tests of homogeneity of the distributions of sub-samples or tests for symmetry of a density against various alternatives. Fattorini, Greco and Naddeo (2002) compared the power of the Kolmogorov-Smirnov test and the chi-square test with permutations for the comparison of the distribution functions of two samples, their simulations studies prove that the Kolmogorov-Smirnov test is the most powerful in all cases.

The invariance property is not generally satisfied, nevertheless the bootstrap tests apply in many cases. The bootstrap is a resampling procedure used in practice to provide approximations of the quantiles and the p-values of the tests with finite sample sizes or with non standard limiting distributions. A bootstrap sample $X^* = (X_1^*, \ldots, X_n^*)$ is drawn with the empirical distribution $\widehat{F}_n$ of the original sample $X_1, \ldots, X_n$ of X from the set $\{X_1, \ldots, X_n\}^{\otimes n}$ and bootstrap estimators and statistics are approximated from a sample $(X^{*b})_{b=1,\ldots,B}$ of the n-dimensional bootstrap variable X^*, with a large size B. The bootstrap estimator of the mean of X is $\bar{X}_{n,B}^* = B^{-1} \sum_{i=1}^{B} \bar{X}^{*b}$, where $\bar{X}^{*b}$ is the empirical mean of the bootstrap sample X^{*b}, for $b = 1, \ldots, B$. The bootstrap estimator of the variance of X is $Var_{n,B}^* X = (B-1)^{-1} \sum_{i=1}^{B} (\bar{X}^{*b} - \bar{X}_{n,B}^*)^2$. The bootstrap estimators of the mean and variance of $\widehat{\theta}_n$ for a parameter θ or $\widehat{\varphi}_n$ for a function φ are calculated in the same way, from the estimators $\widehat{\theta}_n^* b$ and $\widehat{\varphi}_n^* b$.

When the exact or asymptotic variance of a statistic T_n can be calculated and estimated, a pivot statistic $T_n(\widehat{Var}T_n)^{-\frac{1}{2}}$ is considered for a bootstrap test rather than a normalization by bootstrap of the bootstrapped statistic, i.e. $T_{n,B}^*(Var_{n,B}^* T_n)^{-\frac{1}{2}}$. The normalization by bootstrap is however useful for complex variances and small sample sizes, when the asymptotic variance is not sufficiently accurate.

For a statistic T_n with distribution function F_n, $U = F_n(T_n)$ has a uniform distribution $\mathcal{U}_{[0,1]}$, the exact quantiles of a bootstrap test for a statistic T_n are estimated by those of the empirical distribution of T_n. The distribution function of the pivot statistic is approximated by bootstrap and their quantiles provide bootstrap estimators of the critical values and of the power functions of a test. The estimated level and power functions are consistent under the conditions of weak convergence of the statistic and under regularity conditions (Beran 1986, 1988). In multi-sample tests, the size of each sub-sample increases with the total sample size n and the bootstrap mimics the sampling procedure by drawing together $k \geq 2$ independent bootstrap sub-samples with empirical distributions $\widehat{F}_{jn}$, for $j = 1, \ldots, k$ and the bootstrap estimators are calculated similarly in each sub-sample.

The bootstrap distribution is a better approximation of the distribution function of the statistic than the asymptotic normal approximation (Singh 1981, Abramovitch and Singh 1985, Hall 1986): From the Berry-Essen bound for the empirical mean of a sample of a variable X with mean μ, variance σ^2 and third moment μ_3, there exists a constant K such that $\|P(n^{\frac{1}{2}}(\bar{X}_n - \mu) \leq x) - \Phi(\sigma^{-1}x)\|_\infty \leq Kn^{-\frac{1}{2}}\sigma^{-3}(E(|X - \mu|^3)$, hence $\|P_n^*(n^{\frac{1}{2}}(\bar{X}_n^* - \bar{X}_n) \leq x) - \Phi(\widehat{\sigma}^{-1}x)\|_\infty \leq Kn^{-\frac{1}{2}}\widehat{\sigma}^{-3}(E_n^*(|\bar{X}_n^* - \bar{X}_n|^3)$ for the bootstrap sample and

$$\|P_n^*(n^{\frac{1}{2}}(\bar{X}_n^* - \bar{X}_n) \le x) - P(n^{\frac{1}{2}}(\bar{X}_n - \mu) \le x)\|_\infty = O(n^{-\frac{1}{2}}). \qquad (1.9)$$

The quantiles of the distribution function $F_n(x) = P(n^{\frac{1}{2}}(\bar{X}_n - \mu) \le x)$ and of its bootstrap satisfy

$$\|P(n^{\frac{1}{2}}\{\widehat{F}_n^{-1}(t) - F^{-1}(t)\} \le x) - P_n^*(n^{\frac{1}{2}}\{\widehat{F}_n^{*-1}(t) - \widehat{F}_n^{-1}(t)\}$$
$$\le x)\|_\infty = O(n^{-\frac{1}{2}}). \qquad (1.10)$$

This implies the consistency of the level and of the power of bootstrap tests based on statistics $T_n = \psi(\widehat{F}_n)$, with functionals ψ in $C^2(\mathcal{F})$, defined on a metric space $\mathcal{F}$ of distribution functions.

1.8   Relative efficiency of tests

The properties of the tests are described by several notions of efficacy. With finite sample sizes, the relative efficiency of two tests is related to the sample sizes required to obtain the same power. In tests for a simple hypothesis, the asymptotic powers of two tests based on statistics S_{1n} and S_{2n} having asymptotic Gaussian distributions $\mathcal{N}(\mu_1, \sigma_1^2)$ and $\mathcal{N}(\mu_2, \sigma_2^2)$ are $1 - \Phi(c_\alpha - \sigma_1^{-1}\mu_1)$ and $1 - \Phi(c_\alpha - \sigma_2^{-1}\mu_2)$ and the asymptotic relative efficiency of S_{2n} with respect to S_{1n} is defined by the ratio

$$e = \left(\frac{\mu_2\sigma_1}{\mu_1\sigma_2}\right)^2.$$

A test based on S_{1n} is more powerful than a test based on S_{2n} if the relative asymptotic efficiency e is strictly larger than 1.

In a parametric family, a statistic S_θ having an asymptotic Gaussian distribution $\mathcal{N}(\mu_\theta, \sigma_\theta^2)$ yields a test with a Pitman efficiency

$$e_\theta = \dot{\mu}_\theta \sigma_\theta^{-1}.$$

The relative efficiency of two parametric tests defined for the same parameter set Θ is the ratio of their Pitman efficiencies and the most powerful test has the larger Pitman efficiency e_θ. Hodges and Lehmann (1961) compared the Pitman efficiency of several rank tests defined as score tests in parametric models. This notion is only defined for asymptotically Gaussian distributions. The efficiency and the optimality have been defined for parametric and nonparametric tests from the ratio of the bound in Chernov's theorem and the Kullback-Leibler information number (Nikitin 1984, 1987). This notion is also related to the sample size. The Hodges-Lehmann efficiency of a statistic T_{1n} with respect to a statistic T_{2n} is the ratio of the sample sizes $e = n_2 n_1^{-1}$ required for two samples in order that tests performed with the statistics T_{1n_1} and T_{2n_2} have the same power, for finite sample sizes. For asymptotically normal statistics, the power of a statistic depends on the ratio of the difference of its means under the hypothesis and the alternative over its standard deviation under the alternative and on the ratio of its standard deviations. Normal statistics T_{1n} and T_{2n}, have the same power at the level α if

$$\frac{\sigma_{1n_1}^{(0)}}{\sigma_{1n_1}^{(1)}} c_\alpha + \frac{\mu_{1n_1}^{(0)} - \mu_{1n_1}^{(1)}}{\sigma_{1n_1}^{(1)}} = \frac{\sigma_{2n_2}^{(0)}}{\sigma_{2n_2}^{(1)}} c_\alpha + \frac{\mu_{2n_2}^{(0)} - \mu_{2n_2}^{(1)}}{\sigma_{2n_2}^{(1)}}.$$

The parametric theory of statistical tests is presented in the next chapter and it is generalized to the nonparametric tests defined by differentiable functionals of empirical distribution functions and by estimators of the densities under the hypothesis and alternatives. The comparison of the limiting distributions of statistics relies on their Edgeworth expansions depending on their moments, they have been calculated for one and two-sample tests using the expansion for U-statistics (Bickel 1974, Albers, Bickel and van Zwet 1976, Bickel and van Zwet 1978, Serfling 1980) and on their behavior in contiguous alternatives (LeCam, 1956). Applications to the comparison of tests and to their optimality are considered in several parts of the book. In several cases, the asymptotic equivalence of tests is established under the hypothesis and under the local alternatives in neighborhoods of the hypothesis, though they are expressed in different forms. The sequential tests are often presented as more powerful than tests with a single sample and they have been widely used from the first results established by Wald and many others. Several sequential tests are defined and studied in the last chapter.

The local alternatives are mostly defined by small additive perturbations of the functions of the null hypothesis. In models with a change of scale under the alternative, the perturbations are multiplicative. When the functions of H_0 are unknown, the perturbations are chosen with the same order as the order of their estimators. In parametric models and models of cumulative intensities or distribution functions, their order is $n^{\frac{1}{2}}$ and LeCam's theory (1956) apply. For density functions, regression functions or intensities of C^s, their order is $n^{\frac{s}{2s+1}}$ and LeCam's results for the log-likelihood ratios are modified.

The tests presented in this book are robust in the sense that thay are valid in large nonparametric or semi-parametric models. They may be sensitive to outliers such as the kernel estimators, and the optimal local bandwidths constitute a practical improvement of the estimators and tests in reducing their global bias and variance in the presence of outliers so that they have only a local effect. The bias corrections are also robust procedures. Outlier observations can be detected in the least-squares errors of adjustment and their influence is reduced by the classical robustness procedures by truncating the range of the observations. Such procedures modify the distribution of the sample (cf. the censoring models) and introduce a bias. In measurement error models, the outliers are explained by an additional measurement error and the models are modified as convolution models.

Chapter 2

Asymptotic theory

2.1 Parametric tests

Consider a parametric set Θ in a d-dimensional metric space and a family of probabilities $\mathcal{P}_\Theta = \{P_\theta; \theta \in \Theta\}$ in a measurable space $(\Omega, \mathcal{A})$. Let Θ_0 be a subset of Θ and $\Theta_1 = \Theta \setminus \Theta_0$. A parametric hypothesis and its alternative in Θ are expressed as $H_0 : \theta$ belongs to Θ_0 and an alternative $K : \theta$ belongs to Θ_1. For every θ in Θ the mean and the variance of a statistic T_n under a probability P_θ in $\mathcal{P}_\Theta$ are denoted

$$\mu_{n,\theta} = E_\theta T_n, \quad \sigma^2_{n,\theta} = Var_\theta T_n = E_\theta(T_n^2) - \mu^2_{n,\theta}.$$

The asymptotically Gaussian statistics are centered and normalized under the hypothesis, in the form

$$U_{n,\theta} = \sigma^{-1}_{n,\theta}(T_n - \mu_{n,\theta})$$

in order to provide asymptotically free tests. Other free distributions which appear commonly as the asymptotic distributions of test statistics are the χ^2 distributions and the Fisher distributions. The χ^2 variables are not centered since their mean equals their degree of freedom, and they are not normalized. The Fisher variables are the ratio of two χ^2 variables normalized according to their degrees of freedom.

Let $F_{n,\theta}$ be the distribution function of $U_{n,\theta}$ under P_θ, for every θ in Θ. Under H_0, there exists θ in Θ_0 such that P_θ is the probability of the observed variable in the space $(\Omega, \mathcal{A})$. An one-sided test of level α based on a pivot U_n has a rejection domain $D_{n,\alpha} = \{U_{n,\theta} > c_{n,\alpha}\}$, and the level of the test is then $\alpha = 1 - F_{n,\theta}(c_{n,\alpha})$. This test is valid when $\theta_n > \theta$ for every parameter θ_n of the alternative and for every parameter θ of H_0.

For a pivot statistic T_n with distribution function $F_{n,\theta}$ under H_0, we suppose that $F_{n,\theta}(x)$ converges to $F_\theta(x)$ for every x, then the critical value of the test converges to $c_\alpha = F_\theta^{-1}(1 - \alpha)$.

Under a sequence of alternatives $(K_n)_{n\geq 1}$ with parameters $(\theta_n)_{n\geq 1}$ in a neighborhood Θ_{1n} of Θ_0, the mean and the variance of T_n are denoted μ_{n,θ_n} and σ^2_{n,θ_n}, with θ_n converging to θ in Θ_0. The convergence of the power of the one-sided test for the hypothesis H_0 against $(K_n)_{n\geq 1}$ requires conditions for the convergence of the statistic under P_θ, for every θ in Θ_{1n}. The mean and the variance of T_n are supposed to satisfy the following conditions, for every θ in Θ_0 and for every θ_n in Θ_{1n}.

Condition 2.1.

(1) There exist a strictly positive sequence $(a_n)_{n\geq 1}$ and functions μ_θ and σ_θ in $C^1(\Theta)$ such that $\mu_\theta = \lim_{n\to\infty} a_n^{-1}\mu_{n,\theta}$ and $\sigma_\theta = \lim_{n\to\infty} a_n^{-1}\sigma_{n,\theta}$ are finite.
(2) $F_{n,\theta}(x)$ converges to $F_\theta(x)$ for every x, and for every θ in Θ, and $F_\theta(x)$ is continuous in Θ, for every x.
(3) For every θ in Θ_0, $\eta_\theta = \lim_{n\to\infty} a_n(\theta_n - \theta)$ is finite.

Let $\Theta_0 = \{\theta\}$ and let $\Theta_{1,n}$ be such that $\theta_n > \theta$ for every θ_n in $\Theta_{1,n}$.

Definition 2.1. The tangent space of Θ_0 defined by the alternative is

$$H_\theta = \{\eta_\theta > 0; \exists \theta_n \in \Theta_{1,n} : \eta_\theta = \lim_{n\to\infty} a_n(\theta_n - \theta)\}.$$

The transpose of a vector u is denoted u^t.

Proposition 2.1. *Under Condition 2.1, the asymptotic power of the one-sided test for* $\Theta_0 = \{\theta\}$ *against* $\Theta_{1,n}$ *is*

$$\beta_\theta(\alpha) = 1 - \sup_{\eta\in H_\theta} F_\theta(c_\alpha - \eta_\theta^t \dot{\mu}_\theta \sigma_\theta^{-1}),$$

with the critical value $c_\alpha = \lim_{n\to\infty} c_{n,\alpha}$.

Proof. Condition (2) is equivalent to the weak convergence of T_n to a variable T with the distribution function F_θ under P_θ, θ in Θ_0. Under K_n, an expansion of the mean of T_n

$$\mu_{n,\theta_n} - \mu_{n,\theta} = (\theta_n - \theta)^t \dot{\mu}_{n,\theta} + o(1)$$

implies the expansions of U_{n,θ_n} and U_{θ_n} as

$$U_{n,\theta_n} = \frac{\sigma_{n,\theta}}{\sigma_{n,\theta_n}}\left(U_{n,\theta} - \frac{(\theta_n - \theta)^t \dot{\mu}_\theta}{\sigma_{n,\theta}}\right) + o(1),$$

$$U_{n,\theta} = \frac{\sigma_{n,\theta_n}}{\sigma_{n,\theta}}\left(U_{n,\theta_n} + \frac{(\theta_n - \theta)^t \dot{\mu}_\theta}{\sigma_{n,\theta}}\right) + o(1).$$

From this expansion and the conditions, the power of the test in $\Theta_{1,n}$ is the minimum in H_θ of

$$P_{\theta_n}\left(U_{n,\theta} > c_{n,\alpha}\right) = P_{\theta_n}\left(U_{n,\theta_n} > c_{n,\alpha} - \frac{(\theta_n - \theta)^t \dot{\mu}_\theta}{\sigma_{n,\theta}}\right) + o(1)$$

$$= 1 - F_{\theta_n}\left(c_{n,\alpha} - \frac{\eta_\theta^t \dot{\mu}_\theta}{\sigma_\theta}\right) + o(1)$$

$$= 1 - F_\theta\left(c_\alpha - \frac{\eta_\theta^t \dot{\mu}_\theta}{\sigma_\theta}\right) + o(1).$$

$\square$

Let $U_{1,n,\theta}$ and $U_{2,n,\theta}$ be two statistics of one-sided tests of level α for a simple hypothesis $H_0 : \{\theta\}$, against local alternatives $(K_n)_{n\geq 1}$, with limiting mean functions $\mu_{k,\theta}$ and variance functions $\sigma_{k,\theta}$, $k = 1, 2$, in Condition 2.1.(1). The test $\phi_{1,n,\theta}$ based on $U_{1,n,\theta}$ is asymptotically more powerful than the test $\phi_{2,n,\theta}$ based on $U_{2,n,\theta}$ if for all η_θ in H_θ

$$\eta_\theta^t\left(\frac{\dot{\mu}_{1,\theta}}{\sigma_{1,\theta}} - \frac{\dot{\mu}_{2,\theta}}{\sigma_{2,\theta}}\right) < 0,$$

equivalently, for every η in H_θ

$$\frac{\eta_\theta^t \dot{\mu}_{2,\theta}}{\eta_\theta^t \dot{\mu}_{1,\theta}} \frac{\sigma_{2,\theta}}{\sigma_{1,\theta}} > 1. \tag{2.1}$$

In $\mathbb{R}$, this inequality is equivalent to

$$e_1(\theta) < e_2(\theta)$$

for their Pitman efficiency $e(\theta) = \sigma_\theta^{-1}\dot{\mu}_\theta$.

The tangent space of $\Theta_0 = \{\theta\}$ against a two-sided local alternative $\Theta_{1,n}$ is defined by

$$H_\theta = \{\eta; \exists \theta_n \in \Theta_{1,n} : \eta = \lim_{n\to\infty} a_n(\theta_n - \theta)\}. \tag{2.2}$$

The rejection domain of tests is $D_{n,\alpha} = \{|U_{n,\theta}| > c_{n,\alpha}\}$, with the critical value $c_{n,\alpha}$ converging to c_α, it has the asymptotic level

$$\alpha = 1 - F_\theta(c_{\frac{\alpha}{2}}) + F_\theta(-c_{\frac{\alpha}{2}}).$$

Proposition 2.2. *Under Condition 2.1, the asymptotic power of the two-sided test for $\Theta_0 = \{\theta\}$ against $\Theta_{1,n}$ is*

$$\beta_\theta(\alpha) = \inf_{\eta\in H_\theta}\{1 - F_\theta(c_{\frac{\alpha}{2}} - \eta^t \dot{\mu}_\theta\sigma_\theta^{-1}) + F_\theta(-\eta^t\dot{\mu}_\theta\sigma_\theta^{-1} - c_{\frac{\alpha}{2}})\}.$$

The expansion of U_{n,θ_n} is the same as in the proof of Proposition 2.1 and the result follows from the approximation of $P_{\theta_n}(U_{n,\theta} > c)$ which is valid for every real c. The asymptotic relative efficiency of tests based on $U_{1,n,\theta}$ and $U_{2,n,\theta}$ is still given by (2.1).

The parameter of a test of a single hypothesis is known but it has to be estimated for a composite hypothesis H_0 defined by parameters in a set Θ_0. The test statistic is

$$U_{n,\widehat{\theta}_n} = \sigma^{-1}_{n,\widehat{\theta}_n} (T_n - \mu_{n,\widehat{\theta}_n}).$$

Let K_n be a local alternative defined by a parameter set $\Theta_{1,n}$ such that $\theta_n \geq \theta$ for all θ_n in $\Theta_{1,n}$ and θ in Θ_0. The tangent space at θ in Θ_0 defined by the alternative is H_θ given by (2.2).

Proposition 2.3. *Let $\widehat{\mu}_n$ and $\widehat{\sigma}_n$ be consistent estimators of the mean and the variance of T_n under the hypothesis. Under Condition 2.1 and if, under every P_θ of H_0, $\lim_{n\to\infty} a_n^{-1}\widehat{\sigma}_n = \sigma_\theta$ in probability and $a_n(\widehat{\mu}_n - \mu_\theta)$ converges weakly to a variable X_θ, the asymptotic level of the one-sided test of H_0 against K_n based on $U_{n,\widehat{\theta}_n}$ is $\alpha = 1 - \inf_{\theta\in\Theta_0} G_\theta(c_\alpha)$, with the convolution of the distribution functions $G_\theta = F_{\sigma_\theta^{-1}X_\theta} * F_\theta$. The asymptotic local power of the test is $\beta_\theta(\alpha) = 1 - \inf_{\theta\in\Theta_0} \sup_{\eta\in H_\theta} G_\theta(c_\alpha - \eta^t \mu_\theta \sigma_\theta^{-1}).$*

Proof.   Under the conditions,

$$\widehat{\sigma}_n^{-1}(\mu_{n,\widehat{\theta}_n} - \mu_{n,\theta}) = \sigma_\theta^{-1} a_n(\mu_{n,\widehat{\theta}_n} - \mu_{n,\theta}) + o_p(1),$$

it converges weakly to $\sigma_\theta^{-1}X_\theta$ under P_θ, θ in Θ_0. The level of the test is

$$\alpha = \sup_{\theta\in\Theta_0} \lim_{n\to\infty} P_\theta(U_{n,\theta} > c_{n,\alpha} + X_\theta) = 1 - \inf_{\theta\in\Theta_0} F_{Z_\theta} * F_\theta(c_\alpha)$$

where the convolution G_θ of F_θ with the distribution function of the variable $Z_\theta = \sigma_\theta^{-1}X_\theta$ is the limiting distribution under H_0 of the statistic $U_{n,\widehat{\theta}_n}$. The local asymptotic power of the test is deduced like in the proof of Proposition 2.1, with the convolution function $F_{Z_\theta} * F_\theta$ instead of F_θ.   □

Proposition 2.3 is extended to two-sided tests like in Proposition 2.2.

Proposition 2.4. *Under the conditions of Proposition 2.3, the asymptotic power of the two-sided test for Θ_0 against $\Theta_{1,n}$ is*

$$\beta_\theta(\alpha) = \sup_{\theta\in\Theta_0} \inf_{\eta\in H_\theta} \{1 - G_\theta(c_{\frac{\alpha}{2}} - \eta^t \mu_\theta \sigma_\theta^{-1}) + G_\theta(-\eta^t \mu_\theta \sigma_\theta^{-1} - c_{\frac{\alpha}{2}})\}.$$

2.2   Parametric likelihood ratio tests

Let $X_1, \ldots, X_n$ be a sample with distribution function F belonging to a class $\mathcal{G}$ of distribution functions. Let $\mathcal{F} = \{F_\theta, \theta \in \Theta\}$ be a parametric subset of $\mathcal{G}$ indexed by an open Θ set of $\mathbb{R}^{d_1}$, $d_1 \geq 1$. The optimal parametric goodness of fit test of the hypothesis $H_0 : F$ belongs to $\mathcal{F}$ is the likelihood ratio test for the parametric class of functions $\mathcal{F}$. The general alternative of the tests are $K : F$ belongs to $\mathcal{G} \setminus \mathcal{F}$ and a semi-parametric Kolmogorov-Smirnov test is optimal if $\mathcal{G}$ is nonparametric.

Assuming that F has a uniformly continuous density f, the likelihood ratio test for the density of H_0 against an alternative K is defined by maximizing the density ratio of the sample under K and H_0

$$T_n = 2 \log \frac{\sup_{f:F\in\mathcal{G}} \prod_{i=1,\ldots,n} f(X_i)}{\sup_{f:F\in\mathcal{F}} \prod_{i=1,\ldots,n} f(X_i)}.$$

A parametric alternative is defined by a space of parametric distribution functions $\mathcal{H} = \{F_\gamma, \gamma \in \Gamma\}$ in $\mathcal{G}$ indexed by a d_2-dimensional parameter γ belonging to a set Γ and such that $\mathcal{F}$ is a sub-model of $\mathcal{H}$ where components of the parameter γ are zero under the hypothesis and the other components of γ belong to Θ. The alternative is then $K : F$ belongs to $\mathcal{H} \setminus \mathcal{F}$ and the statistic is written

$$T_n = 2 \sum_{i=1}^{n} \{\log f_{\widehat{\gamma}_n}(X_i) - \log f_{\widehat{\theta}_n}(X_i)\},$$

where $\widehat{\gamma}_n = \arg\max_{\gamma\in\Gamma} \sum_{i=1}^{n} \log f_\gamma(X_i)$ is the estimator of the parameter in Γ and $\widehat{\theta}_n$ is the estimator of the parameter in Θ. Under H_0, there exists a parameter value θ_0 belonging to the interior of Θ such that X has the density $f_0 = f_{\theta_0}$.

Let $y^{\otimes 2} = yy^t$ be the tensor product of a vector y with transpose y^t. Under the assumptions that every $F = F_\theta$ of $\mathcal{F}$ has a twice continuously differentiable density f_θ with respect to θ and that the Fisher information matrix $I_\theta = -E_\theta\{f_\theta^{-1}(X)\dot{f}_\theta(X)\}^{\otimes 2}$ is finite and nonsingular, the log-likelihood is locally concave at the true parameter value which implies that $\widehat{\theta}_n$ is a consistent estimator of θ_0 under H_0. The first two derivatives of the log-likelihood at f_0 are written as

$$U_n = n^{-\frac{1}{2}} \sum_{i=1,\ldots,n} \frac{\dot{f}_{\theta_0}}{f_{\theta_0}}(X_i),$$

$$-I_n = n^{-1} \sum_{i=1,\ldots,n} \left\{ \frac{\ddot{f}_{\theta_0}}{f_{\theta_0}}(X_i) - \left(\frac{\dot{f}_{\theta_0}}{f_{\theta_0}}\right)^{\otimes 2}(X_i) \right\},$$

By a Taylor expansion of the derivative $\dot{f}_\theta$ in a neighborhood of θ_0, the estimator of the parameter under H_0 has the expansion

$$n^{\frac{1}{2}}(\widehat{\theta}_n - \theta_0) = I_n^{-1} U_n + o_p(1).$$

Since $\mathcal{H}$ includes $\mathcal{F}$, the true parameter value γ_0 in Γ under H_0 can be written as a vector with components including θ_0 and other parameters with value zero, i.e. $\gamma_0 = (\theta_0^t, 0^t)^t$. Under distributions of $\mathcal{H}$, the notations for the first two derivatives of the log-likelihood are replaced by $\widetilde{U}_n$ and $\widetilde{I}_n$ in $\mathbb{R}^{d_2}$, their d_1 first components are respectively U_n and I_n and they have expansions like U_n and I_n at γ_0, under H_0. Under the same conditions as under H_0, the estimator under the alternative has the expansion $n^{\frac{1}{2}}(\widehat{\gamma}_n - \gamma_0) = \widetilde{I}_n^{-1}\widetilde{U}_n + o_p(1)$, when the variable X has the distribution function F_γ.

For every θ in Θ and γ in Γ, the expansion of the log-likelihood ratio statistic relies on the inversion by blocks of the matrix $\widetilde{I}_n$ and its limit as

$$\widetilde{I} = \begin{pmatrix} \widetilde{I}_{11} & \widetilde{I}_{12} \\ \widetilde{I}_{21} & \widetilde{I}_{22} \end{pmatrix}. \text{ Let } d = \dim \Gamma - \dim \Theta > 0.$$

Proposition 2.5. *Under H_0, $T_n = Y_n^t Y_n + o_p(1)$ where Y_n is a d-dimensional vector of independent and centered variables with variance 1 and it converges weakly to a χ_d^2 variable. Let $(K_n)_{n\geq 1}$ be a sequence of local alternatives indexed by a sequence of parameters $(\gamma_n)_{n\geq 1}$ in sets $(\Gamma_n)_{n\geq 1}$ for which there exist θ in Θ and γ_a in K such that $\gamma_a = \lim_{n\to\infty} n^{\frac{1}{2}}(\gamma_n - \gamma)$ has its first d_1 components zero. Under K_n, the limiting distribution of the statistic T_n is $\chi_d^2 + \gamma_a^t \widetilde{I}_{22}(\gamma)\gamma_a$.*

Proof. From the consistency of the estimator under H_0 and expanding $\log f_{\widehat{\theta}_n} - \log f_{\theta_0} = \log\{1 + f_{\theta_0}^{-1}(f_{\widehat{\theta}_n} - f_{\theta_0})\}$ as n tends to infinity, we obtain

$$\frac{T_n}{2} = n^{\frac{1}{2}}(\widehat{\theta}_n - \theta_0)^t U_n - \frac{n}{2}(\widehat{\theta}_n - \theta_0)^t I_n(\widehat{\theta}_n - \theta_0) + o_p(1)$$

$$= \frac{1}{2} U_n^t I_n^{-1} U_n + o_p(1)$$

and a similar expansion is written in Γ, in a neighborhood of the true parameter value, for $\log f_{\widehat{\gamma}_n} - \log f_{\gamma_0} = \log\{1 + f_{\gamma_0}^{-1}(f_{\widehat{\gamma}_n} - f_{\gamma_0})\}$. The test statistic is then written

$$T_{n,n} = \widetilde{U}_n^t \widetilde{I}_n^{-1} \widetilde{U}_n - U_n^t I_n^{-1} U_n + o_p(1).$$

Let $\widetilde{U}_n = (U_n^t, \widetilde{U}_{n2}^t)^t$, its variance matrix $\widetilde{I}_n$ is split into blocks acccording to the components of the parameter inside or outside Θ (Section A.4). It follows that

$$T_n = (I_{n21} I_n^{-1} U_n - \widetilde{U}_{n2})^t A_n^{-1}(I_n^{-1} U_n - \widetilde{U}_{n2}) + o_p(1)$$

where the variance of the d-dimensional variable $I_{n21}I_n^{-1}U_n - \widetilde{U}_{n2}$ is A_n, hence T_n has an asymptotically free distribution and it converges weakly to a χ_d^2 variable.

Let $\gamma = (\theta^t, 0^t)^t$, under the sequence of local alternatives $(K_n)_{n\geq 1}$ the sequence of parameters of $(K_n)_{n\geq 1}$ is such that there exists a limit γ_a for the sequence $(n^{\frac{1}{2}}(\gamma_n - \gamma))_{n\geq 1}$. Since $f_\gamma = f_\theta$, an expansion of T_n under K_n is obtained from the expansions of $\log\{f_{\widehat{\gamma}_n}f_\theta^{-1}\} = \log\{f_{\gamma_n}f_\gamma^{-1}\} + \log\{1 + f_{\gamma_n}^{-1}(f_{\widehat{\gamma}_n} - f_{\gamma_n})\}$ and $\log\{f_{\widehat{\theta}_n}f_\theta^{-1}\} = \log\{1 + f_\theta^{-1}(f_{\widehat{\theta}_n} - f_\theta)\}$ as n tends to infinity. The statistic is now written

$$T_{n,n} = n\{(\widehat{\gamma}_n - \gamma_n)^t \widetilde{I}_n(\gamma_n)\,(\widehat{\gamma}_n - \gamma_n) + \gamma_a^t \widetilde{I}_n(\gamma_n)\gamma_a$$
$$- (\widehat{\theta}_n - \theta)^t I_n(\theta)\,(\widehat{\theta}_n - \theta)\} + o_p(1)$$
$$= \{I_{n21}(\gamma_n)I_n^{-1}U_n - \widetilde{U}_{n2}(\gamma_n)\}^t A_n^{-1}(\gamma_n)\{I_{n21}(\gamma_n)I_n^{-1}U_n - \widetilde{U}_{n2}(\gamma_n)\}$$
$$+ \gamma_a^t \widetilde{I}_n(\gamma_n)\gamma_a + o_p(1).$$

The variance of $I_{n21}(\gamma_n)I_n^{-1}U_n - \widetilde{U}_{n2}(\gamma_n)$ under the sequence of local alternatives is $A_n(\gamma_n)$ and γ_n tends to γ, so the asymptotic distribution of T_n under the sequence of local alternatives is $\chi_d^2 + \gamma_a^t \widetilde{I}(\gamma)\gamma_a$. Since the first components of γ_a are zero, it is considered as a parameter in $\mathbb{R}^d$ and $\gamma_a^t \widetilde{I}(\gamma)\gamma_a$ is also written $\gamma_a^t \widetilde{I}_{22}(\gamma)\gamma_a$. □

The critical value of the test of level α is c_α such that $\alpha = P(\chi_d^2 > c_\alpha)$ and its asymptotic power under the sequence of local alternatives $(K_n)_{n\geq 1}$ for which there exist θ in Θ and γ_a in K such that $\gamma_a = \lim_{n\to\infty} n^{\frac{1}{2}}(\gamma_n - \gamma)$ is

$$\beta_{\gamma,\gamma_a}(\alpha) = P\{\chi_d^2 > c_\alpha - \gamma_a^t \widetilde{I}_{22}\gamma_a\}.$$

It follows that $\inf_{\gamma,\gamma_a \in \Gamma} \beta_{\gamma,\gamma_a} > \alpha$, the test is therefore unbiased.

Let X be a real variable with a density f_θ indexed by a real parameter set Θ and such that for every x, $f_\theta(x)$ belongs to $C^2(\Theta)$. The Fisher information matrix of X is $I_\theta = \int_{\mathbb{R}} f_\theta^{-2} \dot{f}_\theta^2 \, dF_\theta$, with the first derivative $\dot{f}_\theta$ of f_θ. The mean of X is $\mu_\theta = \int x f_\theta(x)\, dx$ then its derivative μ'_θ equals $-\int(x - \theta)\dot{f}_\theta(x)\, dx$ and the Cauchy-Schwarz inequality entails the Rao-Blackwell inequality

$$\sigma_\theta^{-1}\dot\mu_\theta \leq I_\theta^{\frac{1}{2}},$$

where σ_θ^2 is the variance of X under F_θ.

Let us consider a parameter space defined by a constraint $\psi(\theta) = 0$, where ψ is a nonlinear real function defined $C^1(\mathbb{R}^{d_1})$ which split Θ into disjoint subspaces, $\Theta_1 = \{\theta \in \Theta : \psi(\theta) = 0\}$ and $\Theta_2 = \Theta \setminus \Theta_1$, and let $\mathcal{F} = \{F_\theta, \theta \in \Theta\}$ be the class of the distribution functions. The estimators of the parameters in Θ_1 and the tests for the null hypothesis $H_0 : \theta \in \Theta_1$ against the alternative $K : \theta \in \Theta_2$ are built according to the same arguments with indicators of the parameter sets. In the estimation procedure, the log-likelihood is replaced by the Lagrangian

$$l_n(\theta) = \sum_{i=1}^{n} \log f_\theta(X_i) - \lambda_n \psi(\theta),$$

where $\lambda = \lim_{n\to\infty} n^{-\frac{1}{2}} \lambda_n$ is zero only if the constraint does not apply, hence $l_n(\theta)$ is the log-likelihood in Θ_2.

Proposition 2.6. *Under H_0 and under the assumption of a nonsingular matrix $\dot\psi(\theta_0)^t \dot\psi(\theta_0)$, the LR statistic T_n has the asymptotic distribution*

$$T_0 = [Z - I^{-1}\dot\psi(\theta_0)^t U\{\dot\psi(\theta_0)^t \dot\psi(\theta_0)\}^{-1}]^t [Z - I^{-1}\dot\psi(\theta_0)^t U$$
$$\{\dot\psi(\theta_0)^t \dot\psi(\theta_0)\}^{-1}]1_{\{\psi(Z)=0\}} - Z^t Z 1_{\{\psi(Z)\neq 0\}},$$

where Z is a d-dimensional vector of independent normal variables and $U = IZ$. Let $(K_n)_{n\geq 1}$ be a sequence of local alternatives with parameters θ_n in Θ_2 such that there exist θ in Θ_1 and $a = \lim_{n\to\infty} n^{\frac{1}{2}}(\theta_n - \theta)$, the limiting distribution of the statistic T_n is $T = T_0 + a^t I(\theta)a$.

Proof. Under H_0, the first derivatives of $l_n(\theta)$ are

$$\dot l_n(\theta) = \sum_{i=1}^{n} f_\theta^{-1} \dot f_\theta(X_i) - \lambda_n \dot\psi(\theta) = n^{\frac{1}{2}} U_{n,\theta} - \lambda_n \dot\psi(\theta),$$
$$\ddot l_n(\theta) = n I_{n,\theta} - \lambda_n \ddot\psi(\theta),$$

and $\dot l_n$ is zero at $\widehat\theta_n$. By a first order Taylor expansion of $\dot l_n$ at θ_0 where ψ is zero,

$$n^{\frac{1}{2}}(\widehat\theta_n - \theta_0) = I_n^{-1}\{U_{n,\theta_0} - \lambda\dot\psi(\theta_0)\}1_{\{\psi(\widehat\theta_n)=0\}} + o_p(1).$$

Moreover $\dot l_n(\widehat\theta_n) = 0$ and the continuity of $\dot\psi$ imply

$$\lambda = \{\dot\psi(\theta_0)^t \dot\psi(\theta_0)\}^{-1} U_n^t \dot\psi(\theta_0) + o_p(1),$$

and the expansion of $n^{\frac{1}{2}}(\widehat\theta_n - \theta_0)$ is deduced. Under the alternative, the expansion of $n^{\frac{1}{2}}(\widehat\theta_n - \theta)$ has no constraint $n^{\frac{1}{2}}(\widehat\theta_n - \theta_0) = I_n^{-1}U_n 1_{\{\psi(\widehat\theta_n)\neq 0\}} + o_p(1)$. Let $Z_n = I_n^{-\frac{1}{2}} U_n$, the test statistic for H_0 against K is

$$T_n = [Z_n - I_n^{-1}\dot\psi(\theta_0)^t U_n\{\dot\psi(\theta_0)^t \dot\psi(\theta_0)\}^{-1}]^t [Z_n - I_n^{-1}\dot\psi(\theta_0)^t U_n$$
$$\{\dot\psi(\theta_0)^t \dot\psi(\theta_0)\}^{-1}]1_{\{\psi(Z_n)=0\}} - Z_n^t Z_n 1_{\{\psi(Z_n)\neq 0\}} + o_p(1)$$

and Z_n converges to a normal variable Z in $\mathbb{R}^d$ as n tends to infinity. The limiting distribution under H_0 of the log-likelihood ratio tests with the constraint of Θ_1 is then a difference of the squares of truncated Gaussian variables.

Under the sequence of local alternatives with parameters θ_n in Θ_2, there exist θ in Θ_1 and $a = \lim_{n \to \infty} n^{\frac{1}{2}}(\theta_n - \theta)$. The expansion of $\log\{f_{\theta_n} f_\theta^{-1}\}$, as n tends to infinity, adds a term $a^t I_n(\theta_n) a$ to the likelihood ratio under the alternative and the asymptotic distribution of T_n under the sequence of local alternatives follows. $\qquad\square$

The critical value of the test of level α is c_α such that $\alpha = P(T > c_\alpha)$ and its asymptotic power under the sequence of local alternatives $(K_n)_{n \geq 1}$ is $\beta_{\theta,a}(\alpha) = P\{T > c_\alpha - a^t I(\eta) a\}$ such that

$$\inf_{\theta \in \Theta_1} \inf_{a = \lim_n n^{\frac{1}{2}}(\theta_n - \theta)} \beta_{\eta,a} > \alpha.$$

With linear constraints such as $\sum_k p_k = 1$ for the probabilities of a multinomial distribution, the first derivative of $l_n(p) - \lambda(\sum_k p_k - 1)$ has components $\dot{l}_n(p) - \lambda$ an its second derivative is $\ddot{l}_n(p)$, the estimator $\hat{p}_n$ has the expansion $n^{\frac{1}{2}}(\hat{p}_n - p) = I_n^{-1}(U_n - e\lambda) + o_p(1)$, where e is the vector with components 1. Multiplying by e^t this expression and using the constraint implies $e^t I_n^{-1}(U_n - e\lambda) + o_p(1) = 0$ therefore $\lambda = e^t I_n^{-1} U_n (e^t I_n^{-1} e)^{-1} + o_p(1)$ and

$$n^{\frac{1}{2}}(\hat{p}_n - p) = I_n^{-1}\{U_n - e^t I_n^{-1} U_n e(e^t I_n^{-1} e)^{-1}\} + o_p(1)$$
$$= Z_n - I_n^{-1} e^t Z_n e\, (e^t I_n^{-1} e)^{-1} + o_p(1),$$

with the notation $Z_n = I_n^{-1} U_n$. The variable Z_n converges weakly to a centered Gaussian variable $Z = I^{-1} U$ and the limiting distribution of $\hat{p}_n$ follows.

Under a local alternative K_n such that $\sum_k p_k < 1$, the estimator satisfies the same inequality as n tends to infinity and there is no constraint, $n^{\frac{1}{2}}(\hat{p}_n - p) = I_n^{-1} U_n + o_p(1)$. The log-likelihood statistic T_n for a test of the hypothesis $H_0 : \sum_k p_k = 1$ against the alternative $K : \sum_k p_k < 1$ has the asymptotic distribution

$$T_0 = \{Z - I^{-1} e^t Z e(e^t I^{-1} e)^{-1}\}^t \{Z - I^{-1} e^t Z e(e^t I^{-1} e)^{-1}\} - Z^t Z,$$

under H_0, where Z is a d-dimensional vector of independent normal variables. Under the sequence of local alternatives $(K_n)_{n \geq 1}$ of Proposition 2.6 the limiting distribution of the statistic T_n is $T = T_0 + a^t I(\theta) a$.

2.3   Likelihood ratio tests against local alternatives

In the previous section, the expansions of the parametric likelihood ratio tests under the null hypothesis and the alternative are calculated in the parametric family. When it is unknown, an error for the function f_0 defines a neighborhood of the functional class. Here, we consider neighborhoods for the function and the parameter defining the density under H_0. The local log-likelihood ratio statistics of goodness of fit tests have expansions following Lecam's lemma (1956). In this section, several parametric and nonparametric models are considered. The approximation and the limiting distribution of the nonparametric log-likelihood ratio statistic are similar to the approximation and limits in the parametric models, under H_0 and under the local alternatives.

Let $\mathcal{P} = \{P_\theta, \theta \in \Theta\}$ be a parametric family of probabilities on a measurable space $(\Omega, \mathcal{F})$ and let X be a real variable defined from the family of probability spaces $(\Omega, \mathcal{F}, \mathcal{P})$ into a metric space $\mathbb{X}$. We assume that the probabilities P_θ have respective densities f_θ in $C^2(\Theta)$, a.s. in $\mathbb{X}$, for a subset Θ of $\mathbb{R}^d$. The observations are a n-sample $(X_i)_{i=,...,n}$ of a variable X with density function f_θ, where θ is an unknown parameter of Θ. We consider the hypothesis H_0 of a sampled variable with a known density $f = f_{\theta_0}$ and a sequence of local alternatives K_n of densities f_{n,θ_n} belonging to $C^2(\mathcal{V}_n(\theta_0))$, with a parameter θ_n in a neighborhood $\mathcal{V}_n(\theta_0)$ of θ_0 and $f_n = f_{n,\theta}$ is defined by

$$
K_n : \begin{cases} f_n(t) = f(t)\{1 + n^{-\frac{1}{2}}\gamma_n(t)\}, \\ f_{n,\theta_{n,j}} = f_{\theta_{n,j}}(t)\{1 + n^{-\frac{1}{2}}\gamma_{n,\theta_{n,j}}(t)\}, \\ \theta_n = \theta_0 + n^{-\frac{1}{2}}\rho_n \end{cases} \tag{2.3}
$$

where $(\rho_n)_{n\geq 1}$ is a sequence of positive vectors of $\mathbb{R}^d$ converging to a strictly positive limit ρ and $(\gamma_n)_{n\geq 1}$ is a sequence of uniformly bounded functions of $C^2(\mathbb{R})$, converging uniformly to a nonzero function γ, as n tends to infinity. Second order approximations of the log-likelihood ratio statistic

$$
T_n = \sum_{i=1}^{n}\{\log f_{n,\theta_n}(X_i) - \log f(X_i)\}
$$

and their weak convergence are established under the following conditions.

Condition 2.2.

(1) The parameter set Θ is an open bounded convex of $\mathbb{R}^d$. The functions γ_n belong to $C^1(\mathbb{X})$ and for every x, the functions defined in Θ by $\theta \mapsto \gamma_{n,\theta_n}(x)$ belong to $C^1(\Theta)$.

(2) The functions $f^{-1}\dot{f}$ and γ_n belong to $L^2(\mathbb{X}, f)$, for all n.

The variance $\sigma_K^2 = \int_{\mathbb{X}} (\rho^t f^{-1}\dot{f} + \gamma)^2 f = \rho^t I_0 \rho + \int_{\mathbb{X}} \gamma^2 f$ is then finite. The next results have been established by LeCam (1956).

Theorem 2.1. *Under H_0 and Condition 2.2, the log-likelihood ratio statistic has an uniform expansion $T_n = Y_{0n} - \frac{1}{2} Z_{0n} + o_p(1)$, it converges weakly to $Y_0 - \frac{1}{2}\sigma_K^2$, where Y_0 is a centered Gaussian variable with variance σ_K^2. Under the local alternatives K_n, $T_n = Y_n + \frac{1}{2} Z_n + o_p(1)$ and it converges weakly to $Y + \frac{1}{2}\sigma_K^2$, where Y is a centered Gaussian variable with variance σ_K^2.*

Proof. The logarithm of the density under K_n is approximated using a second order Taylor expansion of $\log f_n$ in the neighborhood of $\log f$ defined by K_n, for every function h_θ of $C^2(\Theta)$

$$\log \frac{h_{\theta_n}}{h_\theta}(x) = (\theta_n - \theta)^t \frac{\dot{h}_\theta}{h_\theta}(x) + \frac{1}{2}(\theta_n - \theta)^t \left\{ \frac{\ddot{h}_\theta}{h_\theta}(x) - \frac{\dot{h}_\theta^{\otimes 2}}{h_\theta}(x) \right\}(\theta_n - \theta) + o(n^{-1}),$$

where $\theta_n - \theta = n^{-\frac{1}{2}}\rho_n$. Deriving f_n, we obtain $f_n^{-1}\dot{f}_n = f^{-1}\dot{f} + n^{-\frac{1}{2}}\dot{\gamma}_n + o(n^{-1})$ and $f_n^{-1}\ddot{f}_n = f^{-1}\ddot{f} + o(1)$. From these approximations and under H_0, the log-likelihood ratio statistic T_n has the following expansion

$$T_n = \sum_{i=1}^n \left\{ \log \frac{f_{n,\theta_n}}{f_n}(X_i) + \log \frac{f_n}{f}(X_i) \right\}$$

$$= \sum_{i=1}^n \left[\log \frac{f_n}{f}(X_i) + (\theta_n - \theta)^t \frac{\dot{f}_n}{f_n}(X_i) \right.$$

$$\left. + \frac{1}{2}(\theta_n - \theta)^t \left\{ \frac{\ddot{f}_n}{f_n}(X_i) - \frac{\dot{f}_n^{\otimes 2}}{f_n^{\otimes 2}}(X_i) \right\}(\theta_n - \theta) \right] + o_p(1)$$

$$= n^{\frac{1}{2}} \sum_{i=1}^n \left\{ \gamma_n(X_i) + \rho_n^t \frac{\dot{f}}{f}(X_i) \right\} - (2n)^{-1} \sum_{i=1}^n \left[\left\{ \rho_n^t \frac{\dot{f}^{\otimes 2}}{f^2}(X_i) - \frac{\ddot{f}}{f}(X_i) \right\} \rho_n \right.$$

$$\left. + \gamma_n^2(X_i) - 2\rho_n^t \dot{\gamma}_n(X_i) \right] + o_p(1)$$

$$= n^{\frac{1}{2}} \sum_{i=1}^n \left\{ \gamma_n(X_i) + \rho_n^t \frac{\dot{f}}{f}(X_i) \right\} - (2n)^{-1} \sum_{i=1}^n \left\{ \rho_n^t \frac{\dot{f}^{\otimes 2}}{f^2}(X_i) \rho_n \right. \qquad (2.4)$$

$$\left. + \gamma_n^2(X_i) - 2\rho_n \dot{\gamma}_n(X_i) \right\} + o_p(1),$$

where the $o_p(1)$ are uniform in Θ under Condition 2.2.1. Under H_0 and K_n, the second sum converges in probability to

$$\mu_0 = -\frac{1}{2} \int_{\mathbb{X}} \left\{ \rho_n^t \frac{\dot{f}^{\otimes 2}}{f^2} \rho_n + \gamma_n^2 - 2\rho_n^t \dot{\gamma}_n \right\} f \, dx.$$

Since f_n is a density for every n, the functions γ_n satisfy $\int_\mathbb{X} \gamma_n f \, dx = 0$, hence $\int_\mathbb{X} \dot{\gamma}_n f \, dx = -\int_\mathbb{X} \gamma_n \dot{f} \, dx$ and $\mu_0 = -\frac{1}{2}\sigma_K^2$. Under H_0, the variable T_n converges weakly to a Gaussian variable with mean μ_0 and variance σ_K^2. According to Lecam's first lemma, the probability distributions of T_n under H_0 and K_n are therefore contiguous and the $o_p(1)$ under H_0 remains an $o_p(1)$ under K_n. Under K_n, the mean of the first two order term of the expansion of T_n is

$$E_{K_n}\left(Y_{0n} - \frac{1}{2}Z_{0n}\right) = n^{\frac{1}{2}} \int_\mathbb{X} \left(\gamma_n + \rho_n^t \frac{\dot{f}}{f}\right) f_{n,\theta_n} \, dx + \mu_0 = \frac{1}{2}\sigma_K^2.$$

The variable $Y_n = Y_{0n} - \sigma_K^2$ is centered under the alternative and it converges weakly to a Gaussian variable with variance σ_K^2.   □

The first order terms Y_{0n} and Y_n of T_n have the same asymptotic distribution, under H_0 and K_n, which is the distribution of a centered Gaussian variable with variance σ_K^2 depending on the limits ρ and γ. The local alternatives are well separated from the null hypothesis by the means of T_n under H_0 and K_n. Theorem 2.1 implies the consistency of the log-likelihood ratio test, the critical value x_α and the asymptotic power at the level α of the one-sided test are

$$\alpha = \lim_{n\to\infty} P_{H_0}(T_n > x_\alpha) = \lim_{n\to\infty} P_{H_0}\left(Y_{0n} - \frac{1}{2}Z_{0n} > x_\alpha\right)$$
$$= 1 - \Phi\left(x_\alpha + \frac{1}{2}\sigma_K\right) + o(1),$$

$$\beta(\alpha) = \lim_{n\to\infty} P_{K_n}(T_n > x_\alpha) = \lim_{n\to\infty} P_{K_n}\left(Y_n + \frac{1}{2}Z_n > x_\alpha\right)$$
$$= 1 - \Phi\left(x_\alpha - \frac{1}{2}\sigma_K\right).$$

The quantile c_α of the normal distribution is related to the asymptotic quantile x_α of T_n by $c_\alpha = x_\alpha + \frac{1}{2}\sigma_K$ and the asymptotic power of the test expressed with c_α as $\beta(\alpha) = 1 - \Phi(c_\alpha - \sigma_K)$. The hypothesis H_0 of the test is also the hypothesis $\rho_n = 0$ and $\gamma_n = 0$.

Theorem 2.1 and the expression of σ_K^2 imply that the Pitman efficiency of the likelihood ratio test is $e = \|I_0\|(\|I_0\| + 1) < 1$, with the derivatives

$$\frac{\partial \sigma_K^2}{\partial \rho} = -2I_0\rho, \quad \frac{\partial \sigma_K^2}{\partial \gamma} = 0$$

and the normalization of ρ and γ to 1.

The score test for H_0 relies on the first derivative of the log-likelihood ratio i.e. on the statistic

$$U_n = n^{-\frac{1}{2}} \sum_{i=1}^{n} \frac{\dot{f}_n}{f_n}(X_i),$$

under the null hypothesis, it is centered and it converges weakly to a Gaussian variable with mean zero and variance $\sigma_U^2 = I_0$. The variance σ_K^2 of the statistic T_n is larger than σ_U^2 and the power of the normalized tests can be compared. Under a separate alternative K of a density in a class $\mathcal{G}_\varepsilon = \{g : f^{-1}g < 1 - \varepsilon\}$, $\varepsilon > 0$, the statistics T_n and U_n tend to infinity, their asymptotic power is therefore 1.

Theorem 2.2. *Under Condition 2.2, the likelihood ratio test of the hypothesis $H_0 : \{f_\theta; \theta \in \Theta\}$ against local alternatives K_n given by (2.3) for every density under H_0 is locally asymptotically more powerful than the score test.*

Proof. The variable U_n converges weakly under H_0 to a centered asymptotically Gaussian centered variable with the variance I_0 and the mean $\mu_U = \int_\mathcal{X} (f^{(1)}\gamma + f\gamma^{(1)}) = 0$. The likelihood ratio test is therefore locally asymptotically more powerful than the score test for $H_0 : f_0$ against local alternatives K_n given by (2.3).

The unknown parameter of the density of X under a composite hypothesis defined by a parametric class of densities is estimated by $\widehat{\theta}_n$ such that $\widehat{\rho}_n = n^{\frac{1}{2}}(\widehat{\theta}_n - \theta_0) = O_p(1)$ and the expansions of Theorem 2.1 apply with this random parameter, the asymptotic distributions of both statistics are modified by a convolution as in Proposition 2.3, which yields the result. $\square$

In a same approach, a two-sample local test of homogeneity aims to compare the density functions of two sub-samples with respective sizes n_1 and n_2 when their difference tends to zero with the rate $n^{-\frac{1}{2}}$, depending on the total sample size $n = n_1 + n_2$. The observations are a n-sample $(X_i)_{i=,...,n}$ of a variable X such that $X_1, \ldots, X_{n_1}$ have a density $f_1 = f_{\theta_1}$ and $X_{n_1+1}, \ldots, X_n$ have the density $f_2 = f_{\theta_2}$, where θ_1 and θ_2 are unknown parameters of the parameter space Θ. We consider the hypothesis H_0 of a common density f_0, with $\theta_1 = \theta_2 = \theta_0$ and a sequence of local alternatives K_n where the densities are f_{n,θ_n} defined by

$$K_n : \begin{cases} f_{1n}(t) = f_{2n}(t)\{1 + n^{-\frac{1}{2}}\eta_n(t)\}, \\ f_{2n}(t) = f_0(t)\{1 + n^{-\frac{1}{2}}\gamma_n(t)\}, \ t \in \mathbb{R}, \end{cases}$$

where $(\eta_n)_{n\geq 1}$ and $(\gamma_n)_{n\geq 1}$ are sequence of uniformly bounded functions converging uniformly to nonzero functions η and γ, as n tends to infinity. The log-likelihood ratio statistic for H_0 against K_n is nonparametric

$$T_n = \sum_{i=1}^{n_1}\{\log f_{1n}(X_i) - \log f_0(X_i)\} + \sum_{i=n_1+1}^{n}\{\log f_{2n}(X_i) - \log f_0(X_i)\}$$

and Theorem 2.1 does not apply. By definition of the density sequences of the sub-samples, $\int_{\mathbb{X}} \eta f_0\, dx = 0$, $\int_{\mathbb{X}} \gamma f_0\, dx = 0$ and $\int_{\mathbb{X}} \eta\gamma f_0\, dx = 0$.

Condition 2.3.

(1) The sample sizes are $n_k = n_{kn}$ such that $n^{-1}n_k$ converge as n tends to infinity, $\lim_{n\to\infty} n^{-1}n_1 = \lambda$ and $\lim_{n\to\infty} n^{-1}n_2 = 1 - \lambda$ in $]0,1[$.
(2) For every n, the functions η_n and γ_n belong to $L^2(\mathbb{X}, f_{\theta_0})$.

The second condition implies that the following variance is finite

$$\sigma_K^2 = \int_{\mathbb{X}}\{\lambda(\eta + \gamma)^2 + (1 - \lambda)\gamma^2\}f_0\, dx = \int_{\mathbb{X}}(\lambda\eta^2 + \gamma^2)f_0\, dx.$$

Theorem 2.3. *Under H_0 and Condition 2.3, the log-likelihood ratio statistic has an expansion $T_n = Y_{0n} - \frac{1}{2}Z_{0n} + o_p(1)$ where $E_0 Y_{0n} = 0$ and it converges weakly to $Y_0 - \frac{1}{2}\sigma_K^2$ where Y_0 is a centered Gaussian variable with variance σ_K^2. Under the local alternative K_n, $T_n = Y_n + \frac{1}{2}Z_n + o_p(1)$ and it converges weakly to $Y + \frac{1}{2}\sigma_K^2$, where Y is a centered Gaussian variable with variance σ_K^2.*

Proof. The logarithm of the density f_{1n} is approximated using second order Taylor expansions for every x

$$\log f_{1n}(x) = \log f_0(x) + \log\{1 + n^{-\frac{1}{2}}(\gamma_n + \eta_n)(x) + n^{-1}(\gamma_n\eta_n)(x)\}$$
$$= \log f_0(x) + n^{-\frac{1}{2}}(\gamma_n + \eta_n)(x) - (2n)^{-1}(\gamma_n^2 + \eta_n^2)(x) + o_p(1).$$

Under H_0, the log-likelihood ratio statistic is expanded as

$$T_n = \sum_{i=1}^{n_1}\log\frac{f_{1n}}{f_0}(X_i) + \sum_{i=n_1+1}^{n_2}\log\frac{f_{2n}}{f_0}(X_i)$$
$$= n^{-\frac{1}{2}}\left\{\sum_{i=1}^{n}\gamma_n(X_i) + \sum_{i=1}^{n_1}\eta_n(X_i)\right\}$$
$$- (2n)^{-1}\left\{\sum_{i=1}^{n_1}\eta_n^2(X_i) + \sum_{i=1}^{n}\gamma_n^2(X_i)\right\} + o_p(1).$$

Under H_0, the first order term of this expansion has the mean zero and the second order term converges in probability to $\mu_K = -\frac{1}{2}\int(\lambda\eta^2 + \gamma^2)f_0$. The variable T_n converges weakly to a Gaussian variable with variance σ_K^2 and mean $\mu_K = -\frac{1}{2}\sigma_K^2$, due to the the integral $\int_{\mathbb{X}}\eta\gamma f_0\,dx = 0$.

Under the alternative, the mean of the expansion of the log-likelihood ratio statistic is modified

$$E_{K_n}T_n = E_0(2n)^{-1}\Big\{\sum_{i=1}^{n_1}\eta_n^2(X_i) + \sum_{i=1}^{n}\gamma_n^2(X_i)\Big\} + o(1) = -\mu_K.$$

The variable T_n is then approximated by the sum of the first order term Y_{0n} and $\frac{1}{2}\sigma_K^2$. $\qquad\square$

Theorem 2.1 extends to a k-sample goodness of fit test, $k \geq 2$. The hypothesis H_0 of a n-sample of a variable X with a known density f has local alternatives K_n defined by k independent sub-samples with respective size $n_j = n_{j,n}$ increasing with n. The density of the observations X_{ji}, $i = 1,\ldots,n_j$ of the jth sub-sample is indexed a parameter $\theta_{n,j}$

$$K_n : \begin{cases} f_n(t) = f(t)\{1 + n^{-\frac{1}{2}}\gamma_n(t)\}, \\ f_{n,\theta_{n,j}} = f_{\theta_{n,j}}(t)\{1 + n^{-\frac{1}{2}}\gamma_{n,\theta_{n,j}}(t)\}, \\ \theta_{n,j} = \theta_j + n^{-\frac{1}{2}}\rho_{n,j} \quad j = 1,\ldots,k. \end{cases}$$

As n tends to infinity, the sequence $(\rho_{n,j})_{n\geq 1}$ of positive vectors of $\mathbb{R}^d$ converges to a strictly positive limit ρ_j, for $j = 1,\ldots,k$, and $(\gamma_n)_{n\geq 1}$ is a sequence of uniformly bounded functions of $C^2(\mathbb{R})$, converging uniformly to a nonzero function γ. In addition to Condition 2.2 we assume that the sample sizes $n_j = n_{n,j}$ converge as n tends to infinity to λ_j in $]0,1[$, for $j = 1,\ldots,k$, and $\sum_{j=1}^{k}\lambda_j = 1$. The variance is denoted

$$\sigma_{K,k}^2 = \sum_{j=1}^{k}\lambda_j\int_{\mathbb{X}}\Big(\rho_j^t\frac{\dot{f}}{f} + \gamma\Big)^2 f.$$

Theorem 2.4. *Under Condition 2.3, the log-likelihood ratio statistic $T_{n,k}$ satisfies Theorem 2.1 with the asymptotic variance $\sigma_{K,k}^2$.*

Proof. Under H_0, the statistic is

$$T_{n,k} = \sum_{j=1}^{k}\sum_{i=1}^{n_j}\Big\{\log\frac{f_{n,\theta_{n,j}}}{f_n}(X_{ji}) + \log\frac{f_n}{f}(X_{ji})\Big\}$$

it is approximated by Taylor expansions of the density under the local alternative with parameter in a neighborhood of the true parameter value

θ_0 and with f_n in a neighborhood of $f = f_{\theta_0}$

$$T_{n,k} = n^{\frac{1}{2}} \sum_{j=1}^{k} \sum_{i=1}^{n_j} \left\{ \gamma_n(X_{ji}) + \rho_{n,j}^t \frac{\dot{f}}{f}(X_{ji}) \right\} - (2n)^{-1} \sum_{j=1}^{k} \sum_{i=1}^{n_j} \Big[\gamma_n^2(X_{ji})$$

$$-2\rho_{n,j}^t \dot{\gamma}_n(X_{ji}) + \rho_{n,j}^t \left\{ \frac{\dot{f}^{\otimes 2}}{f^2}(X_{ji}) - \frac{\ddot{f}}{f}(X_{ji}) \right\} \rho_{n,j} \Big] + o_p(1)$$

and the asymptotic distribution under the null hypothesis is deduced from this expansion. Under K_n, the mean of the first order term of the expansion of the statistic equals $\sigma_{K,k}^2$ and the second order term converges to the same limit as under H_0.   □

Theorem 2.3 also extends to a sample with k independent sub-samples, under the same conditions. We consider the hypothesis H_0 of a common density f_0 and a sequence of local alternatives K_n where the density of X_{ji}, $i = 1, \ldots, n_j$, is f_{jn}, for $j = 1, \ldots, k$. They are defined by

$$K_n : \begin{cases} f_{jn}(t) = f_{kn}(t)\{1 + n^{-\frac{1}{2}}\eta_{n,j}(t)\}, \ j = 1, \ldots, k-1 \\ f_{kn}(t) = f_0(t)\{1 + n^{-\frac{1}{2}}\gamma_n(t)\}, \ t \in \mathbb{R}, \end{cases}$$

where the sequences of functions $(\eta_{n,j})_{n \geq 1}$ and $(\gamma_n)_{n \geq 1}$ converge uniformly to nonzero functions η_j, for $j = 1, \ldots, k$, and γ, as n tends to infinity. The log-likelihood ratio statistic for H_0 against K_n with nonparametric functions is

$$T_n = \sum_{j=1}^{k} \sum_{i=1}^{n_j} \{\log f_{jn}(X_{ji}) - \log f_0(X_{ji})\}.$$

Theorem 2.3 applies under similar conditions with the asymptotic variance $\sigma_K^2 = \sum_{j=1}^{k-1} \int_{\mathbb{X}} (\lambda_j \eta_j^2 + \gamma^2) f_0 \, dx$. Under H_0, the local log-likelihood ratio statistic has the approximation

$$T_{n,k} = \sum_{j=1}^{k} \sum_{i=1}^{n_j} \left\{ \log \frac{f_{jn}}{f_n}(X_{ji}) + \log \frac{f_n}{f}(X_{ji}) \right\}$$

$$= n^{-\frac{1}{2}} \sum_{j=1}^{k} \sum_{i=1}^{n_j} \{\gamma_n(X_{ji}) + \eta_{jn}(X_{ji})\}$$

$$- (2n)^{-1} \sum_{j=1}^{k} \sum_{i=1}^{n_j} \{\gamma_n^2(X_{ji}) + \eta_{jn}^2(X_{ji})\} + o_p(1)$$

and its limits under H_0 and K_n follow LeCam's rules as previously.

2.4 Nonparametric likelihood ratio tests

Consider a real random variable X defined from a probability space $(\Omega, \mathcal{A}, P)$ onto the real metric space $(\mathbb{R}, \mathcal{B}, \|\cdot\|)$, with a density f in $C^s(\mathcal{I}_X)$ where $\mathcal{I}_X$ is a bounded subset of $\mathbb{R}$, the support of f is either bounded or such that $P(X \in \mathcal{I}_X) > 1 - \varepsilon$, with an arbitrary small $\varepsilon > 0$.

A likelihood ratio test of the hypothesis $H_0 : f = f_0$ relies on the comparison of f and f_0 when it is known. The L^2-optimal convergence rate of a nonparametric estimator of a density of $C^s(\mathcal{I}_X)$ from a n-sample is $n^{-\frac{s}{2s+1}}$ and it is reached with a bandwidth $h = h_n = n^{-\frac{1}{2s+1}}$, so we shall consider alternatives converging to hypotheses with this convergence rate. Due to the nonparametric setting, the sample size of the likelihood ratio test is modified as $k_n = n^{\frac{s}{2s+1}}$. Here, we study the asymptotic expansion of the log-likelihood ratio statistic for several hypotheses and local alternatives. The results proved in Section 2.1 for the parametric models differ from the following limiting distributions when f_0 is unknown.

Let f_0 be the density of the variable X under the hypothesis H_0 and let $\Gamma = \{\gamma \in L^2(\mathcal{I}_X); \gamma \text{ uniformly continuous}, \int_{\mathcal{I}_X} \gamma f_0 = 0, \int_{\mathcal{I}_X} \gamma^2 f_0 < \infty\}$. A sequence of local alternatives is defined as

$$K_n : f_n(x) = f_0(x)\{1 + k_n^{-\frac{1}{2}}\gamma_n(x)\}$$

for every x of $\mathcal{I}_X$, where $(\gamma_n)_{n \geq 1}$ is a sequence of uniformly bounded functions converging uniformly to a nonzero function γ of Γ, as n tends to infinity. The log-likelihood ratio statistic for H_0 against K_n is expanded by a Taylor expansion of the logarithm

$$T_{0n} = \sum_{i=1}^{k_n} \log \frac{f_n}{f_0}(X_i) = \sum_{i=1}^{k_n} \log\left\{1 + \frac{f_n - f_0}{f_0}(X_i)\right\}$$

$$= \sum_{i=1}^{k_n} \frac{f_n - f_0}{f_0}(X_i) - \frac{1}{2}\sum_{i=1}^{k_n}\left\{\frac{f_n - f_0}{f_0}(X_i) + o_p((f_n - f_0)(X_i))\right\}^2.$$

Let $\sigma_n^2 = \int (f_n - f_0)^2 f_0^{-1}$ be the nonparametric information of f_n with respect to f_0 and let $\sigma_0^2 = \int \gamma^2 f_0 = E_0\{\gamma^2(X)\}$. By definition of the functions γ_n, $E_0\{\gamma_n(X)\} = \int \gamma f_0 = 0$, for every integer n.

Theorem 2.5. *The log-likelihood ratio has an asymptotic expansion*

$$T_{0n} = Y_{0n} - \frac{1}{2}Z_{0n} + o_p(1)$$

under H_0 and K_n. Under H_0, $E_0 Y_{0n} = 0$, $Var_0 Y_{0n} = E Z_{0n}$ and T_{0n} converges weakly to $Y_0 - \frac{1}{2}\sigma_0^2$ where Y_0 is a centered Gaussian variable with variance σ_0^2. Under K_n, T_{0n} converges weakly to $Y_0 + \frac{1}{2}\sigma_0^2$.

Proof. Under H_0, the log-likelihood ratio statistic is expanded as

$$T_{0n} = k_n^{-\frac{1}{2}} \sum_{i=1}^{k_n} \gamma_n(X_i) - \frac{1}{2k_n} \sum_{i=1}^{k_n} \gamma_n^2(X_i) + o_p(1)$$

and the asymptotic expansion of T_n under the hypothesis is written with the notations $Y_{0n} = k_n^{-\frac{1}{2}} \sum_{i=1}^{k_n} \gamma_n(X_i)$, $Z_{0n} = k_n^{-1} \sum_{i=1}^{k_n} \gamma_n^2(X_i)$. Its asymptotic distribution follows. Under the alternative K_n

$$E_n Y_{0n} = k_n^{\frac{1}{2}} \int \gamma_n f_n = k_n^{\frac{1}{2}} \int \gamma_n \{1 + k_n^{-\frac{1}{2}} \gamma_n\} f_0$$

$$= \int \gamma_n^2 f_0 = \sigma_0^2 + o(1),$$

the variance of Y_{0n} is asymptotically equivalent to $E_n Z_{0n} = \int \gamma_n^2 \{1 + k_n^{-\frac{1}{2}} \gamma_n\} f_0$ and they converges to σ_0^2. $\qquad\square$

In a goodness of fit test to a nonparametric family $\mathcal{F}$ of $C^s(\mathbb{R})$, the unknown density f_0 of the hypothesis is estimated from a n sample of the variable X by its kernel estimator $\widehat{f}_{n,h}$ which converges to f_0 at the optimal rate $n^{\frac{s}{2s+1}}$, with a optimal bandwidth $h = h_n$. The likelihood ratio statistic for a larger sample of size k_n is

$$\widehat{T}_n = \sum_{i=1}^{k_n} \log \frac{f_n}{\widehat{f}_{n,h}}(X_i) = \sum_{i=1}^{k_n} \log \frac{f_n}{f_0}(X_i) - \sum_{i=1}^{k_n} \frac{\widehat{f}_{n,h}}{f_0}(X_i)$$

$$= \sum_{i=1}^{k_n} \frac{f_n - \widehat{f}_{n,h}}{f_0}(X_i) - \frac{1}{2} \sum_{i=1}^{k_n} \frac{(f_n - f_0)^2 - (\widehat{f}_{n,h} - f_0)^2}{f_0^2}(X_i) + o_p(1).$$

Let $\sigma_1^2 = \int (\gamma_s^2 b_{f_0}^2 + \sigma_{f_0}^2) f_0^{-1}$ and $\sigma_Y^2 = \int (f_0^{-1} \gamma_s b_{f_0} - \gamma)^2 f_0 + \int \sigma_{f_0}^2 f_0^{-1}$, with the bias and variance constants of Theorem 1.1.

Theorem 2.6. *Under Condition 1.1 and with a density f_0 having a $(s-1)$th derivative that tends to zero at infinity, the statistic $\widehat{T}_n$ has an asymptotic expansion $Y_n - \frac{1}{2} Z_n + o_p(1)$ under H_0 and K_n. Under H_0, $\widehat{T}_n$ converges weakly to $Y - \frac{1}{2}(\sigma_0^2 - \sigma_1^2)$ where Y is a centered Gaussian variable with variance σ_Y^2. Under K_n, $\widehat{T}_n$ converges weakly to $Y_2 + \frac{1}{2}(\sigma_0^2 + \sigma_1^2)$ where Y_2 is a centered Gaussian variable with variance σ_Y^2.*

Proof. The statistic $\widehat{T}_n$ satisfies the same asymptotic expansion as in Theorem 2.5, with

$$Y_n = k_n^{-\frac{1}{2}} \sum_{i=1}^{k_n} \left\{ \gamma_n(X_i) - k_n^{\frac{1}{2}} \frac{\widehat{f}_{n,h} - f_0}{f_0}(X_i) \right\},$$

$$Z_n = k_n^{-1} \sum_{i=1}^{n} \left\{ \gamma_n^2(X_i) - k_n \frac{(\widehat{f}_{n,h} - f_0)^2}{f_0^2}(X_i) \right\}.$$

Under H_0 and if the $(s-1)$th derivative of the density f_0 tends to zero at infinity, the mean of Y_n is asymptotically equivalent to the integral of the bias constant $\gamma_s b_f$ of $\widehat{f}_{n,h}$ and it tends to zero, the mean of Z_n tends to $\sigma_0^2 - \sigma_1^2$. The variance of Y_n is

$$Var_0 Y_n = \int (\gamma_n^2 - 2\gamma_n f_0^{-1}\gamma_s b_{f_n} + f_0^{-2}\gamma_s^2 b_{f_n}^2 + f_0^{-2}\sigma_{f_n}^2)f_0$$

$$= \int \{(\gamma - f_0^{-1}\gamma_s b_{f_0})^2 + f_0^{-2}\sigma_{f_0}^2\}f_0 + o(1),$$

it converges to σ_Y^2. The moments of Y_n under K_n are

$$E_n Y_n = \sigma_0^2 - \gamma_s \int b_{f_n} f_0^{-1} f_n + o(1) = \sigma_0^2 - \gamma_s \int b_{f_n}\{1 + k_n^{-\frac{1}{2}}\gamma_n\} + o(1),$$

which converges to σ_0^2 and

$$Var_n Y_n = \int \{\gamma_n^2 - 2\gamma_n(f_0^{-1}\gamma_s b_{f_n} + \gamma_n) + (f_0^{-1}\gamma_s b_{f_n} + \gamma_n)^2$$

$$+ f_0^{-2}\sigma_{f_n}^2\}f_n + o(1)$$

$$= \sigma_0^2 + \int (\gamma_s b_{f_0}^2 + \sigma_{f_0}^2)f_0^{-2}f_n + o(1)$$

$$= \sigma_0^2 + \int (\gamma_s b_{f_0}^2 + \sigma_{f_0}^2)f_0^{-1}\{1 + k_n^{-\frac{1}{2}}\gamma_n\} + o(1),$$

it converges to σ_Y^2. The mean of Z_n is

$$E_n Z_n = \int \{\gamma_n^2 - (f_0^{-1}\gamma_s b_{f_0})^2 - \sigma_{f_0}^2)f_0^{-2}\}f_n + o(1),$$

it converges to $\sigma_0^2 - \sigma_1^2$. $\qquad\square$

The likelihood ratio test of H_0 against K_n is asymptotically equivalent to the asymptotically normal test based on $\widehat{\sigma}_{Yn}^{-1}Y_{0n}$, defined with a consistent estimator $\widehat{\sigma}_{Yn}^2$ of the variance of Y_{0n} under H_0. The asymptotic critical value of the likelihood ratio test at the level α is $c = \sigma_Y c_\alpha - \frac{1}{2}(\sigma_0^2 - \sigma_1^2)$, its local asymptotic power is

$$\beta(\alpha) = 1 - \inf_{f_0 \in \mathcal{F}} \sup_{\gamma \in \Gamma} \Phi(c_\alpha - \sigma_Y^{-1}\sigma_0^2),$$

where c_α is the normal critical value.

Consider two independent random variables X_1 and X_2, and a sample of X. Their respective densities f_1 and f_2 are estimated by kernel estimators from independent samples of size n. A nonparametric likelihood ratio test of equality of the densities against local alternatives is defined by sequences of density functions

$$K_n : f_{kn}(t) = f_0(t)\{1 + n^{-\frac{1}{2}}\eta_{kn}(t)\}, \ k = 1, 2,$$

for every real t, where f_0 is the unknown common density belonging to a class $\mathcal{F}$ under the hypothesis, $(\eta_{1n})_{n\geq 1}$ and $(\eta_{2n})_{n\geq 1}$ are sequences of bounded functions converging uniformly to distinct functions η_1 and respectively η_2, as n tends to infinity. The log-likelihood ratio statistic for H_0 against K_n is calculated for k_n-samples of X_1 and X_2

$$\widehat{T}_{1n} = \sum_{i=1}^{k_n} \log \frac{\widehat{f}_{1n,h}}{\widehat{f}_{2n,h}}(X_i)$$

$$= \sum_{i=1}^{k_n} \left\{ \log \frac{\widehat{f}_{1n,h}}{f_1}(X_i) - \log \frac{\widehat{f}_{2n,h}}{f_2}(X_i) + \log \frac{f_1}{f_2}(X_i) \right\}.$$

Under the hypothesis, the statistic $\widehat{T}_{1n}$ is expanded as the difference of the expansions of the independent statistics of Theorem 2.6 defined for each variables. Under the alternatives, the expansion of the last term is added, it is the same as in Theorem 2.5. With the previous notations for each variable, we obtain the following limits, where $\sigma_{0k}^2 = \int_{\mathbb{R}} \eta_k^2 f_0$, $k = 1, 2$.

Theorem 2.7. *Under Condition 1.1 and with a density f_0 having a $(s-1)$th derivative that tends to zero at infinity, $\widehat{T}_{1n}$ has an asymptotic expansion $Y_{1n} - Y_{2n} - \frac{1}{2}(Z_{1n} - Z_{2n}) + o_p(1)$ under H_0 and it converges weakly to $T_1 = Y - \frac{1}{2}(\sigma_{01}^2 - \sigma_{02}^2)$ where Y is a centered Gaussian variable with variance $\sigma_{Y1}^2 + \sigma_{Y2}^2$. Under K_n,*

$$\widehat{T}_{1n} = Y_{01n} + Y_{1n} - Y_{02n} - Y_{2n} - \frac{1}{2}(Z_{01n} + Z_{1n} - Z_{02n} - Z_{2n}) + o_p(1),$$

it converges weakly to $Y + \frac{1}{2}(\sigma_{01}^2 - \sigma_{02}^2)$.

The result under H_0 is straightforward, replacing σ_0^2 by $\sigma_{01}^2 - \sigma_{02}^2$ in Theorem 2.6 and using the same notations. Under K_n, the mean of Y_{kn} tends $\sigma_{01}^2 - \sigma_{02}^2$ and the mean of Z_{kn} has the same limit, this is the difference of two terms obtained in the previous theorem as the mean of Z_n. The variance under K_n of $\widehat{T}_{1n}$ is the sum of the variances of the independent terms. The asymptotic critical value of the likelihood ratio test at the level α is $c_1 = (\sigma_{Y1}^2 + \sigma_{Y2}^2)^{\frac{1}{2}}c_\alpha + \frac{1}{2}(\sigma_{01}^2 - \sigma_{02}^2)$, where c_α is the normal critical value, and its local asymptotic power is

$$\beta_1(\alpha) = 1 - \inf_{f_0 \in \mathcal{F}} \sup_{\eta_1, \eta_2} \Phi\left(c_\alpha - \frac{\sigma_{01}^2 - \sigma_{02}^2}{(\sigma_{Y1}^2 + \sigma_{Y2}^2)^{\frac{1}{2}}}\right).$$

Consider a two-dimensional variable X with dependent components X_1 and X_2, and a sample of X. The density f of X is estimated by a kernel estimator, with a bandwidth h_2 of the optimal order $n^{\frac{1}{2(s+1)}}$, and its marginals f_1 and f_2 are estimated by real kernel estimators. The optimal convergence rate of the bivariate kernel estimator is now $m_n = n^{\frac{1}{s+1}}$. A sequence of density functions is defined in a neighborhood of the independence by

$$K_n : \begin{cases} f_n(x) = f_{1n}(x_1)f_{2n}(x_2)\{1 + k_n^{-\frac{1}{2}}\eta_{3n}(x)\}, \\ f_{kn}(t) = f_k(t)\{1 + k_n^{-\frac{1}{2}}\eta_{kn}(t)\}, \ k = 1,2, \end{cases}$$

for every $x = (x_1, x_2)$ of $\mathbb{X}^2$, where $(\eta_{1n})_{n\geq 1}$, $(\eta_{2n})_{n\geq 1}$ and $(\eta_{3n})_{n\geq 1}$, are sequence of uniformly bounded functions converging uniformly to nonzero functions η_1, η_2 and respectively η_3, as n tends to infinity. The density under H_0 is $f_0(x) = f_1(x_1)f_2(x_2)$, $x = (x_1, x_2)$ of $\mathbb{X}^2$. The log-likelihood ratio test statistic for H_0 against K_n is

$$\widehat{T}_{2n} = \sum_{i=1}^{k_n} \{\log \widehat{f}_{n,h}(X_i) - \log \widehat{f}_{0n,h}(X_i)\}$$

$$= \sum_{i=1}^{k_n} \left\{\log\left(1 + \frac{\widehat{f}_{n,h} - f}{f}(X_i)\right) - \log\left(1 + \frac{\widehat{f}_{0n,h} - f_0}{f_0}(X_i)\right)\right.$$

$$\left. + \log\left(1 + \frac{f - f_0}{f_0}(X_i)\right)\right\}. \tag{2.5}$$

The estimator $\widehat{f}_{0n,h}$ is such that $k_n^{\frac{1}{2}}(\widehat{f}_{0n,h} - f_0) = k_n^{\frac{1}{2}}(\widehat{f}_{1n,h} - f_1)f_2 + k_n^{\frac{1}{2}}(\widehat{f}_{2n,h} - f_2)f_1 + o_p(1)$, its convergence rate to f_0 has the order $k_n^{\frac{1}{2}}$ whereas the order of the estimator $\widehat{f}_{n,h}$ in $\mathbb{R}^2$ is m_n, it follows that the expansion of the first term of the sum (2.5) is a $o_p(1)$ and the likelihood ratio statistic $\widehat{T}_{2n}$ has the same asymptotic properties as $\widehat{T}_n$ in Theorem 2.6, σ_0^2 is replaced by $\sum_{j=1}^{3} \int_{\mathbb{R}} \eta_j^2 f_0$, $\sigma_{f_0}^2$ is replaced by $f_1^2\sigma_{f_{02}}^2 + f_2^2\sigma_{f_{01}}^2$ and b_{f_0} by $f_1 b_{f_{02}} + f_2 b_{f_{01}}$. Under K_n, the expansion of the last two terms yield

$$\widehat{T}_{2n} = \widehat{T}_n + k_n^{-\frac{1}{2}} \sum_{i=1}^{k_n} \frac{k_n^{\frac{1}{2}}(f_n - f_{0n})}{f_{0n}}(X_i)$$

$$- (2k_n)^{-1} \sum_{i=1}^{k_n} \left\{\frac{k_n^{\frac{1}{2}}(f_n - f_{0n})}{f_{0n}}\right\}^2 (X_i) + o_p(1)$$

$$= T_{0n} + \widehat{T}_n + o_p(1)$$

it converges weakly under K_n to a distribution similar to the limit of the statistic $\widehat{T}_n$ given in Theorem 2.6. Its limiting distribution is Gaussian with mean $\mu_2 = \frac{1}{2}(\sigma_1^2 + \sigma_0^2)$ and variance similar to σ_Y^2, with the modified notations. The asymptotic critical value of the likelihood ratio test at the level α is $c_2 = \sigma_Y c_\alpha + \mu_2$, where c_α is the normal critical value, and its local asymptotic power is

$$\beta_2(\alpha) = 1 - \inf_{f_0} \sup_\eta \Phi\left(c_\alpha - \frac{\sigma_0}{\sigma_Y}\right).$$

These tests are generalized in higher dimensions.

2.5   Nonparametric tests for empirical functionals

Let X be a random variable on a probability space $(\Omega, \mathcal{A}, P)$, with values in a metric space $(\mathcal{X}, \mathcal{B}, \|.\|)$ and having the distribution function F_X. Let φ be a real functional defined in a space $\mathcal{P}_\mathcal{X}$ of the distribution functions in $(\mathcal{X}, \mathcal{B})$. A parameter $\theta = \varphi(F_X)$ has the empirical estimator

$$\widehat{\theta}_n = \varphi(\widehat{F}_n),$$

where $\widehat{F}_n$ is the empirical distribution function of a sample $X_1, \ldots, X_n$ of the variable X. Assuming that φ is continuously differentiable, with derivative $\varphi^{(1)}$, the empirical estimator $\widehat{\theta}_n$ is consistent and there exists u in $]0, 1[$ such that

$$n^{\frac{1}{2}}(\widehat{\theta}_n - \theta) = \varphi^{(1)}(F_X + u(\widehat{F}_n - F_X)).\nu_n + o_p(1).$$

The weak convergence of the variable $n^{\frac{1}{2}}(\widehat{\theta}_n - \theta)$ to the transformed Brownian bridge $(\varphi^{(1)} \circ F_X).(B \circ F_X)$ is a consequence of the weak convergence of the empirical process ν_n to a transformed Brownian bridge $B \circ F_X$.

Tests about the distribution function of the variable X are obtained directly from the empirical process ν_n of a sample $X_1, \ldots, X_n$ or from a map of ν_n onto $\mathbb{R}$ which provides statistics of nonparametric tests for hypotheses concerning the distribution function of X. They have the form

$$T_n = \varphi_n(\widehat{F}_n) = \int_\mathcal{X} \varphi_n \, d\widehat{F}_n$$

for a convergent sequence of functionals $(\varphi_n)_{n \geq 1}$ of $C^1(\mathcal{P}_X)$ and square integrable under the hypothesis and the alternative of the test. Under the hypothesis H_0 of a distribution function F_X for the variable X, the mean of T_n is $\mu_{n,X} = \int_\mathcal{X} \varphi_n \, dF_X$ and its variance equals $\sigma_{n,X}^2 = n^{-1} \int_\mathcal{X} \varphi_n^2 \, dF_X - n^{-1}\mu_{n,X}^2$. Pivotal statistic $U_n = \sigma_{n,X}^{-1}(T_n - \mu_{n,X})$ and $\widehat{U}_n = \widehat{\sigma}_{n,X}^{-1}(T_n - \mu_{n,X})$, with an empirical estimators of $\sigma_{n,X}^2$, are then considered, they are asymptotically equivalent.

Proposition 2.7. *Under the condition of the existence of a function φ in $C^1(\mathcal{P}_X)$ such that $\lim_{n\to\infty} \int_{\mathcal{X}} (\varphi_n - \varphi)^2 \, dF_X = 0$, the statistic U_n converges weakly under the hypothesis H_0 to the normal variable*

$$U = \{Var\varphi(X)\}^{-\frac{1}{2}} \int_{\mathcal{X}} \varphi \, d(B \circ F_X).$$

Proof. Under the conditions, the mean and the variance of T_n satisfy Condition 2.1, $\mu_{n,X} - \int_{\mathcal{X}} \varphi \, dF_X = \int_{\mathcal{X}} (\varphi_n - \varphi) \, dF_X$ converges to zero and $n\sigma_{n,X}^2$ is asymptotically equivalent to $\sigma_X^2 = \int_{\mathcal{X}} (\varphi - \int_{\mathcal{X}} \varphi \, dF_X)^2 \, dF_X$. Morever, $E_0 T_n = \varphi_n(F_X)$ is such that $n^{\frac{1}{2}} \{\varphi_n(\widehat{F}_n) - \varphi_n(F_X)\} = \int_{\mathcal{X}} \varphi_n \, d(\nu_n \circ F_X)$ which converges weakly to $\int_{\mathcal{X}} \varphi \, d(B \circ F_X)$ therefore U_n converges weakly to a normal variable $\mathcal{N}(0,1)$. $\qquad\square$

From Proposition 2.1, the asymptotic level of two-sided tests based on statistics U_n, with a sequence of functions $(\varphi_n)_{n\geq 1}$ satisfying the condition of Proposition 2.7 is free and it is expressed from the distribution function Φ of the normal variable as $\alpha = 1 - \Phi(c_{\frac{\alpha}{2}}) + \Phi(-c_{\frac{\alpha}{2}})$. The asymptotic local power of a two-sided test for a simple hypothesis H_0 against a local alternative K_n defined by a distribution function in $\mathcal{P}_X$, $F_n = F + n^{-\frac{1}{2}} H_n$, with a function $H_n = n^{\frac{1}{2}} (F_n - F)$ converging uniformly to a function H in the tangent space H_X to F_X is

$$\inf_{H\in H_X} \{1 - \Phi(c_{\frac{\alpha}{2}} - \Delta_H) + \Phi(-c_{\frac{\alpha}{2}} - \Delta_H)\},$$

with $\Delta_H = \{Var\varphi(X)\}^{-\frac{1}{2}} \int_{\mathcal{X}} \varphi \, dH$, since $n^{\frac{1}{2}} (\mu_{n,F_n} - \mu_{n,F_X}) = \int_{\mathcal{X}} \varphi_n \, dH_n$.

Some tests for two samples rely on a weighted difference of the empirical distribution functions of the two samples. Let φ_n be a sequence of functions defined in $\mathcal{P}_X^{\otimes 2}$ and let $\varphi_n(x) = \varphi_n(x, x)$. The test statistics have the form

$$S_n = \frac{1}{n(n-1)} \left\{ \sum_{i=1}^{n} \sum_{j\neq i} \varphi_n(X_i, X_j) \right.$$

$$= \frac{n}{n-1} \int_{\mathcal{X}^{\otimes 2}} 1_{\{x\neq y\}} \varphi_n(x, y) \, d\widehat{F}_n(x) \, d\widehat{F}_n(y).$$

The variable S_n has the mean

$$\mu_n = \frac{n}{n-1} \left\{ \int_{\mathcal{X}^{\otimes 2}} \varphi_n(x, y) \, dF_X(x) \, dF_X(y) - \int_{\mathcal{X}} \varphi_n(x) \, dF_X(x) \right\}$$

and $n^{\frac{1}{2}} (S_n - \mu_n)$ is asymptotically equivalent to

$$T_n = \int_{\mathcal{X}^{\otimes 2}} \varphi_n(x, y) \, d\nu_n(x) \, dF(y) + \int_{\mathcal{X}^{\otimes 2}} \varphi_n(x, y) \, dF(x) \, d\nu_n(y)$$

$$- \int_{\mathcal{X}} \varphi_n(x) \, d\nu_n(x).$$

As φ_n converges to a function φ of $L^2(\mathcal{P}_X^2)$, the variable $n^{\frac{1}{2}}(S_n - \mu_n)$ converges weakly to a centered variable

$$T = \int_{\mathcal{X}^{\otimes 2}} \varphi(x, y) \, d(B \circ F_X)(x) \, dF_X(y) + \int_{\mathcal{X}^{\otimes 2}} \varphi(x, y) \, dF_X(x) \, d(B \circ F_X)(y)$$
$$- \int_{\mathcal{X}} \varphi(x) \, d(B \circ F_X)(x).$$

Let $\sigma_{1,\varphi}^2$ denote the variance of $\int_{\mathcal{X}} \varphi(x, y) \, d(B \circ F_X)(x) \, dF_X(y)$, let $\sigma_{2,\varphi}^2$ be the variance of $\int_{\mathcal{X}} \varphi(x, y) \, dF_X(x) \, d(B \circ F_X)(y)$, and let $\sigma_{3,\varphi}^2$ be the variance of $\int_{\mathcal{X}} \varphi(x) \, d(B \circ F_X)(x)$, they are written

$$\sigma_{1,\varphi}^2 = \int_{\mathcal{X}^{\otimes 4}} \varphi(x, y) \varphi(u, v) \, d\{F_X(x \wedge u) - F_X(x) F_X(u)\} \, dF_X(y) \, dF_X(v)$$
$$= \int_{\mathcal{X}^{\otimes 4}} \varphi(x, y) \varphi(u, v) \, [1_{\{x \leq u\}} \{1 - F_X(u)\} \, dF_X(x)$$
$$+ 1_{\{u \leq x\}} \{1 - F_X(x)\} \, dF_X(u)] \, dF_X(y) \, dF_X(v),$$

$$\sigma_{2,\varphi}^2 = \int_{\mathcal{X}^{\otimes 4}} \varphi(x, y) \varphi(u, v) \, d\{F_X(y \wedge v) - F_X(x) F_X(v)\} \, dF_X(x) \, dF_X(u)$$
$$= \int_{\mathcal{X}^{\otimes 4}} \varphi(x, y) \varphi(u, v) \, [1_{\{y \leq v\}} \{1 - F_X(v)\} \, dF_X(y)$$
$$+ 1_{\{v \leq y\}} \{1 - F_X(y)\} \, dF_X(v)] \, dF_X(x) \, dF_X(u),$$

$$\sigma_{3,\varphi}^2 = \int_{\mathcal{X}^{\otimes 2}} \varphi(x) \varphi(u) \, [1_{\{x \leq u\}} \{1 - F_X(u)\} \, dF_X(x)$$
$$+ 1_{\{u \leq x\}} \{1 - F_X(x)\} \, dF_X(u)],$$

their covariances are

$$\sigma_{12,\varphi}^2 = \int_{\mathcal{X}^{\otimes 4}} \varphi(x, y) \varphi(u, v) \, d\{F_X(x \wedge v) - F_X(x) F_X(v)\} \, dF_X(y) \, dF_X(u),$$

$$\sigma_{13,\varphi}^2 = \int_{\mathcal{X}^{\otimes 4}} \varphi(x, y) \varphi(u) \, d\{F_X(x \wedge u) - F_X(x) F_X(u)\} \, dF_X(y),$$

$$\sigma_{23,\varphi}^2 = \int_{\mathcal{X}^{\otimes 4}} \varphi(x, y) \varphi(u) \, d\{F_X(y \wedge u) - F_X(y) F_X(u)\} \, dF_X(x),$$

then the variance of the variable T is the sum $\sigma_\varphi^2 = \sigma_{1,\varphi}^2 + \sigma_{2,\varphi}^2 + \sigma_{3,\varphi}^2 + 2(\sigma_{12,\varphi}^2 - \sigma_{13,\varphi}^2 - \sigma_{23,\varphi}^2)$ and its empirical estimator $\widehat{\sigma}_{\varphi,n}^2$ is obtained by plugging the empirical distribution functions in this expression. The normalized statistics $U_n = \widehat{\sigma}_{\varphi,n}^{-1}(S_n - \mu_n)$ provide two-sided tests of hypotheses for the distributions of two samples. Local alternatives K_n are defined by distribution functions in $\mathcal{P}_X$, $F_n = F + n^{-\frac{1}{2}} H_n$, with functions $H_n = n^{\frac{1}{2}}(F_n - F)$ converging uniformly to a function H in the tangent space to F. Let H_0 be the simple hypothesis of a distribution function F_X.

Proposition 2.8. *If there exist functions φ of $C^1(\mathcal{P}_X)$ such that $\lim_{n \to \infty} \int_{\mathcal{X}} (\varphi_n - \varphi)^2 \, dF_X = 0$, the statistic U_n converges weakly under the hypothesis H_0 to a normal variable. The asymptotic local power of a two-sided test for H_0 against the alternative K_n is*

$$\beta_F(\alpha) = \inf_{H \in H_X} \{1 - \Phi(c_{\frac{\alpha}{2}} - \Delta_{\varphi, H}) + \Phi(-c_{\frac{\alpha}{2}} - \Delta_{\varphi, H})\},$$

with

$$\Delta_{\varphi, H} = \sigma_\varphi^{-1} \left[\int_{\mathcal{X}^{\otimes 2}} \varphi(x, y) \, d\{H(x)F(y) + F(x)H(y)\} - \int_{\mathcal{X}} \varphi \, dH \right].$$

Proof. Under K_n, the variable S_n has the mean

$$\mu_{n,n} = \frac{n}{n-1} \left\{ \int_{\mathcal{X}^{\otimes 2}} \varphi_n(x, y) \, dF_n(x) \, dF_n(y) - \int_{\mathcal{X}} \varphi_n(x) \, dF_n(x) \right\}$$

$$= \mu_n + \frac{n^{\frac{1}{2}}}{n-1} \left[\int_{\mathcal{X}^{\otimes 2}} \varphi_n(x, y) \{dH_n(x) \, dF(y) + dH_n(y) \, dF(x)\} \right.$$

$$\left. - \int_{\mathcal{X}} \varphi_n(x) \, dH_n(x) \right] + o(n^{-\frac{1}{2}}).$$

The variable $n^{\frac{1}{2}}(S_n - \mu_{n,n})$ is asymptotically equivalent to a centered variable $T_{n,n}$ such that $T_n = T_{n,n} + \delta_n$, where $\delta_n = n^{\frac{1}{2}}(\mu_{n,n} - \mu_n)$ converges to $\delta_{\varphi, H} = \int_{\mathcal{X}^{\otimes 2}} \varphi(x, y)\{dH(x) \, dF(y) + dF(x) \, dH(y)\} - \int_{\mathcal{X}} \varphi \, dH$ as n tends to infinity. Let $U_{n,n} = \hat{\sigma}_{\varphi,n}^{-1}(S_n - \mu_{n,n})$ and let $\Delta_{\varphi_n, H_n} = \hat{\sigma}_{\varphi,n}^{-1} \delta_n$. The asymptotic local power of the test against a local alternative K_n is deduced from the shift of the asymptotically normal variable $U_{n,n}$

$$\inf_{H \in H_X} \lim_{n \to \infty} P_{K_n}(|U_n| > c_{n,\alpha}) = \inf_{H \in H_X} \lim_{n \to \infty} P_{K_n}(|U_{n,n} + \Delta_{\varphi_n, H_n}| > c_{n,\alpha}).$$

The result follows from the weak convergence of $U_{n,n}$ to a normal variable under K_n and of the convergence of Δ_{φ_n, H_n} to $\Delta_{\varphi, H}$. $\square$

The approximation of a normalized sum by a normal variable is due to the law of large numbers and the approximation is improved by an Egdeworth expansion (Feller, 1966, Abramovitch and Singh, 1985). By an inversion of the expansion, the asymptotic threshold c_α of the test is corrected according to the terms of higher orders in the Egdeworth expansion.

2.6 Tests of homogeneity

Let $(X_1, \ldots, X_n)$ be a sequence of independent variables sampled in a metric space $\mathbb{X}$ with a distribution function F. In a mixture model with k

components, the probability space is split into k subspaces $\Omega = \cup_{1 \leq j \leq k} \Omega_j$ and the distribution F is written as $F = \sum_{j=1}^{k} \lambda_j F_j$, this is the mixture of k distribution functions F_j with respective probabilities λ_j belonging to $]0,1[$, $j = 1, \ldots, k$. The null hypothesis of homogeneity is $H_0 : F_1 = \cdots = F_k$ denoted F_0 and its alternative H_k is a mixture of k distinct distribution functions $F_1, \ldots, F_k$. The sample is supposed to contain k sub-samples of respective size n_j, with empirical distribution functions $\widehat{F}_{jn}$ and mixture probabilities $\widehat{\lambda}_{jn} = n^{-1} n_j$, $j = 1, \ldots, k$, and $\sum_{j=1}^{k} n_j = n$. The empirical estimator of the distribution function F is always denoted $\widehat{F}_n$.

Under H_2 and the conditions of Theorem 2.3, the distribution function of the sample is $F = \lambda F_1 + (1 - \lambda) F_2$ and each distribution function has a continuous density of $C^2(\mathbb{X}) \cap L^2(\mathbb{X})$. The first order term of the expansion of the log-likelihood ratio statistic T_n is

$$U_n = n^{\frac{1}{2}} \widehat{\lambda}_n \int_{\mathbb{X}} \frac{f_1 - f_0}{f_0} \, d\widehat{F}_{1n} + n^{\frac{1}{2}} (1 - \widehat{\lambda}_n) \int_{\mathbb{X}} \frac{f_2 - f_0}{f_0} \, d\widehat{F}_{2n}$$

$$= n^{\frac{1}{2}} \int_{\mathbb{X}} \frac{f_1 - f_2}{f_0} \{ \widehat{\lambda}_n (1 - \lambda) \, d\widehat{F}_{1n} - \lambda (1 - \widehat{\lambda}_n) \, d\widehat{F}_{2n} \}.$$

The mean of U_n is zero under the hypothesis H_0, under the alternative H_2

$$E_2 U_n = n^{\frac{1}{2}} \lambda (1 - \lambda) \int_{\mathbb{X}} \frac{(f_1 - f_2)^2}{f_0} \, dx,$$

it is strictly positive and tends to infinity. The unknown densities of U_n are estimated using a kernel function K and a bandwidth $h = h_n$ which tends to zero as n tends to infinity. For all x in $\mathbb{X}$ and $j = 0, 1, 2$,

$$\widehat{f}_{jn}(x) = \int_{\mathbb{X}} K_h(x - s) \, d\widehat{F}_{jn}(s),$$

with these estimators, an estimator of U_n is proportional to

$$S_n = n^{\frac{1}{2}} \int_{\mathbb{X}} \frac{\widehat{f}_{1n} - \widehat{f}_{2n}}{\widehat{f}_n} \, d(\widehat{F}_{1n} - \widehat{F}_{2n})$$

$$= n^{\frac{1}{2}} \int_{\mathbb{X}} \frac{K_h(x - s) \, d(\widehat{F}_{1n} - \widehat{F}_{2n})(s)}{K_h(x - s) \, d\widehat{F}_n(s)} \, d(\widehat{F}_{1n} - \widehat{F}_{2n})(x)$$

$$= n^{\frac{1}{2}} \int_{\mathbb{X}} \frac{K_h(x - s) \, d(\widehat{F}_{1n} - \widehat{F}_{2n})(s)}{K_h(x - s) \, d\{\widehat{\lambda}_n \widehat{F}_{1n}(s) + (1 - \widehat{\lambda}_n) \widehat{F}_{2n})(s)\}} \, d(\widehat{F}_{1n} - \widehat{F}_{2n})(x).$$

The statistic S_n is asymptotically equivalent to the log-likelihood ratio statistic of Theorem 2.7.

Theorem 2.8. *Under Condition 2.3, a test built with the statistic U_n is asymptotically equivalent to the local log-likelihood ratio test for the homogeneity of two samples. It converges weakly under H_0 to a centered Gaussian variable with variance*

$$\sigma_S^2 = \lambda^2 \left\{ \int_{\mathbb{X}} \phi \, dF_1 - \left(\int_{\mathbb{X}} \phi \, dF_1 \right)^2 \right\} + (1 - \lambda)^2 \left\{ \int_{\mathbb{X}} \phi \, dF_2 - \left(\int_{\mathbb{X}} \phi \, dF_2 \right)^2 \right\},$$

with $\phi = (f_1 - f_2) f_0^{-1}$.

Proof. The statistic U_n is equivalent to the first order term of the expansion of the log-likelihood ratio in Theorem 2.3, under H_0 and it is centered. Under the alternative H_2, it is equivalent to the second order term of its expansion. Under H_0, the statistic U_n converges weakly to a centered Gaussian process and a test based on its normalization is therefore asymptotically equivalent to a Gaussian test based the log-likelihood ratio statistic. Under H_2, U_n is asymptotically equivalent to a Gaussian variable with a mean proportional to its variance, which is the difference of the means of the log-likelihood ratio statistic under the hypothesis and the local alternative.

Under H_0 the difference $\widehat{\lambda}_n (1 - \widehat{\lambda}_n) S_n - U_n$ converges to zero in probability, by the consistency of the density estimators and by the weak convergence of the empirical process $n^{\frac{1}{2}}(\widehat{F}_{1n} - \widehat{F}_{2n})$ to a centered Gaussian process. It also converges to zero under local alternatives of H_2, by contiguity of the models and by the expansion of Theorem 2.3, and it tends to infinity under H_2. $\qquad\square$

We consider local alternatives K_n of densities f_1 and f_2 similar to the alternative of Theorem 2.3, with neighborhoods converging at the rate $n^{-\frac{1}{4}}$ which is larger than the nonparametric convergence rate of the kernel estimators. The statistic S_n converges weakly under the local alternatives to a Gaussian variable with the variance σ_S^2 and its asymptotic mean is $\mu_\eta = \lambda (1 - \lambda) \int_{\mathbb{X}} \eta^2 f_0 \, dx$.

Let $\widehat{\sigma}_n^2$ be the empirical estimator of the asymptotic variance σ_S^2. The critical value at the level α of the test based on $T_n = \widehat{\sigma}_n^{-1} \widehat{\lambda}_n (1 - \widehat{\lambda}_n) S_n$ is the normal quantile c_α and its asymptotic power for H_0 against the local alternatives K_n is $\beta(\alpha) = \inf_{\eta \neq 0} \{ 1 - \Phi(c_\alpha - \sigma_S^{-1} \mu_\eta) \}$.

Under the alternative and when the change of distribution function occurs after an unknown sample size, the mixture proportions are unknown and they cannot be estimated directly from the sample, this alternative is denoted K_2. A test of homogeneity against the alternative K_2 is performed with a varying sample size m for the first sub-sample. The statistic S_n is replaced by

$$S_n^* = n^{-\frac{1}{2}} \min_{m=1,\ldots,n} S_{n,m}$$

such that $S_{n,m}$ is the statistic S_n performed with the value m for the sample size n_1. It tends to infinity under K_2 and it converges weakly to a centered distribution under H_0. The sample size n_1 is consistently estimated under the alternative K_2 by

$$\widehat{m}_n = \arg \min_{m=1,\ldots,n} S_{n,m}$$

since the estimated value is uniquely defined. This problem is the estimation of a change-point in a sample and the asymptotic distribution of the maximum likelihood estimators have been studied by Pons (2009). With parametric densities, the convergence rate of $\widehat{m}_n$ to the change-point value m is n and the parameter estimators are $n^{\frac{1}{2}}$-consistent.

Under H_k and the conditions of Theorem 2.3, the distribution function of the sample is $F = \sum_{j=1}^{k} \lambda_j F_j$ and each distribution function has a continuous density of $C^2(\mathbb{X}) \cap L^2(\mathbb{X})$. The first order term of the expansion of the log-likelihood ratio statistic T_n is

$$Y_{kn} = n^{\frac{1}{2}} \sum_{j=1}^{k} \widehat{\lambda}_{jn} \int_{\mathbb{X}} \frac{f_j - f_0}{f_0} \, d\widehat{F}_{jn}.$$

The mean of Y_{kn} in the mixture model with k-density components is

$$E_k Y_{kn} = n^{\frac{1}{2}} \sum_{j=1}^{k} \lambda_j \int_{\mathbb{X}} \frac{f_j - f_0}{f_0} \, dF_j$$

and it is zero under H_0, therefore it also equals

$$n^{\frac{1}{2}} \sum_{j=1}^{k} \lambda_j \int_{\mathbb{X}} \left(\frac{f_j}{f_0} - 1 \right) d(F_j - F_0).$$

The modified variable

$$U_{kn} = n^{\frac{1}{2}} \sum_{j=1}^{k} \widehat{\lambda}_{jn} \int_{\mathbb{X}} \frac{f_j - f_0}{f_0} \, d(\widehat{F}_{jn} - \widehat{F}_n)$$

is always positive and it tends to infinity under the alternative H_k. The kernel estimators of unknown densities are used for the estimation of U_n by plugging, which yields

$$S_n = n^{\frac{1}{2}} \sum_{j=1}^{k} \widehat{\lambda}_{jn} \int_{\mathbb{X}} \frac{\widehat{f}_{jn} - \widehat{f}_n}{\widehat{f}_n} \, d(\widehat{F}_{jn} - \widehat{F}_n).$$

The asymptotic variance of S_n is defined with the notation $\phi_j = f_j f_0^{-1} - 1$ as the sum of the variances in the sub-samples

$$\sigma_S^2 = \sum_{j=1}^{k} \lambda_j^2 \left\{ \int_{\mathbb{X}} \phi_j \, dF_j - \left(\int_{\mathbb{X}} \phi_j \, dF_j \right)^2 \right\}.$$

Theorem 2.9. *Under Condition 2.3, the test built on the statistic S_n is asymptotically equivalent to the log-likelihood ratio test of homogeneity of k sub-samples. It converges weakly under H_0 to a centered Gaussian variable with variance σ_S^2.*

Under local alternatives of the densities f_j similar to K_n for Theorem 2.3 with the convergence rate $n^{-\frac{1}{4}}$, the statistic S_n converges weakly under the local alternatives to a Gaussian variable with the variance σ_S^2 and the asymptotic mean $\mu = \sum_{j=1}^{k} \lambda_j \int_{\mathbb{X}} (\eta_j + \gamma)^2 f_0 \, dx$. Let $\hat{\sigma}_n^2$ be the empirical estimator of the asymptotic variance σ_S^2. The critical level at the level α of the test based on the statistic $T_n = \hat{\sigma}_n^{-1} \hat{\lambda}_n (1 - \hat{\lambda}_n) S_n$ is the normal quantile c_α and its asymptotic power for H_0 against the local alternatives K_n is $\beta(\alpha) = \inf_{\eta 1, \ldots, \eta_k, \gamma \neq 0} \{1 - \Phi(c_\alpha - \sigma_S^{-1} \mu_{\eta,\gamma})\}$.

Optimal goodness of fit tests for parametric families of densities are defined from the likelihood ratio. Many tests have been specifically defined as asymptotically locally optimal tests in parametric families. The density under a change of location of a density f in $\mathbb{R}$ is $f(\cdot - \theta)$, its score under H_0 is $-f^{-1} f^{(1)}$ and it is $-f^{-1} f^{(1)}(\cdot - \theta)$ under the probability P_θ. Let f_θ be the Gaussian density with mean θ and variance σ^2. Consider local alternatives of the hypothesis H_0 of the normal density f, defined by K_n : $f(\cdot - \theta_n)$, $\theta_n = n^{-\frac{1}{2}} \rho_n$ and bounded ρ_n converging to $\rho > 0$. Under K_n, the variable $n^{-\frac{1}{2}} \sum_{i=1}^{n} (X_i - \theta_n) = n^{-\frac{1}{2}} \sum_{i=1}^{n} X_i - \rho_n$ has a bounded translation, it is denoted $Y_n - \rho_n$. The log-likelihood ratio depends on θ_n through $\sum_{i=1}^{n} X_i^2 - \sum_{i=1}^{n} (X_i - \theta_n)^2 = 2\rho_n Y_n - \theta_n^2$ and the variable $\hat{\sigma}_n^{-1} Y_n$ is sufficient for a test of H_0 against K_n.

A change of scale of a density f has the density $f_\sigma = \sigma^{-1} f(\sigma^{-1} \cdot)$ under a Gaussian probability $P_{0,\sigma}$. Consider local alternatives of the hypothesis H_0 of the normal density f, defined by K_n : f_{σ_n}, $\sigma_n^2 = 1 + n^{-\frac{1}{2}} \zeta_n$, with ζ_n converging to $\zeta > 0$. Under K_n and up to an additive constant, the log-likelihood ratio is

$$\frac{1}{2} \left\{ \left(\sum_{i=1}^{n} X_i^2 \right) (1 - \sigma_n^{-2}) - n \log \sigma_n \right\} = \frac{\zeta_n}{2} n^{-\frac{1}{2}} \left(\sum_{i=1}^{n} X_i^2 - 1 \right)$$

and the statistic $Y_{2n} = n^{-\frac{1}{2}} \sum_{i=1}^{n}(X_i^2 - 1)$ is sufficient for an asymptotically locally optimal test of H_0 against K_n. The asymptotically locally optimal test of the normal density against changes of location and scale are based on the sufficient statistic (Y_n, Y_{2n}).

With a density f, the log-likelihood ratio for a local change of location has the expansion

$$l_n = -\rho n^{-\frac{1}{2}} \sum_{i=1}^{n} f^{-1} f^{(1)}(X_i) - \frac{\rho^2}{2} \int f^{-1} f^{(1)2} + o_p(1).$$

Let $I_n(f) = n^{-1} \sum_{i=1}^{n} \{f^{-1} f^{(1)}(X_i)\}^{\otimes 2}$ be the empirical information matrix, the normalized score statistic $T_{n,f} = I_n^{-1}(f) n^{-\frac{1}{2}} \sum_{i=1}^{n} f^{-1} f^{(1)}(X_i)$ defines an asymptotically optimal local test, for every density f under the conditions of the expansion. The log-likelihood ratio for a local change of scale $\lambda_n = 1 + n^{-\frac{1}{2}} \zeta_n$, such that $\lim_{n \to \infty} \zeta_n = \zeta$, has the expansion

$$l_n = \zeta_n n^{-\frac{1}{2}} \sum_{i=1}^{n} f^{-1} f^{(1)}(X_i) - \frac{1}{2} \zeta_n^2 I_n(f) + o_p(1)$$

and the normalized score statistic $T_{n,f}$ still defines an asymptotically optimal local test.

2.7   Mixtures of exponential distributions

Let X be an exponential variable with a random parameter $\theta > 0$ and let G be the distribution function of the variable θ in $\mathbb{R}_+$. The distribution function F_X of the variable X is the continuous mixtures of exponential distributions

$$F_X(x) = 1 - \int_0^{\infty} e^{-\theta x} \, dG(\theta). \tag{2.6}$$

Let $(X_i)_{i=1,\ldots,n}$ be a sample of X, an empirical estimator of the distribution function G is written in the form

$$G_n(\theta) = \sum_{k=1}^{K_n} p_{kn} 1_{\{\theta_{kn} \leq \theta\}}, \; \theta > 0,$$

assuming that the jumps of the estimator of G have an amplitude p_{kn} in at most $K_n = O(n)$ values θ_{kn} in $\mathbb{R}_+$. The values p_{kn} are the probabilities at θ_{kn} of a discrete version of the mixing distribution function G and their sum is $\sum_{k=1}^{K_n} p_{kn} = 1$. The survival function $\bar{F}_X = 1 - F_X$ of the variable X is then approximated by $\bar{F}_{Xn}(x) = \sum_{k=1}^{K_n} e^{-\theta_{kn} x} p_{kn}$ as n tends to infinity.

It is differentiable and we assume that G has a density, then the density f_X of X is approximated by

$$f_n(x) = \sum_{k=1}^{K_n} \theta_{kn} e^{-\theta_{kn}x} p_{kn}. \tag{2.7}$$

An empirical maximum likelihood estimator of the unknown distribution function G consists in estimating the probabilities p_{kn} and of their locations θ_{kn} by maximization of the approximated likelihood of the sample under the constraint $\sum_{k=1}^{K_n} p_{kn} = 1$. From (2.7), the likelihood $\prod_{i=1}^{n} f(X_i)$ is approximated by

$$L_n(\theta, p) = \prod_{i=1}^{n} \left\{ \sum_{k=1}^{K_n} \theta_{kn} e^{-\theta_{kn}X_i} p_{kn} \right\}$$

and the maximum likelihood estimators of the vector $(\theta_{kn}, p_{kn})_{k=1,\ldots,K_n}$ provide a nonparametric estimator of the distribution function G. Let $\widehat{F}_n$ be the empirical distribution function of the sample, an empirical maximum likelihood estimator of the density is $\widehat{f}_{n,K_n}$ which estimates the approximated density f_n given by (2.7), using the maximum likelihood estimators of the parameters.

For $k = 1, \ldots, K_n$, the maximum likelihood estimators of the parameters θ_{kn} and p_{kn}, $k = 1, \ldots, K_n$, are solutions of the score equations defined as the first derivatives of the log-likelihood $\log L_n(\theta, p)$ with respect to the components of θ and p, under the constraint $\sum_{k=1}^{K_n} p_{kn} = 1$. It follows that the estimators of the parameters θ_{kn} and $p_{kn} = g(\theta_{kn})$ maximizing $L_n(\theta, p)$ are solutions of the equations

$$\int_0^\infty \frac{(1 - \theta_{kn}x)e^{-\theta_{kn}x}}{f_n(x)} d\widehat{F}_n(x) = 0, \tag{2.8}$$

$$\int_0^\infty \theta_{kn} e^{-\theta_{kn}x} f_n^{-1}(x) d\widehat{F}_n(x) = \int_0^\infty \theta_{K_n n} e^{-x\theta_{K_n n}} f_n^{-1}(x) d\widehat{F}_n(x).$$

Proposition 2.9. *For $k = 1, \ldots, K_n$, the maximum likelihood estimators of the parameters θ_{kn} satisfy, for $k = 1, \ldots, K_n$*

$$\frac{1}{\widehat{\theta}_{kn}} = \int_0^\infty x \frac{f_{X|\widehat{\theta}_{kn}}}{\widehat{f}_n(x)} d\widehat{F}_n(x),$$

$$\widehat{p}_{kn} = \int_0^\infty \frac{f_{\widehat{\theta}_{kn}|X}}{\sum_{j=1}^{K_n} f_{\widehat{\theta}_{jn}|X}} d\widehat{F}_n.$$

Proof.   Multiplying the second equation of (2.8) by p_{kn} and summing them up to $K_n - 1$ implies that for $k = 1, \ldots, K_n$

$$\int_0^\infty \widehat{\theta}_{kn} e^{-x\widehat{\theta}_{kn}} \widehat{f}_n^{-1}(x) \, d\widehat{F}_n(x) = 1,$$

then the first equation of (2.8) becomes

$$\frac{1}{\widehat{\theta}_{kn}} = \int_0^\infty \frac{\widehat{\theta}_{kn} x e^{-\widehat{\theta}_{kn} x}}{\widehat{f}_n(x)} \, d\widehat{F}_n(x) = \int_0^\infty x \frac{f_{X|\widehat{\theta}_{kn}}}{\widehat{f}_n(x)} \, d\widehat{F}_n(x).$$

From the Bayes rule, the second equation of (2.8) is rewritten in the equivalent form

$$\int_0^\infty f_{\theta|X}(\theta_{kn}; x) p_{kn}^{-1} \, d\widehat{F}_n(x) = \int_0^\infty f_{\theta|X}(\theta_{K_n n}; x) p_{K_n n}^{-1}(x) \, d\widehat{F}_n(x),$$

due to the expression of $p_{kn} = g(\theta_{kn})$ and the expression of the estimator of p_{kn} follows by summing over k. □

From the joint distribution function of (X, θ), the mean of the variable X^p is $EX^p = \int_0^\infty \theta^{-p} dG(\theta)$ and by the expression of G_n, it is approximated by $\sum_{k=1}^{K_n} p_{kn} \theta_{kn}^{-p}$ and estimated by the empirical mean $(\bar{X}^p)_n$ of the variables $(X_i^p)_{i=1,\ldots,n}$. Using the constraint on the probabilities and the moment estimators, Equations (2.8) can be solved numerically using recursively an EM algorithm based on the density of θ conditionally on X

$$f_{\theta|X}(\theta, x) = \frac{e^{-x\theta} g(\theta)}{\int_0^\infty e^{-x\theta} g(\theta) \, d\theta}.$$

Bootstrap tests on the number of components K_n sufficient for the approximation of G by G_n, with a sum of K terms, can be performed as in Proposition 2.7, using the statistics $T_n(K) = n^{-1} \sum_{i=1}^n S_n^2(X_i, K)$, with the processes

$$S_n(x, K) = n^{-1} \sum_{i=1}^n 1_{\{X_i \geq x\}} - \sum_{k=1}^K \widehat{p}_{kn} e^{-x\widehat{\theta}_{kn}}. \tag{2.9}$$

In a generalized exponential model, the conditional density $f_{X|Y}$ of the variable X, given a dependent real variable Y, is expressed as the exponential of a sufficient statistic $T(X, Y)$, $f_{X|Y}(x; Y) = \exp\{T(x, Y) - b(Y)\}$, with the normalization function $b(Y) = \log \int \exp\{T(x, Y)\} \, dx$. More generally, the variable X is supposed to have a semi-parametric density

$$f_{X|Y,\eta}(x; Y) = \exp\{\eta^t T(x, Y) - b(\eta, Y)\}$$

with $b(\eta, Y) = \log \int \exp\{\eta^t T(x, Y)\} \, dx$, and the density of $\log X$ conditionally on Y is the linear function $\eta^t T(x, Y) - b(\eta, Y)$. When Y has the distribution function F_Y, the variable X has the density $f_\eta(x) = E_Y f_{X|Y,\eta}(x; Y)$ with an expectation with respect to F_Y.

Let $E_{X|Y}$ be the conditional expectation with respect to the probability distribution of X conditionally on Y. When the distribution function F_Y is known, the distribution function of X is parametric with parameter η. The derivative with respect to η of $\log f_\eta$ is $f_\eta^{-1} \dot{f}_\eta$ defined by

$$\dot{f}_\eta(x) = E_Y[\{T(x, Y) - E_{X|Y} T(X, Y)\} f_{X|Y,\eta}(x; Y)]$$

with $\dot{b}_\eta(\eta, Y) = E_{X|Y} T(X, Y)$ and the information matrix of the sample is

$$I_\eta = E\{f_\eta^{-1}(X) \dot{f}_\eta(X)\}^{\otimes 2} = \int f_\eta^{-1}(x) \dot{f}_\eta^{\otimes 2}(x) \, dx.$$

For the sample $(X_i)_{1 \leq i \leq n}$, the maximum likelihood estimator of η is solution of the score equation $\dot{l}_n(\eta) = \sum_{i=1}^n f^{-1}(X_i; \eta) \dot{f}_\eta(X_i) = 0$ and its asymptotic behavior follows the classical theory of the parametric maximum likelihood. In particular

$$\widehat{\eta}_n = \eta_0 + I_{n,\eta_0}^{-1} n^{-1/2} \sum_{i=1}^n f^{-1}(X_i; \eta_0) \dot{f}_{\eta_0}(X_i; \eta_0) + o_p(n^{-1/2}),$$

where the matrix $I_{n,\eta} = n^{-1} \sum_{i=1}^n f_\eta^{-2}(X_i) \dot{f}_\eta^{\otimes 2}(X_i)$ converges to I_η, uniformly in a neighborhood of the true parameter value η_0. The normalized variable $n^{-1/2} \sum_{i=1}^n f^{-1}(X_i; \eta_0) \dot{f}_{\eta_0}(X_i)$ converges weakly to a Gaussian variable with mean zero and variance $I_0 = I_{\eta_0}$, then $n^{1/2}(\widehat{\eta}_n - \eta_0)$ converges weakly to a Gaussian variable with distribution $\mathcal{N}(0, I_0^{-1})$. Tests about the parameter η are built on the score statistic $n^{-1} \dot{l}_{n,\eta}(X_i)$, it converges to a centered Gaussian variable with variance I_η, estimated by $I_{n,\eta}$.

Assuming that the mixing distribution function F_Y of the latent variable Y is unknown and belongs to a regular parametric family $\mathcal{G} = \{G_\gamma, \gamma \in \Gamma\}$ for a k-dimensional real set Γ, the parameter γ is a nuisance parameter for the estimation of η. The density of X is written $f_{\eta,\gamma}(x) = E_\gamma f_{X|Y,\eta}(x; Y)$, the score function for η is the vector $f_{\eta,\gamma}^{-1} \dot{f}_{\eta,\gamma}(x; \eta) = f_{\eta,\gamma}^{-1} E_\gamma \dot{f}_{X|Y,\eta}(x; Y)$ and the score function for γ is $f_{\eta,\gamma}^{-1} \dot{f}_{\eta,\gamma}(x; \gamma)$ where the derivative of $f_{\eta,\gamma}$ with respect to γ is $\dot{f}_{\eta,\gamma}(x; \gamma) = \int f_{X|Y,\eta}(x; y) \dot{g}_\gamma(y) \, dy$. The asymptotic results for the maximum likelihood estimators of parametric densities with nuisance parameters apply, the score function and the estimators are asymptotically Gaussian and asymptotically normal statistics can be used.

With a nonparametric mixing distribution, the distribution function F_Y is approximated by a countable function

$$F_{Yn}(y) = \sum_{k=1}^{K_n} p_{kn} 1_{\{y_{kn} \leq y\}},$$

where the probabilities

$$p_{kn} = f_{Yn}(y_{kn}) \tag{2.10}$$

have a sum $\sum_{k=1}^{K_n} f_{Yn}(y_{kn}) = \int_0^\infty dF_{Yn} = 1$. The empirical distribution function of the observed variable X is denoted $\widehat{F}_{Xn}$. Let $f_{X|Y,\eta}(x; y)$ be the density of $F_{X|Y,\eta}$, it is supposed to be twice continuously differentiable with respect to the parameter η and with respect to y, with first derivative with respect to η, $\dot{f}_{X|Y,\eta}$ and $f^{(1)}_{X|Y,\eta}(x; y)$, with respect to y. Under the constraint $\sum_{k=1}^{K_n} p_{kn} = 1$, the maximum likelihood estimators of the probabilities p_{kn} are

$$\widehat{p}_{nk} = \frac{\int f_{Y|X,\widehat{\eta}_n}(\widehat{y}_{nk}; x)\, d\widehat{F}_n(x)}{\sum_{k=1}^{K_n} \int f_{Y|X,\widehat{\eta}_n}(\widehat{y}_{nk}; x)\, d\widehat{F}_n(x)}, \quad k = 1, \ldots, K_n,$$

where $\widehat{\eta}_n$ and $(\widehat{y}_{nk})_{k \leq K_n}$ are the maximum likelihood estimators of the parameters η and $(y_{kn})_{k \leq K_n}$. They are solutions of the equations

$$0 = \sum_{i=1}^n \frac{\sum_{k=1}^{K_n} \widehat{p}_{nk} \dot{f}_{X|Y,\eta}(X_i; y_{K_n,n})}{f_{X,\eta}(X_i)},$$

$$0 = \sum_{i=1}^n \frac{\widehat{p}_{nk} f^{(1)}_{X|Y,\eta,k}(X_i; y_{kn})}{f_{X,\eta}(X_i)}, \quad k = 1, \ldots, K_n.$$

The first equation is written

$$\int f_{X|\eta}^{-1}(x) \sum_{k=1}^{K_n} \widehat{p}_{nk}\{T(x, y_{kn}) - E_{X|Y} T(X, y_{kn})\} f_{X|Y,\eta}(x, y_{kn})\, d\widehat{F}_n(x) = 0$$

and the second equation is equivalent to

$$\int f_{X|\eta}^{-1}(x) \sum_{k=1}^{K_n} \widehat{p}_{nk}\{T_y^{(1)}(x, y_{kn}) - E_{X|Y} T_y^{(1)}(X, y_{kn})\} f_{X|Y,\eta}(x, y_{kn})\, dx = 0.$$

If η is a vector of dimension d, $K_n + d$ parameters are estimated by these equations, they converge to the true parameter values as n tends to infinity if $K_n = o(n)$, by the classical theory of the maximum likelihood estimation. Tests about the number of point which can be estimated for the distribution of the unobserved variable Y can be performed with processes similar to (2.9) defined for mixtures of exponential distributions.

2.8 Nonparametric bootstrap tests

Let F be a distribution function F of $C^1(\mathbb{R})$, let $\widehat{F}_n$ be its empirical distribution function for a n-sample and let $\nu_n = n^{\frac{1}{2}}(\widehat{F}_n - F)$ be the related empirical process. At $y_{n,i} = \widehat{F}_n(X_i)$, x in $\mathbb{R}$, the quantiles of $\widehat{F}_n$ satisfy by differentiability of the inverse function F^{-1}, the next equalities

$$F^{-1}(y_{n,i}) = F^{-1}\{F(X_i) + n^{-\frac{1}{2}}\nu_n(X_i)\} = X_i + \frac{\nu_n}{f}(X_i) + o(1),$$

$$n^{\frac{1}{2}}\{\widehat{F}_n^{-1}(y_{n,i}) - F^{-1}(y_{n,i})\} = -\frac{\nu_n}{f}(X_i) + o(1),$$

$$n^{\frac{1}{2}}|\widehat{F}_n^{-1}(y) - F^{-1}(y)| \le \left|\frac{\nu_n}{f} \circ F^{-1}(y)\right| + o(1)$$

for every real y. Its bootstrap $\nu_n^* = n^{\frac{1}{2}}(\widehat{F}_n^* - \widehat{F}_n)$ satisfies the similar inequalities at $y_{n,i}^* = \widehat{F}_n^*(X_i^*)$

$$n^{\frac{1}{2}}\{\widehat{F}_n^{*-1}(y_{n,i}^*) - \widehat{F}_n^{-1}(y_{n,i}^*)\} = -\frac{\nu_n^*}{f}(X_i^*) + o(1),$$

$$n^{\frac{1}{2}}|\widehat{F}_n^{*-1}(y) - \widehat{F}_n^{-1}(y)| \le \left|\frac{\nu_n^*}{f} \circ F^{-1}(y)\right| + o(1),$$

Let P_n^* be the probability in the bootstrap space $\{X_1, \ldots, X_n\}^{\otimes n}$, conditionally on the sample.

Proposition 2.10. *If the distribution function F belongs to $C^1(\mathbb{R})$, the maximum*

$$\|P(n^{\frac{1}{2}}\{\widehat{F}_n^{-1}(t) - F^{-1}(t)\} \le x) - P_n^*(n^{\frac{1}{2}}\{\widehat{F}_n^{*-1}(t) - \widehat{F}_n^{-1}(t)\} \le x)\|_\infty$$

converges in probability to zero as n tends to infinity.

Proof. Let M_n be the maximum of this variable, it is reached at an observation $X_i^* = X_j$ where it equals

$$\|P(\nu_n(X_j) \ge -xf(X_j)) - P_n^*(\nu_n^*(X_j) \ge -xf(X_j))\|_\infty + o(1).$$

Let $\nu_n = W_n \circ F$ and $\nu_n^* = W_n^* \circ \widehat{F}_n$, the empirical processes W_n and W_n^* converge weakly to processes having both the distribution of a transformed Brownian Bridge $W \circ F$, due to the convergence of their covariance functions to the same limit as n tends to infinity. It follows that for every real a, $P(\nu_n(t) \ge a) = P(W_n \circ F(t) \ge a)$ and $P_n^*(\nu_n^*(t) \ge a) = P(W_n^* \circ \widehat{F}_n(t) \ge a)$ have the same limit and $M_n = o(1)$. $\square$

When the function F is replaced by functions F_n such that $n^{\frac{1}{2}}\{F_n - F\}$ converges to a function H, the empirical process $\nu_{n,n} = n^{\frac{1}{2}}(\widehat{F}_n - F_n)$ is asymptotically equivalent to $\nu_n - H$ and its variance under F_n converges to the variance under F, the convergence of Proposition 2.10 is still satisfied. This implies the consistency of the level and of the power of bootstrap tests based on statistics $T_n = \psi(\widehat{F}_n)$, with functionals ψ of $C^1(\mathcal{F}_{\mathbb{R}})$, for the class $\mathcal{F}_{\mathbb{R}}$ of the distribution functions in $\mathbb{R}$.

2.9   Exercises

2.9.1. Let $(\Omega, \mathcal{A}, P_n)_{n \geq 1}$ and $(\Omega, \mathcal{A}, Q_n)_{n \geq 1}$ be two sequences of probability spaces with respective densities $dP_n/dP_0 = f_n$ and $dQ_n/dQ_0 = g_n$ for every integer n. Prove that if probability densities f_n and g_n have a log-likelihood ratio $l_n = \sum_{i=1}^{n}\{\log g_n(X_i) - \log f_n(X_i)\}$ converging weakly to a Gaussian distribution with $\mathcal{N}(-\frac{1}{2}\sigma^2, \sigma^2)$, then the sequences $(P_n)_{n \geq 1}$ and $(Q_n)_{n \geq 1}$ are contiguous, i.e. for every sequence of variables $(T_n)_n$ converging in probability to zero under $(P_n)_{n \geq 1}$, $(T_n)_n$ converges to zero under $(Q_n)_{n \geq 1}$.

Proof. Let $\varepsilon > 0$ and $\delta > 0$. For every sequence $(T_n)_n$ converging in probability to zero under $(P_n)_{n \geq 1}$, let $\phi_n = 1_{\{T_n > \delta\}}$. Let $F_n(x) = P_n(L_n \leq x)$ be the distribution function of the likelihood ratio $L_n = \prod_{i=1}^{n}(dQ_n/dP_n)(X_i)$, such that $\log L_n$ has a normal limiting distribution $\mathcal{N}(-\frac{1}{2}\sigma^2, \sigma^2)$. This implies that the mean of L_n is 1. Let $y > 0$ be the smallest real value such that $1 - \int 1_{\{x \leq y\}} x\, dF_n(x) \leq \frac{1}{2}\varepsilon$ and let δ be such that $yP_n(T_n > \delta) \leq \frac{1}{2}\varepsilon$. Writing $dQ_n = L_n\, dP_n$ we obtain, for n large enough

$$\int \phi_n\, dQ_n = \int_{\{L_n \leq y\}} \phi_n\, dQ_n + \int_{\{L_n > y\}} \phi_n\, dQ_n$$

$$\leq y\int \phi_n\, dP_n + \int 1_{\{L_n > y\}} dQ_n$$

$$= y\int \phi_n\, dP_n + 1 - \int 1_{\{L_n \leq y\}} dQ_n$$

$$= y\int \phi_n\, dP_n + 1 - \int 1_{\{x \leq y\}} x\, dF_n(x) \leq \varepsilon.$$

2.9.2. Let $(\Omega, \mathcal{A}, P_\theta)_{\theta \in \Theta}$ be a sequence of probabilities with parametric densities indexed at parameter θ in an open and bounded subset Θ of $\mathbb{R}^d$, such that $\theta \mapsto f_\theta$ belongs to $C^2(\Theta)$, f_θ and $f_{\theta_0}^{-2}\ddot{f}$ belong to $L^2(\mathbb{R}, P_{\theta_0})$ for every θ, with θ_0 in Θ. Let $(P_n)_{n \geq 1}$ be a sub-sequence of probabilities defined as $P_n = P_{\theta_n}$ defined as $\theta_n = \theta_0 + n^{-\frac{1}{2}}\gamma_n$ where $\lim_{n \to \infty} \gamma_n = \gamma \neq 0$ is finite. Prove the contiguity of the sequence $(P_n)_{n \geq 1}$ to P_{θ_0}.

Proof. Let $f_0 = f_{\theta_0}$ and $f_n = f_{\theta_n}$. From Exercise 2.9.1, it is sufficient to prove that the log-likelihood $l_n = \sum_{i=1}^{n} \log(dP_n/dP_0)(X_i)$ has a Gaussian limiting distribution $\mathcal{N}(-\frac{1}{2}\sigma^2, \sigma^2)$. This is proved by a Talor expansion of f_n in a neighborhood of f_0, for n large enough

$$f_n = f_0 + n^{-\frac{1}{2}}\gamma_n^t \dot{f}_0 + \frac{1}{2}n^{-1}\gamma_n^t \ddot{f}_0 \gamma_n + o(n^{-1})$$

and by an expansion of the logarithm in $\log(f_n/f_0)(x) = \log(1 + (f_n - f_0)/f_0)(x)$ as in the proof of Proposition 2.5

$$\log \frac{f_n}{f_0} = \frac{f_n - f_0}{f_0} - \frac{1}{2}\frac{(f_n - f_0)^2}{f_0^2} + o(n^{-1})$$

$$= n^{-\frac{1}{2}}\gamma_n^t \frac{\dot{f}_0}{f_0} + \frac{1}{2}n^{-1}\gamma_n^t \left\{ \frac{\ddot{f}_0}{f_0} - \frac{\dot{f}_0^2}{f_0^2} \right\} \gamma_n + o(n^{-1}).$$

Under the integrability conditions and by the central limit theorem, it follows that $l_n = \sum_{i=1}^{n} \log(f_n/f_0)(X_i)$ converges weakly to Gaussian variable with variance $\sigma^2 = \gamma^t(\int \dot{f}_0^2 f_0^{-1})\gamma$, since $\int \dot{f}_0 = 0$, and with mean is $-\frac{1}{2}\sigma^2$.

2.9.3. Let $S_n = \psi(\bar{X}_n)$ be a statistic defined by a $C^1(\mathbb{R})$ function ψ from the empirical mean. Prove the consistency of the level and the power of parametric bootstrap tests, conditionally on the observed sample, using equations (1.9)-(1.10).

Proof. Expanding $S_n - \psi(\mu)$, the variable $n^{\frac{1}{2}}\{S_n - \psi(\mu)\} = n^{\frac{1}{2}}(\bar{X}_n - \mu)\dot{\psi}(\mu)$ converges weakly to a Gaussian variable with mean zero and variance $\sigma^2\dot{\psi}^2(\mu)$, where σ^2 is the variance of the observed variable X. The bootstrap statistic $S_n^* = \psi(\bar{X}_n^*)$ has the expansion $n^{\frac{1}{2}}(S_n^* - S_n) = n^{\frac{1}{2}}(\bar{X}_n^* - \bar{X}_n)\dot{\psi}(\bar{X}_n)$. Applying (1.9), with $\dot{\psi}(\bar{X}_n) = \dot{\psi}(\mu) + o(1)$, for every real x

$$\|P_n^*(n^{\frac{1}{2}}(S_n^* - S_n) \leq x) - P(n^{\frac{1}{2}}(S_n - \psi(\mu)) \leq x)\|_\infty = O(n^{-\frac{1}{2}}).$$

The critical value of a test based on S_n is a quantile of the distribution $F_n(x) = P(n^{\frac{1}{2}}(S_n - \psi(\mu)) \leq x)$, for the bootstrap statistic, this is a quantile of the conditional distribution $F_n^*(x) = P_n^*(n^{\frac{1}{2}}(S_n^* - S_n) \leq x)$. They are continuous and their quantiles have the same limits, like in (1.10).

2.9.4. Let $S_n = \hat{\sigma}_n^{-1}\bar{X}_n$ be a Student statistic related to a n-sample of a variable X. Prove the consistency of the level and the power of parametric bootstrap tests, with statistic $S_n^* = \hat{\sigma}_n^{*-1}\bar{X}_n^*$, conditionally on the observed sample.

Proof. Conditionally on the sample $\mathbb{X}_n = (X_1, \ldots, X_n)$, the bootstrap sample is drawn with the distribution function $\widehat{F}_n$, where $\widehat{F}_n$ is empirical distribution of $\mathbb{X}_n$. A bootstrap mean $\bar{X}_n^{*b}$ has the conditional mean $\int x \, d\widehat{F}_n(x) = \bar{X}_n = \varphi_1(\widehat{F}_n)$ tending to zero under the hypothesis $H_0 : E_0 X = 0$. Its conditional variance is $\varphi_2(\widehat{F}_n) = n^{-1} \sum_{i=1}^n X_i^2 - \bar{X}_n^2$, it is asymptotically equivalent to the empirical variance $\widehat{\sigma}_n^2$ of the sample. Under H_0, the statistic is written as

$$S_n = \psi(\widehat{F}_n) = \frac{\int x \, d\nu_n(x)}{n^{\frac{1}{2}} \widehat{\sigma}_n},$$

with a $C^1(\mathcal{F}([0,1])$ function ψ. Under local alternatives defined by distribution functions F_n such that $H_n = n^{\frac{1}{2}}(F_n - F)$ converges to a limit H, let $\nu_{n,n} = n^{\frac{1}{2}}(\widehat{F}_n - F_n)$, then $S_n - \psi(F_n)$ is asymptotically equivalent to

$$\frac{\int x \, d\nu_{n,n}(x)}{n^{\frac{1}{2}} \widehat{\sigma}_n} - \frac{\int x^2 \, d\nu_{n,n}(x) - \{\int x \, d\nu_{n,n}(x)\}^2}{n \widehat{\sigma}_n^2} \int x \, dF_n(x).$$

The bootstrap statistic S_n^* has similar approximations with ν_n^* and $\widehat{\sigma}_n^*$ and the proof ends as in Proposition 2.10.

Chapter 3

Nonparametric tests for one sample

3.1 Introduction

A large part of this chapter is devoted to goodness of fit tests for a distribution function, a regression function and other functions in models for partial observations, including the observation of the regression variable by intervals. Several forms of the statistics are considered according to the alternatives. Most statistics are asymptotically free, they are Kolmogorov-Smirnov type statistics or normalized squared L^2-distances between the estimator under general conditions and under the conditions of the hypotheses of the tests, like the Cramer-von Mises statistics, or χ^2 statistics.

Tests about the form of a density are tests of monotony, unimodality, concavity or convexity in sub-intervals of its domain. They rely on smooth estimators of the density under the regularity Condition 1.1. The asymptotic properties of the statistics are deduced from those of the density estimators under the constraints defining the hypotheses of the tests and from Theorem 1.1 for the kernel density estimator. The estimator of the mode of a unimodal density by (1.7) and the isotonization of the density estimator are used, as in Lemma 1.1. For densities and regression curves with a finite number of jumps, the estimation of the jump points allows us to extend the results to sub-intervals where the functions are regular. If the functions are C^1 with a Lipschitz derivative, the order of the bias of their estimators is still h^2, its exact expression cannot be used and we have only a bound for their bias, with the L^2-optimal bandwidths h. The consistency of the tests is preserved but the conditions provide only a bound for their asymptotic power against local alternatives.

3.2   Kolmogorov-Smirnov tests for a distribution function

Let F be a distribution function of an observed random variable X and let $X_1, \ldots, X_n$ be a sample of X. A statistic for a test of the adequacy hypothesis $H_0 : F = F_0$ against the alternative $K : F \neq F_0$ is the Kolmogorov-Smirnov statistic

$$\sup_{x \in \mathbb{R}} |W_n(x)| = n^{\frac{1}{2}} \sup_{\mathbb{R}} |\widehat{F}_n - F_0|.$$

Its rejection domain is $D_n(\alpha, F_0) = \{\sup_{t \in \mathbb{R}} |W_n(t)| > c_{\alpha, n}\}$, at the nominal level α, it is defined by a critical value $c_{\alpha, n}$ converging to the $(1 - \alpha)$th quantile c_α of the distribution of the supremum of the transformed Brownian bridge $\sup_{t \in \mathbb{R}} |B \circ F_0(t)|$, $\alpha = P\{\sup_{t \in \mathbb{R}} |B \circ F_0(t)| > c_\alpha\}$ (see Section 1.4).

A distribution free Kolmogorov-Smirnov test is defined by the rejection domain $D_n(\alpha) = \{\sup_{x \in [0,1]} |W_n \circ F_0^{-1}(x)| > c_{\alpha, n}\}$, it has the asymptotic level $\alpha = P\{\sup_{x \in [0,1]} |B(x)| > c_\alpha\}$. The tests based on the processes W_n and $W_n \circ F_0^{-1}$ are asymptotically equivalent since

$$\sup_{x \in \mathbb{R}} |W_n(x)| = \sup_{x \in \mathbb{R}} |B_n \circ F_0(x)| = \sup_{t \in [0,1]} |B_n(t)|.$$

Let Θ be a parameter subset of $\mathbb{R}^d$ and let $\mathcal{F} = \{F_\theta, \theta \in \Theta\}$ be a parametric class of distribution functions. A semi-parametric goodness of fit test for the hypothesis $H_0 : F$ belongs to the parametric class $\mathcal{F}$ against an alternative $K : F$ does not belong to $\mathcal{F}$ is built on the process

$$W_n = n^{\frac{1}{2}}(\widehat{F}_n - F_{\widehat{\theta}_n}), \qquad (3.1)$$

where $\widehat{\theta}_n$ is an estimator of the parameter of the distribution of the observed variable under H_0. Assuming that for every real x, the functions of $\mathcal{F}$ have a twice continuously differentiable density in Θ, the optimal estimator $\widehat{\theta}_n$ is the maximum likelihood estimator, otherwise it may be the estimator of the maximum of probability $\widehat{\theta}_n = \arg\max_{\theta \in \Theta} \prod_{i=1}^n F_\theta(X_i)$ or an estimator of the minimum of the distance of the distribution function F to $\mathcal{F}$. Let $\nu_{n,F} = n^{\frac{1}{2}}(\widehat{F}_n - F)$ be the empirical process under a distribution function F, a test of H_0 relies on the process

$$W_n = n^{\frac{1}{2}}(\widehat{F}_n - F_{\widehat{\theta}_n}) = \nu_{n,F} - n^{\frac{1}{2}}(F_{\widehat{\theta}_n} - F_{\theta_0}),$$

for the parameter value θ_0 in Θ such that $F = F_{\theta_0}$. Let $\dot{F}_\theta$ denote the partial derivative of F_θ with respect to θ, for large n

$$W_n = \nu_{n,F} - n^{\frac{1}{2}}(\widehat{\theta}_n - \theta_0)^t \dot{F}_{\theta_n},$$

where θ_n is between θ_0 and $\widehat{\theta}_n$ which converges to θ_0 in probability. The process W_n converges weakly to $B \circ F - Z^t \dot{F}_{\theta_0}$, where B is the standard Brownian bridge and Z is a centered Gaussian variable with the variance $I_0 = [E_0 \{ f_{\theta_0}^{-2} \dot{f}_{\theta_0}^{\otimes 2} \}(X)]^{-1}$, which is the limiting distribution of $n^{\frac{1}{2}}(\widehat{\theta}_n - \theta_0)$. Using the maximum likelihood estimator $\widehat{\theta}_n$ of θ_0, the covariance function of $B \circ F$ and Z, at x, is

$$C(x) = E\{n^{-\frac{1}{2}} \ddot{i}^{-1}(\theta_0) \dot{l}_n(\theta_0) \nu_{n,F}(x)\}$$

$$= \ddot{i}^{-1}(\theta_0) E_0 \frac{\dot{f}_{\theta_0}}{f_{\theta_0}}(X) 1_{\{X \leq x\}} = \ddot{i}^{-1}(\theta_0) \int_{-\infty}^{x} \dot{f}_{\theta_0}(y) \, dy.$$

The variance of $W_n(x)$ is therefore

$$v(x) = F_{\theta_0}(x) - F_{\theta_0}^2(x) + \dot{F}_{\theta_0}^t(x) \ddot{i}^{-1}(\theta_0) \dot{F}_{\theta_0}(x) - 2\dot{F}_{\theta_0}(x)$$

and it is estimated using the estimators $\widehat{F}_n$ and $\widehat{\theta}_n$. Under a fixed alternative such that $\|G - F\| > 0$ for every F in $\mathcal{F}$ and G in a set of distribution functions $\mathcal{G}$ disjoint of $\mathcal{F}$, $\sup_{F \in \mathcal{F}} \inf_{G \in \mathcal{G}} \|W_n\|$ tends to infinity and the power of the test based on the uniform norm of W_n converges to 1.

Let Ξ be a set of measurable functions defined from $\mathbb{R}$ to $\mathbb{R}^*$ and let $\mathcal{G}_{n,\Xi}$ be a class of local alternatives such that for every F_n of $\mathcal{G}_{n,\Xi}$, there exist F in $\mathcal{F}$ and a sequence $(\xi_n)_{n \geq 1}$ in Ξ such that the distribution function of X is $F_n = F + n^{-\frac{1}{2}} \xi_n$ where F belongs to $\mathcal{F}$ and ξ_n converges to a limit $\xi = \lim_{n \to \infty} n^{\frac{1}{2}}(F_n - F)$. In $\mathcal{G}_{n,\Xi}$, the limiting distribution under $P_{F_{n,\xi}}$ of the process W_n is $B \circ F - Z^t \dot{F}_{\theta_0} + \xi$, where $F = F_\theta$ belongs to $\mathcal{F}$.

Proposition 3.1. *The test defined by the statistic* $\sup_{t \in \mathbb{R}} |W_n(t)|$ *has the asymptotic level* $\alpha = \sup_{F_\theta \in \mathcal{F}} P_{F_\theta} \{ \|B \circ F_\theta - Z^t \dot{F}_\theta\|_\infty > c_\alpha \}$ *and its asymptotic local power is*

$$\beta_{\alpha,\mathcal{F},\Xi} = \lim_{n \to \infty} \sup_{F_\theta \in \mathcal{F}} \inf_{F_n = F_\theta + n^{-\frac{1}{2}} \xi_n} P_{F_n}(D_{n,\alpha})$$

$$= \sup_{F_\theta \in \mathcal{F}} \inf_{\xi \in \Xi} [1 - P\{\|B \circ F_\theta - Z^t \dot{F}_\theta + \xi\|_\infty \leq c_\alpha\}]$$

and $\beta_{\alpha,\mathcal{F},\Xi} > \alpha$.

Proof. The expression of c_α is a consequence of the asymptotic distribution of the process W_n in $\mathcal{F}$ and the asymptotic local power for ξ in Ξ and F_θ in $\mathcal{F}$ is $\beta_{F_\theta,\xi} = P\{\|B \circ F_\theta - Z^t \dot{F}_\theta + \xi\| > c_\alpha\}$. For every F_θ in $\mathcal{F}$, $\sup_{\xi \in \Xi} P_\theta \{\|B \circ F_\theta - Z^t \dot{F}_\theta + \xi\|_\infty \leq c_\alpha\}$ is larger than $P_\theta \{\|B \circ F_\theta - Z^t \dot{F}_\theta\|_\infty \leq c_\alpha\}$ because Z has a Gaussian centered density and the function ξ is not zero. Then $\inf_{\xi \in \Xi} \beta_{F,\xi} > \alpha$ for every F in $\mathcal{F}$ and the test is locally asymptotically consistent. $\square$

The distribution of this test depends on the classes $\mathcal{F}$ and $\mathcal{G}_{n,\Xi}$ and we consider tests based on transformed variables. Under H_0, their exists a function $F = F_\theta$ in $\mathcal{F}$ such that the variables $U_{\theta,i} = F_\theta(X_i)$ have a uniform distribution in $[0,1]$, $i = 1, \ldots, n$.

The transformed empirical process is

$$\nu_{\theta,n}(x) = n^{-\frac{1}{2}} \sum_{i=1}^{n} \{1_{\{U_{\theta,i} \leq x\}} - F \circ F_\theta^{-1}(X_i)\}$$

and under H_0, $\nu_{\theta,n}(x) = n^{-\frac{1}{2}} \sum_{i=1}^{n} \{1_{\{U_{\theta,i} \leq x\}} - x\}$. Estimating the unknown distribution function of the distribution leads to replace $\nu_{\theta,n}(x)$ by $\nu_{\widehat{\theta}_n,n}(x)$. The statistic reduces to the transformed empirical process $\nu_{\widehat{\theta}_n,n} = \nu_n \circ F_{\widehat{\theta}_n}^{-1}$. The behavior of a test based on $\nu_{\widehat{\theta}_n,n}$ is a consequence of its limiting distribution.

Proposition 3.2. *The test of H_0 against $\mathcal{G}_{n,\Xi}$ has rejection domain*

$$D_{n,\alpha} = \left\{ \sup_{t \in [0,1]} |\nu_n \circ F_{\widehat{\theta}_n}^{-1}(t)| > c_{n,\alpha} \right\},$$

such that $c_{n,\alpha}$ converges to the $(1-\alpha)$th quantile c_α of the distribution of $\sup_{t \in [0,1]} |B(t)|$. Its asymptotic local power is $\sup_{F_\theta \in \mathcal{F}} \inf_{\xi \in \Xi} \beta_{F_\theta,\xi}$, where

$$\beta_{F_\theta,\xi} = \lim_{n \to \infty} P_{F_n,\xi}(D_{n,\alpha}) = 1 - P\left\{ \sup_{[0,1]} |B + \xi \circ F_\theta^{-1}| \leq c_\alpha \right\},$$

under a sequence of alternatives $(F_n)_{n \geq 1}$ such that there exist θ in Θ and $\xi = \lim_{n \to \infty} n^{\frac{1}{2}}(F_n - F_\theta)$ in Ξ. Then $\sup_{F \in \mathcal{F}} \inf_{\xi \in \Xi} \beta_{F,\xi} > \alpha$.

Proof. Under a sequence of alternatives $(F_n)_{n \geq 1}$, $\sup_{t \in [0,1]} |F_{\widehat{\theta}_n} - F_\theta|$ converges in probability to zero and the process $\nu_n = B_n \circ F_\theta + n^{\frac{1}{2}}(F_n - F_\theta)$ converges weakly to $B \circ F_\theta + \xi$, hence $\sup_{[0,1]} |\nu_n \circ F_{\widehat{\theta}_n}^{-1}|$ converges weakly to $\sup_{[0,1]} |B + \xi \circ F_\theta^{-1}|$. The test is asymptotically unbiased like in Proposition 3.1, since B is a centered process. □

The level of the test of Proposition 3.2 is free from the class of the distribution functions $\mathcal{F}$, so the statistic $\sup_{t \in [0,1]} |\nu_n \circ F_{\widehat{\theta}_n}^{-1}(t)|$ should be prefered to the statistic of Proposition 3.1.

3.3   Tests for symmetry of a density

Smirnov proposed a test for symmetry of the density of a variable with mean zero which has an asymptotically free distribution. The Cramer-von Mises statistic related to this problem has also been studied Rothman and Woodroofe (1972), with its asymptotic power under the alternative of a local translation. These statistics are generalized to tests of symmetry with an unknown center of symmetry and the local alternatives are changes of location and scale. The asymptotic properties of rank tests of symmetry are considered in the same models, with pararametric densities. Most of them have asymptotically free distributions.

3.3.1 *Kolmogorov-Smirnov tests for symmetry*

On a probability space $(\Omega, \mathcal{F}, P)$, let X be a real random variable with a symmetric density f centered at zero. Its distribution F satisfies the hypothesis of symmetry $H_0 : F(x) + F(-x) = 1$, for every real x. The hypothesis of symmetry is also expressed as the equality of the left and right tails of the distribution function

$$H_0 : P(X \leq -x) = P(X \geq x), \quad \text{for every } x > 0. \tag{3.2}$$

By the means of the survival function $\bar{F}(x) = 1 - F(x) = P(X > x)$, this is equivalent to $H_0 : \bar{F}(x^-) = F(-x)$, for every real x. Tests for symmetry are based on the empirical distribution $\widehat{F}_n$ of a n-sample of variables with distribution F, $(X_1, \ldots, X_n)$. Let $\nu_n = \sqrt{n}(\widehat{F}_n - F)$ be the empirical process of the sample and let $\bar{\nu}_n = \sqrt{n}(\widehat{\bar{F}}_n - \bar{F})$. Tests of H_0 are defined from the process

$$U_n(x) = \sqrt{n}\{\widehat{F}_n(-x) + \widehat{F}_n(x^-) - 1\}, \tag{3.3}$$

under the hypothesis H_0, $U_n(x) = \nu_n(-x) - \bar{\nu}_n(x^-)$. Otherwise it is not centered and for every distribution function F, $U_n(x) = \nu_n(-x) + \nu_n(x^-) + \sqrt{n}\{F(-x) - \bar{F}(x^-)\}$, where $\sup_{x>0} |\sqrt{n}\{F(-x) - \bar{F}(x^-)\}|$ tends to infinity with n.

A test for symmetry of the density of independent and identically distributed variables is defined from the Kolmogorov-Smirnov type statistics

$$T_n = \sup_{x \in \mathbb{R}} U_n(x),$$
$$T_n^+ = \sup_{x \in \mathbb{R}} |U_n(x)|. \tag{3.4}$$

Under H_0, for real every x, the variance of $U_n(x)$ is

$$Var\, U_n(x) = F(x) - F^2(x) + F(-x) - F^2(-x)$$
$$+ 2F(x \wedge (-x)) - 2F(x)F(-x)$$
$$= F(x) + F(-x) + 2F(x \wedge (-x)) - \{F(x) + F(-x)\}^2$$
$$= 2F(x \wedge (-x)) = 2F(-|x|),$$

it is also written as

$$VarU_n(x) = 1 + F(-|x|) - F(|x|) = 1 - P(|X| \leq |x|).$$

By the weak convergence of the empirical process to the standard Brownian bridge B transformed by F, under the hypothesis of symmetry (3.2), the statistics T_n and T_n^+ converge weakly to

$$T = \sup_{x \in \mathbb{R}}\{B \circ F(x) + B \circ F(-x)\} = \sup_{x \in \mathbb{R}}\{B \circ F(x) + B \circ \bar{F}(x)\}$$
$$= \sup_{t \in [0,1]}\{B(t) + B(1 - t)\},$$
$$T^+ = \sup_{x \in \mathbb{R}}|B \circ F(x) + B \circ F(-x)| = \sup_{t \in [0,1]}|B(t) + B(1 - t)|.$$

By symmetry of the distribution of B and of the density f, the suprema in $\mathbb{R}$ and in $\mathbb{R}^+$ or $\mathbb{R}^-$ have the same distribution. The variance of the Gaussian variable $B(t) + B(1 - t)$ is

$$E\{B^2(t)\} + E\{B^2(1 - t)\} + 2E\{B[0,t]\,B[t,1]\} = 2t(1 - t)$$

for every t in $[0,1]$. It is maximum at $t = \frac{1}{2}$ where the variance equals $\frac{1}{2}$. For every $c > 0$

$$\lim_{n \to \infty} P(T_n > c) = P(\sup_{t \in [0,1]}\{B(t) + B(1 - t)\} > c)$$
$$\leq \frac{1}{c^2}E \sup_{t \in [0,1]}\{B(t) + B(1 - t)\}^2 \leq \frac{1}{2c^2}.$$

The thresholds of the tests for H_0 based on T_n or T_n^+, with an asymptotic level α, have therefore the bound $c_\alpha \leq (2\alpha)^{-\frac{1}{2}}$.

The asymptotic power of the tests is calculated under the alternatives of a variable X with density $f(x)$ satisfying an assumption of a local change of location

$$K_l : f(x + \theta) = f(\theta - x), \ \theta \neq 0, \ x \in \mathbb{R}, \tag{3.5}$$

and under the alternative of a change of location and scale

$$K_s : f(x + \theta) = \frac{1}{\sigma}f\left(\frac{\theta - x}{\sigma}\right), \ \theta \neq 0, \ \sigma \notin \{0,1\}, \ x \in \mathbb{R}. \tag{3.6}$$

A density satisfying the assumption K_l is a density with a center of symmetry at some nonzero θ, K_l is equivalent to the existence of $\theta \neq 0$ such that

$$F(x + \theta) + F(\theta - x) = 1, \quad \text{for every } x > 0.$$

The assumption K_s is equivalent to the existence of $\theta \neq 0$ and σ different from zero and one, such that

$$F(x + \theta) + F\left(\frac{\theta - x}{\sigma}\right) = 1, \quad \text{for every } x > 0$$

so the density is not symmetric under K_s.

Under the alternative K_l, there exists $\theta \neq 0$ such that for every $x > 0$, $F(x) + F(2\theta - x) = 1$, i.e. $\bar{F}(-x) = F(2\theta + x)$. Let

$$P_\theta(x) = F(2\theta + x) - F(x) = P(x \leq X \leq 2\theta + x), \ \theta > 0,$$

the mean $E_\theta U_n(x) = \sqrt{n}\{F(x) - \bar{F}(-x)\}$ of the process U_n defined by (3.3) is also written as

$$E_\theta U_n(x) = \sqrt{n}\{F(x) - F(2\theta + x)\} = -\sqrt{n}P_\theta(x).$$

Its variance is

$$\begin{aligned}
Var_\theta U_n(x) &= F(x) + 3F(-x) - \{F(x) + F(-x)\}^2 \\
&= 1 - P_\theta(x) + 2F(-x) - \{1 - P_\theta(x)\}^2 \\
&= P_\theta(x)\{1 - P_\theta(x)\} + 2F(-x).
\end{aligned}$$

They differ from the mean and the variance of $U_n(x)$ under H_0 and $\lim_{\theta \to 0} \sup_{x \in \mathbb{R}} P_\theta(x) = 0$.

Proposition 3.3. *Let $(K_{\theta_n})_{n \geq 1}$ be a sequence of local alternative K_l with $(\theta_n)_{n \geq 1}$ converging to zero and such that $\lim_{n \to \infty} \sqrt{n}\theta_n = \theta_0$ different from zero. If the density is bounded by a constant M, the asymptotic power of the tests of level α based on the statistics T_n and T_n^+, against the sequence of alternatives K_{θ_n} have the bound $\beta(\theta_0) \leq (2c_\alpha^2)^{-1}(8\theta_0^2 M^2 + 1)$.*

Proof. For every $x > 0$, the limits of the mean and variance of $U_n(x)$ are $\lim_{n \to \infty} E_{\theta_n} U_n(x) = -2\theta_0 f(x)$ and $\lim_{n \to \infty} Var_{\theta_n} U_n(x) = Var_0 U_n(x)$. A bound for the asymptotic power of the test is deduced from the a uniform bound for the variance of $U_n(x)$

$$\beta(\theta_0) = \lim_{n \to \infty} P_{\theta_n}(T_n \geq c_\alpha) \leq c_\alpha^{-2} \lim_{n \to \infty} E_{\theta_n} \sup_{x > 0} U_n^2(x),$$

and the result follows from the inequality

$$\lim_{n \to \infty} E_{\theta_n} U_n(x)^2 \leq 2(2\theta_0^2 M^2 + 1), \ x > 0.$$

The proof is similar for T_n^+. $\qquad\square$

Under the alternative K_s, there exist $\theta \neq 0$ and σ different from 0 and 1, such that for every $x > 0$, $P_{\theta,\sigma}(X \leq x) = F(x)$ and $F(x) + F(\frac{2\theta - x}{\sigma}) = 1$. The mean $E_{\theta,\sigma} U_n(x) = \sqrt{n} \{ F(x) + F(-x) - 1 \}$ is

$$E_{\theta,\sigma} U_n(x) = \sqrt{n} \{ F(x) - F(\sigma^{-1}(2\theta + x)) \} = -\sqrt{n} P(x \leq X \leq \sigma^{-1}(2\theta + x)).$$

The variance $Var_{\theta,\sigma} U_n(x)$ is expressed as

$$Var_{\theta,\sigma} U_n(x) = \varphi_{\theta,\sigma}(x) - \varphi_{\theta,\sigma}^2(x) + 2F(-|x|),$$
$$\varphi_{\theta,\sigma}(x) = F(\sigma^{-1}(2\theta + x)) - F(x),$$

and $E_{\theta,\sigma} U_n(x) = -\sqrt{n} \varphi_{\theta,\sigma}(x)$. Let $\theta_0 \neq 0$ and $\sigma_0 > 0$.

Proposition 3.4. *Let $(K_{\theta_n,\sigma_n})_{n \geq 1}$ be a sequence of local alternatives of K_s with sequences $(\theta_n)_{n \geq 1}$ converging to zero and $(\sigma_n)_{n \geq 1}$ strictly positive converging to 1, such that $\lim_{n \to \infty} \sqrt{n} \theta_n = \theta_0$ and $\lim_{n \to \infty} \sqrt{n}(\sigma_n - 1) = \sigma_0$. If there exists a bound M for the density, the asymptotic power of the tests of level α based on the statistics T_n and T_n^+, against the sequence of alternatives K_{θ_n,σ_n} is one.*

Proof. For every $x > 0$, as $\lim_{n \to \infty} \sqrt{n} \theta_n = \theta_0$ and $\lim_{n \to \infty} \sqrt{n}(\sigma_n - 1) = \sigma_0$, the mean of $U_n(x)$ develops as

$$E_{\theta_n,\sigma_n} U_n(x) = \sqrt{n} \{ -2\theta_n f(x) + \sigma_n^{-1}(\sigma_n - 1) x f(x) \} \{ 1 + o(1) \},$$
$$= \{ -2\theta_0 + \sigma_0 x \} f(x) + o(1)$$

hence $\lim_{n \to \infty} E_{\theta_n,\sigma_n} U_n(x) \leq \{ -2\theta_0 + \sigma_0 x \} M$. Since $\lim_{n \to \infty} \varphi_{\theta_n,\sigma_n}(x)$ is 1, the variance of $U_n(x)$ converges to $\lim_{n \to \infty} Var_0 U_n(x) = 2F(-|x|)$. The asymptotic power of the test cannot be lower than 1 since $\sup_{x \in \mathbb{R}} \lim_{n \to \infty} E_{\theta_n,\sigma_n} U_n(x)$ is infinite. $\square$

3.3.2   Semi-parametric tests, with an unknown center

On a probability space $(\Omega, \mathcal{F}, P)$, let X be a real random variable with mean θ_0 and with a symmetric continuous distribution. The distribution function F of $X - \theta_0$ is centered and $F_{\theta_0}(x) = P(X - \theta_0 \leq x) = F(x + \theta_0)$ is the distribution function of the variable X.

The symmetry of the distribution is X is expressed by the hypothesis

$$H_0 : F_{\theta_0}(x^-) + F_{\theta_0}(-x) = 1,$$

i.e. $P(X - \theta_0 < x) + P(X - \theta_0 \leq -x) = 1$, for every real x. The center of symmetry of the distribution of the variable X is its mean and it is estimated by the empirical mean $\widehat{\theta}_n$ of a n-sample $(X_1, \ldots, X_n)$ of X. Test for symmetry when θ_0 is unknown rely on

$$U_n(x) = n^{\frac{1}{2}}\{\widehat{F}_n^-(\widehat{\theta}_n + x^-) + \widehat{F}_n(\widehat{\theta}_n - x) - 1\}. \tag{3.7}$$

Under the hypothesis H_0, the process $U_n(x) = \nu_n^-(\widehat{\theta}_n + x^-) - \bar{\nu}_n(\widehat{\theta}_n - x)$ is centered.

Lemma 3.1. *If the variable X has a continuous and bounded density f_{θ_0}, then under H_0, the mean of $U_n(x)$ conditionally on $\widehat{\theta}_n$ is asymptotically equivalent to the random variable $Z_n(x) = 2n^{\frac{1}{2}}(\widehat{\theta}_n - \theta_0)f_{\theta_0}(x)$ and the conditional variance of $U_n(x)$ converges in probability to $\sigma^2(x) = 2F_{\theta_0}(-|x|)$.*

Proof. Under H_0, $X - \widehat{\theta}_n = (X - \theta_0) - (\widehat{\theta}_n - \theta_0) = X - \theta_0 + o_p(1)$ and $P_{\theta_0}(X - \widehat{\theta}_n < x|\widehat{\theta}_n) = F_{\theta_0}(x + \widehat{\theta}_n - \theta_0)$. The probability $P_{\theta_0}(\widehat{\theta}_n - X \geq x|\widehat{\theta}_n)$ is written in the same way as $F_{\theta_0}(\widehat{\theta}_n - \theta_0 - x)$ therefore the mean of $U_n(x)$ conditionally on $\widehat{\theta}_n$ is

$$\begin{aligned}
E_{\theta_0}\{U_n(x)|\widehat{\theta}_n\} &= n^{\frac{1}{2}}\{F_{\theta_0}(x + \widehat{\theta}_n - \theta_0) + F_{\theta_0}(\widehat{\theta}_n - \theta_0 - x) - 1\} \\
&= -n^{\frac{1}{2}}\{F_{\theta_0}(x) + F_{\theta_0}(-x) - 1\} \\
&\quad + n^{\frac{1}{2}}(\widehat{\theta}_n - \theta_0)\{f_{\theta_0}(x) + f_{\theta_0}(-x)\} + o_p(1).
\end{aligned}$$

Since $f_{\theta_0}(x) = f_{\theta_0}(-x)$, it follows that $E_{\theta_0}\{U_n(x)|\widehat{\theta}_n\} = 2n^{\frac{1}{2}}(\widehat{\theta}_n - \theta_0)f_{\theta_0}(x) + o_p(1)$ and $U_n(x)$ converges weakly to a centered Gaussian variable Z with variance $2f_{\theta_0}^2(x)Var_0X$, as n tends to infinity. For $x > 0$, its conditional variance is approximated from

$$\begin{aligned}
Var_{\theta_0}\{\nu_n(\widehat{\theta}_n + x)|\widehat{\theta}_n\} &= F_{\theta_0}(x + \widehat{\theta}_n - \theta_0) + F_{\theta_0}^2(x + \widehat{\theta}_n - \theta_0) - 1, \\
&= F_{\theta_0}(x) + F_{\theta_0}^2(x) + o_p(1),
\end{aligned}$$

This approximation entails that $Var_{\theta_0}\{\nu_n(\widehat{\theta}_n + x)|\widehat{\theta}_n\} = 2F_{\theta_0}(-|x|) + o_p(1)$, as in Proposition 3.3. $\qquad\square$

Let $\sigma_{\theta_0}^2$ be the variance of the variable X under P_{θ_0}.

Proposition 3.5. *Under the condition of a bounded density f_{θ_0}, the process U_n converges weakly under H_0 to the process $B \circ F_{\theta_0}(x) - B \circ \bar{F}_{\theta_0}(-x) + Z(x)$ where Z is a centered Gaussian process with variance $4f_{\theta_0}^2(x)\sigma_{\theta_0}^2$.*

Since the process process Z is centered, tests of symmetry of the distribution of X with an unknown center of symmetry are defined from the statistics

T_n and T_n^+, based on the process U_n (3.7), where θ_0 is estimated. Under H_0, the statistics T_n and T_n^+ converge weakly to

$$T = \sup_{x>0}\{B \circ F_{\theta_0}(x) + B \circ F_{\theta_0}(-x) + Z(x)\}$$

$$= \sup_{t\in[0,1]} \{B(t) + B(1-t) + Z \circ F_{\theta_0}^{-1}(t)\},$$

$$T^+ = \sup_{x>0}|B \circ F(x) + B \circ F(-x) + Z(x)|$$

$$= \sup_{t\in[0,1]} |B(t) + B(1-t) + Z \circ F_{\theta_0}^{-1}(t)|,$$

they do not depend directly on the center of symmetry of X but on the symmetric density f_{θ_0}. The asymptotic levels of the tests differ from those tests with a known center of symmetry considered in Section 3.3.1, with $\theta_0 = 0$. The empirical process and $n^{\frac{1}{2}}(\bar{X}_n - \theta_0)$ are asymptotically independent, this entails that for every $c > 0$

$$\lim_{n\to\infty} P(T_n > c) = P\left(\sup_{t\in[0,1]} \{B(t) + B(1-t) + Z \circ F_{\theta_0}^{-1}(t)\} > c \right)$$

$$\leq \frac{1}{c^2}E \sup_{t\in[0,1]} \{B(t) + B(1-t) + Z \circ F_{\theta_0}^{-1}(t)\}^2$$

$$\leq \frac{1}{2c^2}(1 + 8M^2\sigma_{\theta_0}^2)$$

if the density is bounded by a constant M. Its asymptotic power against a sequence of local asymmetric alternatives tends to 1.

Proposition 3.6. *Let $(K_{\theta_n})_{n\geq 1}$ be a sequence of local alternative K_{ln} with $(\theta_n)_{n\geq 1}$ converging to zero and such that $\sqrt{n}(\theta_n - \theta_0) = \xi_n$ converges to a limit ξ different from zero. If the density is bounded by a constant M, the asymptotic power of the tests of level α based on the statistics T_n and T_n^+, against the alternatives K_{ln} have the bound $\beta(\theta_0) \leq (2c_\alpha^2)^{-1}\{8(\xi^2 + \sigma_0^2)M^2 + 1\}$.*

As in the proof of Proposition 3.3, $\lim_{n\to\infty} E_{\theta_n} U_n(x) = -2\xi f_{\theta_0}(x)$ and $\lim_{n\to\infty} Var_{\theta_n} U_n(x) = Var_{\theta_0} U_n(x)$, the limiting process $Z(x)$ is only translated by the function $-2\xi f_{\theta_0}(x)$ under K_{ln}.

3.3.3   Rank test for symmetry

Rank test for symmetry depend on the parametrization of the density f. According to Hájek and Sidák (1967, Section 2.4.3), the locally most powerful rank test for symmetry at zero of a parametric and continuous density

f against an alternative of another center of symmetry is the score test. With a normal density, it is based on the statistic

$$S_n = \sum_{i=1}^{n} \text{sign} X_i \, a(R_n(|X_i|)),$$

$$a(i) = -\frac{f'}{f}(X_{n:i}),$$

where $R_n(|X_i|)$ is the rank of $|X_i| = X_i \, \text{sign} X_i$ in a sample $(X_1, \ldots, X_n)$. The statistic S_n is used for other densities. As $\text{sign} X_i = \pm 1$ with probability .5, the mean of $\text{sign} X_i$ is zero under H_0, for $i = 1, \ldots, n$. The density of $|X|$ is $f_{|X|} = 2 f_X$ and $P(R_n(|X_i|) = r) = (n!)^{-1}$, then the variables $Y_i = \text{sign} X_i \, a(R_n(|X_i|))$ are independent and identically distributed, for $i = 1, \ldots, n$. The mean of S_n is zero, by independence of $|X|$ and $\text{sign} X$, and the variance σ_n^2 of S_n is such that

$$n^{-1} \sigma_n^2 = n^{-1} \sum_{i=1}^{n} (f' f^{-1})^2 (X_i) \tag{3.8}$$

converges in probability to $\sigma^2 = E\{f' f^{-1})^2(X)\}$. Then the normalized statistic $U_n = \sigma_n^{-1} S_n$ converges to a normal variable and the critical values of the test is approximated by the normal quantiles.

The local asymptotic power $\beta(\alpha, \theta_0)$ of the test of symmetry at zero against a sequence of alternatives $(K_{\theta_n})_{n \geq 1}$ with centers of symmetry θ_n such that $\lim_{n \to \infty} \sqrt{n} \theta_n = \theta_0$, is calculated using the signs and the ranks of the sample $(X_i)_{n \geq 1}$, where the variables X_i are symmetric at θ_n. Under this assumption, $\text{sign}(X_i - \theta_n) = \pm 1$ with equal probabilities

$$P_{\theta_n}(\text{sign}(X_i - \theta_n) = 1) = P_{\theta_n}(X_i > \theta_n) = 1 - F(\theta_n) = \frac{1}{2}$$

$$= P_{\theta_n}(\text{sign}(X_i - \theta_n) = -1)$$

and $\text{sign}(X_i - \theta_n) = 0$ has the probability zero. The density of $|X_i - \theta_n|$ is twice the density $f(\theta_n + x)$ of $X_i - \theta_n$ and $P_{\theta_n}(R_n(|X_i - \theta_n|) = r) = (n!)^{-1}$. Under K_n, $P_{\theta_n}(\text{sign} X_i = -1) = F(0) = 1 - F(2\theta_n)$, where $F(\theta_n) = \frac{1}{2}$ and $\lim_{n \to \infty} P_{\theta_n}(\text{sign} X_i = -1) = \frac{1}{2}$.

Proposition 3.7. *Let $(K_{\theta_n})_{n \geq 1}$ be a sequence of local alternative with densities having centers of symmetry $(\theta_n)_{n \geq 1}$ converging to zero and such that $\lim_{n \to \infty} \sqrt{n} \theta_n = \theta_0$. Let $\mu = \lim_{n \to \infty} n^{-1} E_{\theta_n} \sum_{i=1}^{n} a(R_n(|X_i|))$, then μ is not zero and the asymptotic power of the tests of level α based on the statistic S_n given by (3.8) against the sequence of alternatives K_{θ_n} is*

$$\beta(\alpha, \theta_0) = \lim_{n \to \infty} P_{\theta_n}(\sigma_n^{-1} S_n > c_\alpha) = 1 - \Phi\left(c_\alpha - \frac{m}{\sigma}\right),$$

where $m = \lim_{n \to \infty} n^{-\frac{1}{2}} E_{\theta_n} S_n = 2\mu \theta_0 f(0)$.

Proof. Under K_n, $\lim_{n\to\infty} E_{\theta_n}\operatorname{sign}X_i = 0$ hence $E_{\theta_n}n^{-1}S_n$ tends to zero, and the variance of $n^{-\frac{1}{2}}S_n$ under K_n is identical its variance under H_0. The mean of $\sigma_n^{-1}S_n$ under P_{θ_n} is calculated using the independence of the sign and the absolute value of the variables, first

$$n^{\frac{1}{2}}\left\{P_{\theta_n}(\operatorname{sign}X_i = 1) - \frac{1}{2}\right\} = n^{\frac{1}{2}}\left\{\frac{1}{2} - F(0)\right\} = n^{\frac{1}{2}}\{F(\theta_n) - F(0)\}$$

$$= n^{\frac{1}{2}}\theta_n f(0)\{1 + o(1)\},$$

$$n^{\frac{1}{2}}\left\{P_{\theta_n}(\operatorname{sign}X_i = -1) - \frac{1}{2}\right\} = n^{\frac{1}{2}}\left\{F(0) - \frac{1}{2}\right\} = n^{\frac{1}{2}}\{F(0) - F(\theta_n)\}$$

$$= -n^{\frac{1}{2}}\theta_n f(0)\{1 + o(1)\},$$

where $n^{\frac{1}{2}}\theta_n$ converges to θ_0. It follows that

$$\lim_{n\to\infty} n^{\frac{1}{2}} E_{\theta_n}\operatorname{sign}X_i = 2\theta_0 f(0), \quad i = 1,\ldots,n.$$

For independent variables X_i having a density f with a center of symmetry θ_n, the event $\{|X_i - \theta_n| = |X - \theta_n|_{n:r}\}$ has the probability $(n!)^{-1}$ and it is equivalent to $\{|X - \theta_n|_{n:r-1} < |X_i - \theta_n| < |X - \theta_n|_{n:r+1}\}$, then for every $j = 1,\ldots,n$, $\{|X_i - \theta_n| < |X - \theta_n|_{n:r+1}\}$ is equivalent to

$$\{X_j < X_i < 2\theta_n - X_j\}, \quad \text{or} \quad \{2\theta_n - X_j < X_i < X_j\},$$

with the probability $P(R(|X_j - \theta_n|) = r + 1) = \{(n-1)!\}^{-1}$. The ranks of the variables $(X_i)_{i=1,\ldots,n}$ are therefore not uniformly distributed in $\{1,\ldots,n\}$ and the limit μ of $\mu_n = n^{-1}E_{\theta_n}\sum_{i=1}^{n} a(R_n(|X_i|))$ is not zero. Under the sequence of alternatives $(K_{\theta_n})_{n\geq 1}$, the sequence of variables $U_n = n^{-\frac{1}{2}}S_n$ converges weakly to a Gaussian distribution with mean $m = 2\mu\theta_0 f(0)$ and variance σ^2. Then

$$\beta(\alpha, \theta_0) = \lim_{n\to\infty} P_{\theta_n}(n^{\frac{1}{2}}\sigma_n^{-1}(n^{-\frac{1}{2}}S_n - m) > c_\alpha - n^{\frac{1}{2}}\sigma_n^{-1}m)$$

$$= 1 - \Phi\left(c_\alpha - \frac{m}{\sigma}\right). \qquad \square$$

Under an alternative of fixed shift of the center of symmetry K_l, $n^{\frac{1}{2}}\theta$ diverges for every θ of the alternative and the asymptotic power of the test based on $\sigma_n^{-1}S_n$ is 1. When the center of symmetry of the density of X is unknown, the locally most powerful rank test for symmetry at zero of a parametric and continuous density f having a center of symmetry is the score test based on the statistic

$$\widehat{S}_n = \sum_{i=1}^{n} \operatorname{sign}(X_i - \bar{X}_n)\, a(R_n(|X_i - \bar{X}_n|)) := \sum_{i=1}^{n} \operatorname{sign}(X_i - \bar{X}_n)\widehat{a}_n(i),$$

$$\widehat{a}_n(i) = -\frac{f'}{f}(X_{n:i} - \bar{X}_n),$$

Its mean is still zero and its asymptotic distribution is modified, we have

$$\widehat{S}_n = S_n + \sum_{i=1}^{n} [\{\operatorname{sign}(X_i - \bar{X}_n) - \operatorname{sign}(X_i - \theta_0)\}a(i)$$

$$+ \sum_{i=1}^{n} \{\operatorname{sign}(X_i - \bar{X}_n)\{\widehat{a}_n(i) - a(i)\}$$

$$= S_n + \sum_{i=1}^{n} [\{\operatorname{sign}(X_i - \bar{X}_n) - \operatorname{sign}(X_i - \theta_0)\}a(i)$$

$$+ \sum_{i=1}^{n} \operatorname{sign}(X_i - \bar{X}_n)(\bar{X}_n - \theta_0)\Big\{\frac{f''}{f} - \Big(\frac{f'}{f}\Big)^2\Big\}(X_{n:i} - \theta_0) + o_p(1).$$

and the mean number of observations between $\bar{X}_n$ and θ_0 has the order $2(\bar{X}_n - \theta_0)f(\theta_0)$.

Proposition 3.8. *The rank statistic $n^{-\frac{1}{2}}\widehat{S}_n$ converges under the hypothesis of a center of symmetry at θ_0 to a centered Gaussian variable with a variance larger than the variance of $n^{-\frac{1}{2}}S_n$.*

The limiting variance of the statistic can be estimated empirically and the asymptotic critical value of the normalized statistic is a normal quantile. By a similar expansion, under the alternative of a non symmetric density with a change of location (3.5), the mean of $n^{-\frac{1}{2}}\widehat{S}_n$ is shifted and its asymptotic variance in unchanged, as in Proposition 3.3.

3.4 Tests about the form of a density

Let X be a real random variable with density f in $\mathcal{I}_X$ and let $\mathcal{H}$ be the class of unimodal densities in $C^s(\mathbb{R})$, $s \geq 2$. A density f is estimated from a sample $X_1, \ldots, X_n$ of X by smoothing its distribution function

$$\widehat{f}_{n,h}(x) = \int K_h(x - s)\, d\widehat{F}_n(s) = \frac{1}{n}\sum_{i=1}^{n} K_h(x - X_i),$$

its asymptotic behavior in a sub-interval $\mathcal{I}_{X,h} = \{s \in \mathcal{I}_X; [y-h, y+h] \in \mathcal{I}_X\}$ of the interior of $\mathcal{I}_X$ is detailed in Theorem 1.1. The estimator of the mode M_f of a density f of $\mathcal{H}$ defined by (1.7) is $\widehat{M}_{n,f} = M_{\widehat{f}_{n,h}}$, it has the same convergence rate as the estimator of the first derivative of the density. As proved in Pons (2011), under Condition 1.1 for a density of class C^s, $s \geq 2$, the variable $(nh^3)^{1/2}(\widehat{M}_{f,n,h} - M_f)$ converges weakly to a Gaussian variable

with mean zero and variance $\sigma_{M_f}^2 = f(M_f)\{f^{(2)}(M_f)\}^{-2} \int K^{(1)2}(z)\, dz$.
With the optimal bandwidth of the density, the convergence rate of the
estimator of the mode of the density is $(nh^3)^{1/2} = n^{\frac{s-1}{2s+1}}$.

A density is monotone in an interval $\mathcal{I}$ if the sign of its derivative is
constant in $\mathcal{I}$. Its derivative is estimated by

$$\widehat{f}_{n,h}^{(1)} = n^{-1} \sum_{i=1}^{n} K_h^{(1)}(x - X_i),$$

with a kernel in $C^1([0,1])$. A test of monotony can be performed with the
sequence of variables $Z_{i,h}(x) = K_h^{(1)}(x - X_i)$. For every x in the interior
of the interval $\mathcal{I}$, their mean and variance have the next expansions as n
tends to infinity

$$EZ_{i,h}(x) = \int_{\mathcal{I}_X} K_h^{(1)}(x-y)\, dF_X(y) = \int_{\mathcal{I}_X} K_h(x-y) f_X^{(1)}(y)\, dy$$

$$= f_X^{(1)}(x) + \frac{1}{2} m_{2K} h^2 f_X^{(3)}(x) + o(h^2)$$

and $VarZ_{i,h}(x) = \int_{\mathcal{I}_X} K_h^{(1)2}(x-y)\, dF_X(y)$ converges to $f(x) \int K^{(1)2}(s)\, ds$.
A test statistic for an increasing density in $\mathcal{I}$ is

$$S_n = \inf_{x \in \mathcal{I}} \widehat{f}_{n,h}^{(1)}(x),$$

it converges to $S = \inf_{x \in \mathcal{I}} f^{(1)}(x)$ which is positive under the hypothesis
and strictly negative under the alternative. The test based on S_n has then
the critical value zero. The hypothesis and the alternative K of a non
increasing density are well separated since $P_0(S_n < 0)$ tends to zero and
$P_K(S_n < 0)$ tends to one, as n tends to infinity. The asymptotic power of
the test against local alternatives is also zero.

The comparison of a kernel estimator and a piecewise constant isotonic
estimator of the density through an integrated difference is not optimal
since the later is a histogram type estimator. Let $\widehat{f}_{n,h}$ be a kernel estimator
of a density f of $C^2(\mathbb{R})$ under Condition 1.1. The difference between the
integral estimator $A_{n,h}(t) = \int_{-\infty}^{t} \widehat{f}_{n,h}(x)\, dx$ and the distribution function
$F(t) = \int_0^t f(x)\, dx$ is

$$A_{n,h}(t) - F(t) = \int_{-\infty}^{t} (\widehat{f}_{n,h} - f)(s)\, ds,$$

its mean is the bias of the integrated kernel estimator for the distribution
function F. Integrating the asymptotic expansions of the bias and the

variance of the kernel estimator of the density, the process $A_{n,h}$ satisfies

$$EA_{n,h}(t) - F(t) = \int_{-\infty}^{t} E(\widehat{f}_{n,h} - f)(x)\,dx$$

$$= \frac{1}{2}\int_{-\infty}^{t}\left[\int_{\mathbb{R}} K(u)\{h^2 f^{(2)}(x) + o(\|h\|_2^2)\}\,du\right]dx$$

$$= \frac{m_{2K}h^2}{2}f^{(1)}(t) + o(h^2), \tag{3.9}$$

$$VarA_{n,h}(t) = \int_{-\infty}^{t} E(\widehat{f}_{n,h} - E\widehat{f}_{n,h})^2(s)\,ds$$

$$+ 2E\int_{-\infty}^{t}\int_{-\infty}^{y} E\{(\widehat{f}_{n,h} - E\widehat{f}_{n,h})(s)(\widehat{f}_{n,h} - E\widehat{f}_{n,h})(u)\}\,du\,ds$$

$$= (nh)^{-1}\kappa_2 F(t) + o((nh)^{-1}), \tag{3.10}$$

by the asymptotic independence of $\widehat{f}_{n,h}(t) - E\widehat{f}_{n,h}(t)$ and $\widehat{f}_{n,h}(s) - E\widehat{f}_{n,h}(s)$ for all distinct s and t. It follows that the optimal convergence rate of the process $A_{n,h} - F$ is $n^{\frac{2}{5}}$ as the order of the bandwidth is $n^{-\frac{1}{5}}$.

The isotonic estimator of a monotone decreasing density is defined by (1.8) from the integrated kernel estimator, $f_{n,h}^* = f_{D,n}^*$ for a decreasing density and $f_{n,h}^* = f_{I,n}^*$ for an increasing density, where

$$f_{D,n}^*(x) = \sup_{u \le x} \inf_{v \ge x} \frac{1}{v - u}\{A_{n,h}(v) - A_{n,h}(u)\},$$

$$f_{I,n}^*(x) = \inf_{v \ge x} \sup_{u \le x} \frac{1}{v - u}\{A_{n,h}(v) - A_{n,h}(u)\}.$$

With a monotone kernel estimator, the estimator $f_{n,h}^*$ is identical to the kernel estimator so their convergence rates are the same and they cannot be improved. However, the variations in the sample induce variations in the kernel estimator $\widehat{f}_{n,h}$ and it is generally not monotone for small n. The following lemma is a consequence of the weak convergence of the kernel estimator.

Lemma 3.2. *The estimator $f_{n,h}^*$ of a real monotone density of $C^s(\mathbb{R})$ converges to a monotone density with the rate $O_p(n^{\frac{s}{2s+1}})$. This is also the optimal convergence rate of the process $A_{n,h}$.*

A statistic for a test of a unimodal density f of $C^s(\mathcal{I}_X)$, $s > 1$, is defined as a weighted integrated squared mean

$$T_n = n^{\frac{2s}{2s+1}}\left\{\int_{]-\infty,\widehat{M}_{n,f}]\cap\mathcal{I}_{X,h}} (f_{n,h}^* - \widehat{f}_{n,h})^2(x)w(x)\,dx\right.$$

$$\left. + \int_{[\widehat{M}_{n,f},+\infty[\cap\mathcal{I}_{X,h}} (f_{n,h}^* - \widehat{f}_{n,h})^2(x)w(x)\,dx\right\},$$

where the bandwidth h_n tends to zero and w is a positive weight function with integral 1. A single term of the sum defining the statistic T_n is used for a test of monotony.

For every x in a compact sub-interval $\mathcal{I}_{X,h_n}$ of the support of the density f, let $B^*_{n,f}(x) = n^{\frac{s}{2s+1}}(f^*_{n,h} - f^*)(x)$ be the empirical process of the isotonic estimator and let $B_{f,n} = n^{\frac{s}{2s+1}}(\widehat{f}_{n,h} - f)$ be the empirical process for the kernel estimator. The bias b^*_f of the isotonic estimator $h^{-s}f^*_{n,h}$ is obtained by an application of the isotonization Lemma 1.1 on $]-\infty, \widehat{M}_{n,h}[$ and $]\widehat{M}_{n,h}, \infty[$ to the functions $Ef^*_{n,h}$ and f^*. The weak convergence of the process $B^*_{n,m}$ to a process $B^* + \gamma_s b^*_f$, where B^* is a centered Gaussian process, is a consequence of Theorem 1.1 and Lemma 3.2.

Under Condition 1.1 for the kernel estimator of a density f of $C^s(\mathcal{I}_X)$, $s \geq 2$, the statistic T_n converges weakly to the variable $T = \int_{\mathcal{I}_{X,h}} \{B^*(x) - B(x) - \gamma_s b^*_f(x) + \gamma_s b_f(x)\}^2 w(x)\, dx$ and it tends in probability to infinity under a fixed alternative.

Proposition 3.9. *Under Condition 1.1 and H_0, the statistic T_n converges weakly to the variable $T = \int_{\mathcal{I}_{X,h}} \{B^*(x) - B(x)\}^2 w(x)\, dx$. Under a local alternative of a nonmonotone density $f_n = f + n^{-\frac{s}{2s+1}} g_n$ where f is a monotone function and $(g_n)_{n\geq 1}$ is a sequence of nonmonotone functions of $L^2(w(x)\, dx)$, converging uniformly to a function g as n tends to infinity, the statistic T_n converges weakly to $\int_{\mathcal{I}_{X,h}} \{B^*(x) - B(x) - g(x)\}^2 w(x)\, dx$.*

Proof. Under Condition 1.1 and H_0, the bias functions b^*_f and b_f are identical. Under the alternative, the isotonic estimator $f^*_{n,h}$ is unchanged and it remains an estimator of an isotonic function. The test statistic is expanded as

$$T_n = n^{\frac{2s}{2s+1}} \int_{\mathcal{I}_{X,h}} \{(f^*_{n,h} - f)(x) - (\widehat{f}_{n,h} - f_n)(x) - n^{-\frac{s}{2s+1}} g_n(x)\}^2 w(x)\, dx,$$

it is asymptotically equivalent in probability to $\int_{\mathcal{I}_{X,h}} \{B^*_{f,n} - B_{f_n,n} - g_n(x)\}^2 w(x)\, dx$. $\qquad\square$

Under the hypothesis H_0, the limit T is the weighted integral of a squared centered Gaussian process. Its mean depends on the variances and on the covariance of $f^*_{n,h}$ and $\widehat{f}_{n,h}$. Since $Cov\{A_{n,h}, \widehat{f}_{n,h}\}(x) = Var A_{n,h}(x)$, this covariance is the variance of $B^*_{f,n}$ that is smaller than the variance of $B_{f,n}$ under H_0. The asymptotic mean of T_n is $\int_{\mathcal{I}_{X,h}} (\sigma^2_f - \sigma^{*2}_f) w(x)\, dx$ and its variance depends on the fourth order moment of $B^*_{f,n} - B_{f,n}$, this is a $O(nh)^{-1}$ and $Var T_n$ has a strictly positive and finite limit.

Under alternatives, the existence of the difference of the bias terms $(b_f^* - b_f)$ reduces the power of the test. A bias correction of the statistic is performed using the estimator $\widehat{f}_{n,h} - \gamma_s \widehat{b}_{f,n,h}$ instead of $\widehat{f}_{n,h}$ and $\widehat{f}_{n,h}^*$ is replaced by $\widehat{f}_{n,h}^* - \gamma_s \widehat{b}_{f,n,h}^*$ in Proposition 3.9. The asymptotic behavior of the statistic is not modified under H_0 and local alternatives, under fixed alternatives the statistic T_n is asymptotically equivalent to $T = \int_{\mathcal{I}_{X,h}} \{n^{\frac{2s}{2s+1}} (f^* - f)(x) + B^*(x) - B(x)\}^2 w(x) \, dx$ and its asymptotic mean is infinite so the power of the test tends to 1. Under the local alternative of Proposition 3.9, the asymptotic mean of T_n has the form $ET = E_0 T + \int_{\mathbb{R}} g^2(x) w(x) \, dx$ and it is larger than its mean $E_0 T$ under H_0, which implies the consistency of the test.

An adaptive kernel estimator of the density is defined with a local bandwidth $h_n(x) = O(n^{-\frac{1}{2s+1}})$ minimizing the L^2-risk $E(\widehat{f}_{n,h_n(x)} - f)^2(x)$ at every x, it improves the previous estimator where the global bandwidth minimizes the integrated L^2-risk. The empirical process for the adaptative kernel estimator is $B_n(x) = \{nh_n(x)\}^{\frac{1}{2}} \{\widehat{f}_{n,h_n(x)}(x) - f(x)\}$ for every x in $\mathcal{I}_{X,\|h_n\|}$, it converges weakly to a Gaussian process with mean $\gamma_s b_f(x)$ and covariance $\sigma_f^2(x) \delta_{\{x,x'\}}$ at x and x', under a condition for the uniform convergence of the bandwidth sequence $\lim_{n \to \infty} \|nh_n^{2s+1} - h\| = 0$ (Pons, 2011). The varying bandwidth kernel estimator is integrated and an isotonic estimator of the density is defined from this integral, which defines a varying bandwidth isotonic estimator of the density. A statistic T_n is defined as above from the squared difference between both estimators and its asymptotic behavior is similar.

A test for monotony of the density in an interval $\mathcal{I}$ is performed in the same way, with a comparison of the adaptive kernel estimator of the density and its isotonic version. The existence of several modes in a density is a hint for a mixture of several subpopulations with shifted modes in the sample. Tests for unimodality against nonparametric alternatives of k local modes, $k > 1$, can be considered as tests for a single unimodal density against alternatives of mixtures of k unimodal densities in parametric families or in nonparametric models. Several test statistics of this hypothesis have been studied according to the models in Pons (2009).

The hypothesis of convexity of a density f having a continuous second order derivative $f^{(2)}$ in an interval $\mathcal{I}$ may be considered as the hypothesis $H_0 : f^{(2)} > 0$ against the alternative $H_1 :$ there exist sub-intervals of $\mathcal{I}$

where $f^{(2)} \leq 0$. An estimator $\widehat{f}^{(2)}_{n,h}$ of the second derivative of the density is defined from the second order derivative of kernel K

$$K_h^{(k)}(x) = h^{-(k+1)} K^{(k)}(h^{-1}(x - X_i)), \; k \geq 1.$$

Under Condition 1.1, the estimator $\widehat{f}^{(k)}_{n,h}$ of the k-order derivative of a density in class C^s has a bias $O(h^s)$ and a variance $O((nh^{2k+1})^{-1})$, its L^2-optimal bandwidth is $O(n^{-1/(2k+2s+1)})$, (Pons, 2011). This estimator is uniformly consistent and, with the optimal bandwidth, the empirical process $B^{(2)}_{n,h} = n^{\frac{s}{2s+5}}(\widehat{f}^{(2)}_{n,h} - f^{(2)})$ converges weakly to a Gaussian process with mean function $\mu_{(2),s} = \frac{1}{s!}\gamma_{s,2}^{\frac{1}{2}} m_{sK} f^{(4)}$ and with variance function $\sigma^2_{(2)} = f(x) \int K^{(2)2}(z) \, dz$, where $\gamma_{s,2} = \lim_{n\to\infty} nh_n^{2s+5}$.

Under H_0, the statistic $T_{(2),n} = \inf_{x\in\mathcal{I}} \widehat{f}^{(2)}_{n,h}(x)$ is positive and $T_{(2),n} < 0$ under the alternative. The threshold zero is therefore the critical value of the test based on $T_{(2),n}$, the hypothesis and the alternative are well separated since the level of this test is zero and its power is 1, for every n. For a test of a local mode at the unknown value M_f, the statistic $T_{(2),n} = \widehat{f}^{(2)}_{n,h}(M_{\widehat{f}_{n,h}})$ yields a test with the same properties. A bias correction of $\widehat{f}^{(2)}_{n,h}$ and $M_{\widehat{f}_{n,h}}$ improves the test.

Removing the assumption of differentiability of the density, the hypothesis of a convex density in an interval $\mathcal{I}$ is expressed by an inequality for the density in $\mathcal{I}$

$$D_\lambda(x,y) = f(\lambda x + (1-\lambda)y) \geq \lambda f(x) + (1-\lambda)f(y)$$

for every λ in $]0,1[$ and for all x and y in $\mathcal{I}$. Under the alternative, there exist λ in $]0,1[$, x and y in $\mathcal{I}$ where the inequality is reversed. The function $D_\lambda(x,y)$ is estimated by $\widehat{D}_{\lambda,n,h}(x,y) = \widehat{f}_{n,h}(\lambda x + (1-\lambda)y) - \lambda\widehat{f}_{n,h}(x) - (1-\lambda)\widehat{f}_{n,h}(y)$ and the statistic

$$S_n = \inf_{(x,y)\in\mathcal{I}^2} \inf_{\lambda\in]0,1[} n^{\frac{s}{2s+1}} \widehat{D}_{\lambda,n,h}(x,y)$$

is positive under H_0, it is strictly negative under the alternative and it tends to $-\infty$ as n tends to infinity. The critical value for the test based on S_n is therefore zero, and it satisfies $P_0(S_n \geq 0) = 1$ and for every alternative K, $P_K(S_n < 0) = 1$, for every n.

3.5   Goodness of fit test in biased length models

On a probability space $(\Omega, \mathcal{F}, P)$, let $Y > 0$ be a random variable with values in a separable and complete metric space $(\mathcal{X}, \mathcal{B})$. A uniform biased length variable X related to Y is the sampled variable Y up to an uniform and independent random time variable U on $[0, 1]$. The variable Y is not directly observed but only a biased length variable $X = YU$, therefore $F_Y \leq F_X$ and $EX = \frac{1}{2}EY$. By definition of X, the uniform sampling variable is written as $U = (Y^{-1}X) \wedge 1$, if $Y > 0$ and 0 if $Y = 0$. The following expressions of the distributions functions of X and Y, and of the mean lifetime function of the variable X

$$m_X(x) = E(X1_{\{X \leq x\}}),\tag{3.11}$$

are established in Pons (2011), with the weak convergence of their empirical estimators. From the observation of a n-sample $X_1, \ldots, X_n$ of X, we define the empirical distribution function $\widehat{F}_{X,n}(x) = n^{-1}\sum_{i=1}^{n} 1_{\{X_i \leq x\}}$ of F_X and the empirical version of the function m_X

$$\widehat{m}_{X,n}(x) = n^{-1}\sum_{i=1}^{n} X_i 1_{\{X_i \leq x\}}.$$

The distribution functions of X and Y satisfy

$$F_X(x) = E(xY^{-1} \wedge 1) = F_Y(x) + x\int_x^{\infty} y^{-1}\, dF_Y(y),\tag{3.12}$$

$$F_Y(y) = 1 - E(Xy^{-1} \wedge 1) = F_X(y) - y^{-1}\int_0^y x\, dF_X(x),\tag{3.13}$$

for all $x > 0$ and $yx > 0$. The expected mean lifetime function m_X is related to the distribution functions of the variables X and Y by (3.13)

$$m_X(x) = x\{F_X(x) - F_Y(x)\}, \quad x > 0,\tag{3.14}$$

and the densities of X and Y are deduced from (3.13) and (3.13)

$$f_X(x) = \frac{m_X(x)}{x}, \quad f_Y(y) = \frac{m_X(y)}{y^2}.\tag{3.15}$$

By plugging in (3.13), $\widehat{F}_{X,n}$ and $\widehat{m}_n$ define the empirical distribution function $\widehat{F}_{Y,n}$ of the unobserved variable Y

$$\widehat{F}_{Y,n}(y) = n^{-1}\sum_{i=1}^{n}\left(1 - \frac{X_i}{y}\right)1_{\{X_i \leq y\}}.$$

The variance of the empirical process $\nu_{Y,n} = n^{\frac{1}{2}}(\widehat{F}_{Y,n} - F_Y)$ is

$$\sigma_Y^2(y) = \{F_Y(1 - F_Y)\}(y) + E\left(\frac{X^2}{y^2}1_{\{X \leq y\}}\right) - \frac{m_X(y)}{y} \leq \{F_Y(1 - F_Y)\}(y).$$

The estimator $\widehat{F}_{Y,n}$ converges uniformly to F_Y in probability and the empirical process $\nu_{Y,n}$ converges weakly to a centered Gaussian process ν_Y with variance function σ_Y^2.

A goodness of fit tests for the distribution function F_Y are now defined like in Section 3.2. Let $X_1, \ldots, X_n$ be a sample with distribution function F_X. A test of the hypothesis $H_0 : F_Y = F_0$ against the alternative $K : F_Y \neq F_0$ relies on the process $\nu_{Y,n}$. Its asymptotic variance under H_0 is known and denoted $\sigma_{Y,0}^2$. A statistic for a test of H_0 against K is

$$T_n = \sup_{\mathbb{R}} |\sigma_{Y,0}^{-1} \nu_{Y,n}|.$$

Under H_0, F_0 is the distribution of the variable Y and the process $\sigma_{Y,0}^{-1}\nu_{Y,n}$ converges weakly to a centered Gaussian process with variance 1 and covariance $\{\sigma_{Y,0}(x)\sigma_{Y,0}(y)\}^{-1} E_0\{(1 - x^{-1}X)(1 - y^{-1}X)1_{\{X \leq x\}} 1_{\{X \leq y\}}\}$ at x and y in $]0, 1[$. It is not a standard process and the statistic T_n is not asymptotically free, a bootstrap test based on T_n can be performed.

Consider a sequence of local alternatives K_n defined by distribution functions $F_{Y,n} = F_Y + n^{-\frac{1}{2}} H_n$ such that $\lim_{n\to\infty} \sup_{\mathbb{R}} |H_n - H| = 0$, for a function H. Under the alternative K_n, the empirical process for F_Y equals $n^{\frac{1}{2}}(\widehat{F}_{Y,n} - F_{Y,n}) = \nu_{Y,n} + H_n$ and it converges weakly to the centered Gaussian process $\nu_Y + H$. The local asymptotic power of the test with a level α is then $\lim_{n\to\infty} P_n(|\sup_{\mathbb{R}} \sigma_{Y,0}^{-1}|\nu_{Y,n} + H_n| > c_\alpha) = P(|\sup_{\mathbb{R}} \sigma_{Y,0}^{-1}|\nu_Y + H| > c_\alpha)$ for every H and the test is unbiased.

A goodness of fit test for a parametric hypothesis $H_0 : F_Y$ belongs to $\mathcal{F}_\Theta = \{F_\theta, \theta \in \Theta\}$ against the alternative $K : F_Y$ does not belong to $\mathcal{F}_\Theta$ relies on the process

$$W_n = n^{\frac{1}{2}}(\widehat{F}_{Y,n} - F_{\widehat{\theta}_n}), \tag{3.16}$$

where $\widehat{\theta}_n$ is an estimator of the unknown value θ_0 of the parameter. The likelihood of the sample $X_1, \ldots, X_n$ is $L_n(\theta) = \prod_{i=1}^n f_{X,\theta}(X_i, \theta)$ and the maximum likelihood estimator $\widehat{\theta}_n$ is deduced. Under H_0, $n^{\frac{1}{2}}(\widehat{\theta}_n - \theta_0)$ converges weakly to a centered normal variable with variance I_{X0}^{-1} where I_{X0} is the information matrix for θ_0 under H_0, defined as $I_{X0} = -\int_0^\infty f_{X,\theta}^{-1} \dot{f}_{X,\theta}^2 \, dx$, with $\dot{f}_{X,\theta}(x) = x\dot{f}_{Y,\theta}$, by (3.15). Let $\nu_{Y,\widehat{\theta}_n, n} = n^{\frac{1}{2}}(F_{Y,\widehat{\theta}_n} - F_{Y,\theta})$, then $W_n = \nu_{Y,n} - \nu_{Y,\widehat{\theta}_n}$ and a test statistic for H_0 against K is

$$T_n = \sup_{\mathbb{R}} |\widehat{\sigma}_n^{-1}(\nu_{Y,n} - \nu_{Y,\widehat{\theta}_n})|,$$

where $\widehat{\sigma}_n$ is a consistent estimator of the variance of the process W_n. It is approximated as $W_n = \nu_{Y,n} - n^{\frac{1}{2}}(\widehat{\theta}_n - \theta)^t \dot{F}_{Y,\theta} + o_p(1)$ and it converges weakly to a centered Gaussian process with variance function

$$\sigma_0^2 = \sigma_Y^2 + \dot{F}_{Y,\theta}^t I_{X0}^{-1} \dot{F}_{Y,\theta} - 2C_{Y,\theta}$$

where $C_{Y,\theta}$ is the asymptotic covariance of $\nu_{Y,n}$ and $n^{\frac{1}{2}}(\widehat{\theta}_n - \theta)^t \dot{F}_{Y,\theta}$

$$C_{Y,\theta}(x) = E_\theta \left[\{1_{\{Y \leq x\}} - F_Y(x)\} \frac{\dot{f}_{X,\theta}}{f_{X,\theta}}(X) \right]^t I_{X0}^{-1} \dot{F}_{Y,\theta}(x)$$

$$= E_\theta [\{1_{\{Y \leq x\}} \frac{\dot{f}_{X,\theta}}{f_{X,\theta}}(X)]^t I_{X0}^{-1} \dot{F}_{Y,\theta}(x)$$

$$= \dot{F}_{Y,\theta}(x) I_{X0}^{-1} \dot{F}_{Y,\theta}(x),$$

using (3.15). A consistent estimator of the variance $\sigma_0^2(x)$ is

$$\widehat{\sigma}_n^2(x) = \widehat{\sigma}_{Y,n}^2(x) - \{\dot{F}_{Y,\widehat{\theta}_n}(x)\}^t \widehat{I}_{X,n}^{-1}(x) \dot{F}_{Y,\widehat{\theta}_n}(x)$$

where the variance of the empirical process $\nu_{Y,n}$ is estimated by

$$\widehat{\sigma}_{Y,n}^2(x) = \{\widehat{F}_{Y,n}(1 - \widehat{F}_{Y,n})\}(y) + n^{-1} \sum_{i=1}^n \frac{X_i^2}{y^2} 1_{\{X_i \leq y\}}) - \frac{\widehat{m}_{X,n}(y)}{y},$$

$\widehat{I}_{X,n}$ is the empirical estimator of I_{X0}, $\dot{F}_{Y,\theta}$ is defined from the derivative $\dot{f}_{X,\theta}$ in (3.13), $\dot{F}_{Y,\theta}(x) = \dot{F}_{X,\theta}(x) - x^{-1} \int_0^x y \dot{f}_{X,\theta}(y)\, dy$ and it is estimated using $\widehat{\theta}_n$ for θ. Under H_0, the variable W_n converges weakly to a centered Gaussian process with variance 1 and nonzero covariances, and the statistic T_n converges to the supremum of this process. A bootstrap test based on T_n based on T_n can be performed.

Consider a sequence of local alternatives K_n defined by distribution functions $F_{Y,n} = F_{Y,\theta} + n^{-\frac{1}{2}} H_n$, θ in Θ, such that H_n converges uniformly to a function H. Under the alternative K_n, the empirical process for F_Y is $\nu_{Y,n,n} = n^{\frac{1}{2}}(\widehat{F}_{Y,n} - F_{Y,n}) = \nu_{Y,n} + H_n$. Since H_n does not depend on the parameter and by the relationship $f_{X,n}(x) = x f_{Y,n}(x)$, we get $\dot{f}_{X,n} = \dot{f}_{X,\theta}$. Under the alternative, there exists θ_0 in Θ such that for the maximum likelihood estimator of the parameter, $n^{\frac{1}{2}}(\widehat{\theta}_n - \theta_0)$ has the same behavior under K_n and H_0. The process $n^{\frac{1}{2}}(F_{\widehat{\theta}_n} - F_{Y,n})$ has the expansion $\nu_{Y,\widehat{\theta}_n,n} - H_n$ and the process

$$W_n = n^{\frac{1}{2}}(\widehat{F}_{Y,n} - F_{\widehat{\theta}_n}), \nu_{Y,n,n} - \nu_{Y,\widehat{\theta}_n,n} + H_n$$

converges weakly to the noncentered Gaussian process $\nu_Y + H$. The local asymptotic power of the test with a level α is then $\lim_{n \to \infty} P_n(|\sup_{\mathbb{R}} |\nu_{Y,n} + H_n| > c_\alpha) = P(|\sup_{\mathbb{R}} |\nu_Y + H| > c_\alpha)$ for every H and the test is unbiased.

3.6 Goodness of fit tests for a regression function

Let (X, Y) be a two-dimensional random variable defined on a probability space $(\Omega, \mathcal{A}, P)$, with values in $\mathbb{R}^2$. Let f_X and $f_{X,Y}$ be the continuous densities of X and (X, Y) respectively, and let F_X and F_{XY} be their respective distribution functions. The conditional mean function of Y given $X = x$ is the nonparametric regression function defined for every x inside the support of X by

$$m(x) = E(Y|X = x) = \int y \frac{f_{X,Y}(x, y)}{f_X(x)} \, dy,$$

it is continuous when the density $f_{X,Y}$ is continuous with respect to its first component. Let K be a kernel satisfying Condition 1.1, Watson's kernel estimator (1964) of a smooth function m is

$$\widehat{m}_{n,h}(x) = \int y \frac{K_h(x - s)\widehat{F}_{XY,n}(ds, dy)}{\widehat{f}_{X,n,h}(x)} = \frac{\sum_{i=1}^n Y_i K_h(x - X_i)}{\sum_{i=1}^n K_h(x - X_i)}. \quad (3.17)$$

Its denominator is $\widehat{f}_{X,n,h}(x)$ and its numerator is denoted

$$\widehat{\mu}_{n,h}(x) = \frac{1}{n} \sum_{i=1}^n Y_i K_h(x - X_i).$$

Its mean is $\mu_{n,h}(x) = \int \int y K_h(x-s) \, dF_{XY}(s, y)$ and its limit is the product $\mu(x) = f_X(x)m(x)$, the mean of $\widehat{m}_{n,h}(x)$ is denoted $m_{n,h}(x)$. The variance of Y is supposed to be finite and its conditional variance is denoted

$$\sigma^2(x) = E(Y^2|X = x) - m^2(x) = f_X^{-1}(x)w_2(x) - m^2(x),$$

where $w_2(x) = \int y^2 f_{XY}(x, y) \, dy = f_X(x) \int y^2 f_{Y|X}(y; x) \, dy$. One of the next additional conditions is necessary for the weak convergence of $\widehat{m}_{n,h}$.

Condition 3.1.

(1) The functions f_X, m and μ are twice continuously differentiable on $\mathcal{I}_X$, with bounded second order derivatives; f_X is strictly positive on $\mathcal{I}_X$;
(2) The functions f_X, m and σ belong to the class $C^s(\mathcal{I}_X)$, with $s \geq 2$; $f_X > 0$ on $\mathcal{I}_X$.

As n tends to infinity, the bias and the variance of the estimator of m have the expansions $b_{m,n,h}(x) = E\widehat{m}_{n,h}(x) - m(x) = h^s b_m(x) + o(h^s)$ and $v_{m,n,h}(x) = (nh)^{-1}\{\sigma_m^2(x) + o(1)\}$, where the constants are

$$b_m(x) = \frac{m_s K}{s!} f_X^{-1}(x)\{\mu^{(s)}(x) - m(x)f_X^{(s)}(x)\},$$

$$\sigma_m^2(x) = \kappa_2 f_X^{-1}(x)\sigma^2(x).$$

Under Conditions 1.1 and 3.1 and with the L^2-optimal bandwidth of the estimator $\widehat{m}_{n,h}$, $h_n = O(n^{-\frac{1}{2s+1}})$, the process $U_{n,h} = n^{\frac{s}{2s+1}}\{\widehat{m}_{n,h} - m\}\mathcal{I}_{\{\mathcal{I}_{X,h}\}}$

converges in distribution to $\sigma_m W_1 + \gamma_s b_m$ where W_1 is a centered Gaussian process on $\mathcal{I}_X$ with variance function 1 and covariance function zero, and γ_s is defined in Condition 1.1 (Pons, 2011).

A test for the hypothesis H_0 in a class of parametric regression models $\mathcal{M} = \{m_\theta, \theta \in \Theta\}$, with a bounded parameter space Θ, against the alternative $K : m$ does not belong to $\mathcal{M}$, relies on the integrated squared difference between the nonparametric estimator of the regression function and a minimum of distance estimator of the curve in $\mathcal{M}$. Let d be a weighted L^2-distance of functions of $\mathcal{M}$

$$d(m_1, m_2) = [E\{w(X)(m_1 - m_2)^2(X)\}]^{\frac{1}{2}}.$$

A sample $(X_i, Y_i)_{i=1,\ldots,n}$ of the variable (X, Y) has a regression function with an unknown parameter value θ_0 under H_0, it is estimated by

$$\widehat{\theta}_n = \arg\min_{\theta \in \Theta} d_n(m_\theta, \widehat{m}_{n,h}),$$

where $d_n(m_\theta, \widehat{m}_{n,h}) = n^{-1} \sum_{i=1}^{n} w_n(X_i)(m_\theta - \widehat{m}_{n,h})^2(X_i)$ is defined with a sequence of positive random weighting functions $(w_n)_{n \geq 1}$.

Condition 3.2.

(1) The sequence $(w_n)_{n \geq 1}$ is a sequence of uniformly convergent functions in $\mathcal{I}_X$, with limit $w > 0$;
(2) The variance Σ_0 of $w(X)\dot{m}_{\theta_0}(X)n^{\frac{s}{2s+1}}(\widehat{m}_{n,h} - m_{n,h})(X)$ under F_0 and $v_0 = E_0\{w(X)\dot{m}_{\theta_0}^{\otimes 2}(X)\}$ are finite matrices.

Proposition 3.10. *Let $\mathcal{M} = \{m_\theta, \theta \in \Theta\}$ be a space of real functions defined in $\mathcal{I}_X \times \Theta$, such that for every x, $m(x, \cdot)$ belongs to $C^2(\Theta)$. Under Conditions 1.1, 3.1 and 3.2, and with a bandwidth $h_n = O(n^{-\frac{1}{2s+1}})$*

$$\sqrt{nh}(\widehat{\theta}_n - \theta_0) = -(nh)^{\frac{1}{2}}\ddot{d}_n^{-1}(m_{\theta_0}, \widehat{m}_{n,h})\dot{d}_n(m_{\theta_0}, \widehat{m}_{n,h}) + o_p(1), \quad (3.18)$$

it converges weakly under H_0 to the centered variable

$$G_0 = v_0^{-1} \int_{\mathcal{I}_X} w(\sigma_{m_0} W_1 + \gamma_s b_{m_0})\dot{m}_{\theta_0} \, dF_X$$

with variance $V_0 = v_0^{-1} \Sigma_0 v_0^{-1}$.

Proof. Let m_{θ_0} be the regression function in $\mathcal{M}$ of (X, Y) under H_0. The L^2-consistency of the kernel estimator implies that for every m_θ, $\lim_{n \to \infty} d_n(m_\theta, \widehat{m}_{n,h}) = E_0\{w^2(X)(m_\theta - \widehat{m}_{n,h})^2(X)\} = d(m_\theta, m_0) + o(1)$

and it is zero if and only if $\theta = \theta_0$. It follows that $\widehat{\theta}_n$ is a consistent estimator of θ_0. The first two derivatives of $d(m_\theta, \widehat{m}_{n,h})$ in Θ are

$$\dot{d}_n(m_\theta, \widehat{m}_{n,h}) = 2n^{-1} \sum_{i=1}^{n} w_{n,i} \dot{m}_\theta(X_i)(m_\theta - \widehat{m}_{n,h})(X_i)$$

$$= 2 \int w_n(x) \dot{m}_\theta(x)(m_\theta - \widehat{m}_{n,h})(x) \, d\widehat{F}_{X,n}(x),$$

$$\ddot{d}_n(m_\theta, \widehat{m}_{n,h}) = 2n^{-1} \sum_{i=1}^{n} w_{n,i} \{\ddot{m}_\theta(X_i)(m_\theta - \widehat{m}_{n,h})(X_i) + \dot{m}_\theta^{\otimes 2}(X_i)\},$$

where $w_{n,i} = w_n(X_i)$. Let $\mathcal{V}_n$ be a neighborhood of θ_0 converging to θ_0, then $\sup_{\theta \in \mathcal{V}_n} |\ddot{d}_n(m_\theta, \widehat{m}_{n,h}) - 2v_0|$ tends to zero as n tends to infinity. Since $2v_0$ is strictly positive $d_n(m_\theta, \widehat{m}_{n,h})$ is a sequence of locally convex functions in $\mathcal{V}_n$. Moreover, by the weak law of large numbers for a triangular array, the variable $(nh)^{\frac{1}{2}} \dot{d}_n(m_{\theta_0}, \widehat{m}_{n,h})$ converges weakly to the Gaussian variable $-2 \int_{\mathcal{I}_X} w \dot{m}_{\theta_0}(W_1 + \gamma_s b_{m_0}) \, dF_X$ with variance matrix Σ_0. A Taylor expansion of $\dot{d}_n(m_{\widehat{\theta}_n}, \widehat{m}_{n,h})$ yields

$$\dot{d}_n(m_{\widehat{\theta}_n}, \widehat{m}_{n,h}) = \dot{d}_n(m_{\theta_0}, \widehat{m}_{n,h}) + \ddot{d}_n(m_{\theta_n}, \widehat{m}_{n,h})(\widehat{\theta}_n - \theta_0),$$

and it is zero by definition of $\widehat{\theta}_n$. It follows that

$$(nh)^{\frac{1}{2}}(\widehat{\theta}_n - \theta_0) = -(nh)^{\frac{1}{2}} \ddot{d}_n^{-1}(m_{\theta_n}, \widehat{m}_{n,h}) \dot{d}_n(m_{\theta_0}, \widehat{m}_{n,h})$$

where θ_n is between θ_0 and $\widehat{\theta}_n$, and it converges to a Gaussian variable, its mean and its variance are deduced from the limiting distribution of $(nh)^{\frac{1}{2}} \dot{d}_n(m_{\widehat{\theta}_n}, \widehat{m}_{n,h})$. □

The asymptotic behavior of the test statistic is a consequence of the weak convergence of the estimator of the regression function and of the previous proposition.

Proposition 3.11. *Under H_0 and under the conditions of Proposition 3.10, the process $\sqrt{nh}(\widehat{m}_{n,h_n} - m_{\widehat{\theta}_n})$ converges weakly to a centered Gaussian process*

$$W = \sigma_m W_1 + \gamma_s b_m - G_0^t \dot{m}_{\theta_0} \tag{3.19}$$

and $S_n = nh_n d_n^2(m_{\widehat{\theta}_n}, \widehat{m}_{n,h_n})$ converges weakly to $E\{w(X)W^2(X)\}$.

Proof. By an expansion, the function $m_{\widehat{\theta}_n}$ is approximated in the form $m_{\widehat{\theta}_n} = m_{\theta_0} + \dot{m}_{\theta_n}(\widehat{\theta}_n - \theta_0)$, where $\dot{m}_{\theta_n}$ converges to $\dot{m}_{\theta_0}$. It follows that

$$\sqrt{nh}(\widehat{m}_{n,h_n} - m_{\widehat{\theta}_n}) = \sqrt{nh}(\widehat{m}_{n,h_n} - m_{\theta_0}) - \sqrt{nh}(\widehat{\theta}_n - \theta_0)^t \dot{m}_{\theta_0} + o_p(1).$$

Its asymptotic variance is obtained from the expansion (3.18) and it is bounded by $2\sigma_{m_0}^2 + 2\dot{m}_{\theta_0}^t V_0 \dot{m}_{\theta_0}$ which is finite. □

The test of a single hypothesis with regression function m_0 is performed with a statistic $S_n = nh_n d_n^2(m_0, \widehat{m}_{n,h_n})$, using for w_n the inverse of a consistent estimator of the asymptotic variance $\sigma_{m_0}^2$ of the estimator of the regression function m_0. With a bandwidth $h_n = o(n^{-\frac{1}{2s+1}})$, the statistic S_n converges weakly under H_0 to $E_0\{W_1^2(X)\}$ so the critical value of the test can be computed if the distribution F_X is known, or it is estimated from $\widehat{F}_{X,n}$. The bias term can be estimated by bootstrapping kernel estimators. For every separate alternative K, the asymptotic power of the test is one.

Proposition 3.12. *Let $(K_n)_{n\geq 1}$ be a sequence of local alternatives defined by regression functions $m_n = m_0 + (nh_n)^{-\frac{1}{2}}r_n$ such that $(r_n)_{n\geq 1}$ converges uniformly to a function r. Under the conditions of Proposition 3.11, the statistic S_n converges weakly under $(K_n)_{n\geq 1}$ to $\int_{\mathcal{I}_X} w(\sigma_{m_0}W_1 + \gamma_s b_{m_0} - r)^2\, dF$.*

Proof. Under K_n, $\widehat{m}_{n,h_n}$ is an estimator of m_n and the statistic S_n is written

$$S_n = nh_n d_n^2(m_n - (nh_n)^{-\frac{1}{2}}r_n, \widehat{m}_{n,h_n})$$

$$= nh_n \int_{\mathcal{I}_X} w_n\{(\widehat{m}_{n,h_n} - m_n)^2 + (nh_n)^{-1}r_n^2$$

$$- 2(\widehat{m}_{n,h_n} - m_n)(nh_n)^{-\frac{1}{2}}r_n\}\, d\widehat{F}_n.$$

With a bandwidth $h_n = O(n^{-\frac{1}{2s+1}})$, $\widehat{m}_{n,h_n} - m_n$ converges weakly to $\sigma_{m_0}W_1 + \gamma_s b_{m_0}$. $\square$

If $h_n = o(n^{-\frac{1}{2s+1}})$, the limiting distributions of S_n under H_0 and $(K_n)_{n\geq 1}$ are identical and the asymptotic local power of the test is its asymptotic level. A bias correction must be performed using $\widehat{m}_{n,h} - \widehat{b}_{m,n,h}$ instead of $\widehat{m}_{n,h}$ in the statistic S_n, its limit is then a weighted integral of squared centered Gaussian process under H_0 and modified as $\int_{\mathcal{I}_X} w(\sigma_{m_0}W_1 - r)^2\, dF$ under K_n.

The test of a smooth parametric model $\mathcal{M}$ is performed with a statistic $S_n = nh_n d_n^2(m_{\widehat{\theta}_n}, \widehat{m}_{n,h_n})$, using for w_n the inverse of a consistent estimator of the asymptotic variance $\sigma_{m_0}^2$ of the estimator of the regression function in $\mathcal{M}$. Applying Propositions 3.10 and 3.11, for every separate alternative K, the asymptotic power of the test is one. The asymptotic local power is established in Proposition 3.12.

Proposition 3.13. *Let $(K_n)_{n\geq 1}$ be a sequence of local alternatives defined by regression functions $m_n = m + (nh_n)^{-\frac{1}{2}}r_n$ such that m belongs to $\mathcal{M}$*

and $(r_n)_{n \geq 1}$ *converges uniformly to a function* r. *Under the conditions of Proposition 3.11 and with the bias correction, the statistic* S_n *converges weakly under* $(K_n)_{n \geq 1}$ *to* $\int_{\mathcal{I}_X} w(W - r)^2 \, dF$, *with* W *defined by (3.19)*.

3.7   Tests about the form of a regression function

The mode of a real regression function m in $\mathcal{I}_X$ is $M_m = \sup_{x \in \mathcal{I}_X} m(x)$. For a regular regression function, it is estimated by the mode of a regular estimator of the function, $\widehat{M}_{m,n,h} = M_{\widehat{m}_{n,h}}$. Under Conditions 1.1, 3.1 and 3.2, $(nh^3)^{1/2}(\widehat{M}_{m,n,h} - M_m)$ converges weakly to a centered Gaussian variable with finite variance $\{m^{(2)}(M_m)\}^{-2} f_X^{-1}(M_m) E(Y^2 | X = M_m)\{\int K^{(1)2}\}$, with a kernel in $C^1(\mathcal{I}_X)$ (Pons, 2011). A test unimodality is then deduced from tests of monotony in sub-intervals defined by the mode.

Several estimators of monotone estimators of regression functions have been proposed. The isotonization Lemma 1.1 applies to the integrated kernel estimator of the curve $\int_{\mathcal{I}_X} 1_{s \leq x} \widehat{m}_{n,h}(s) \, ds$ as for a density. A decreasing kernel estimator $\widehat{m}_{n,h}^*$ of a regression function is obtained as $m_{D,n,h}^*$ and an increasing kernel estimator as $\widehat{m}_{I,n,h}^*$. If the kernel estimator $\widehat{m}_{n,h}$ is monotone, it is identical to $m_{n,h}^*$ and they have the same convergence rate, as in Lemma 3.2. Due to the weak convergence of $n^{\frac{s}{2s+1}}\{\widehat{m}_{n,h}(x) - m(x)\}$, at every x in $\mathcal{I}_{X,h}$, $\widehat{m}_{n,h}^*$ still has the convergence rate $n^{\frac{s}{2s+1}}$.

A statistic for a test of unimodality of a density f of $C^s(\mathcal{I}_X)$ is defined as a weighted integrated squared mean

$$T_n = n^{\frac{2s}{2s+1}} \left\{ \int_{]-\infty, \widehat{M}_{n,m}] \cap \mathcal{I}_{X,h}} (m_{n,h}^* - \widehat{m}_{n,h})^2(x) w_n(x) \, dx \right.$$
$$\left. + \int_{[\widehat{M}_{n,m}, +\infty[\cap \mathcal{I}_{X,h}} (m_{n,h}^* - \widehat{m}_{n,h})^2(x) w_n(x) \, dx \right\},$$

where w_n is a sequence of positive weight functions converging uniformly to a limiting function w.

The notations and the following convergences are similar to those of Section 3.4 for densities. For every x in a compact sub-interval $\mathcal{I}_{X,h_n}$ of the support of the density m, let $B_{n,m}^*(x) = n^{\frac{s}{2s+1}}(m_{n,h}^* - m)(x)$ be the empirical process of the isotonic estimator and let $B_{n,m} = n^{\frac{s}{2s+1}}(\widehat{m}_{n,h} - m)$ be the empirical process for the kernel estimator. The weak convergence of the process $B_{n,m}^*$ to a process $B_m^* + \gamma_s b_m^*$, where B_m^* is centered, is a consequence of the weak convergence of the process $B_{n,m}$. The bias of $B_{m,n}^*$ is denoted b_m^* and its variance σ_m^{*2}. Under Conditions 1.1, 3.1

and 3.2, for the kernel estimator of a curve m of $C^s(\mathcal{I}_X)$, $s \geq 2$, the statistic T_n converges weakly to $T = \int_{\mathcal{I}_{X,h}} \{B_m^*(x) - B_m(x)\}^2 w(x) \, dx$ under the hypothesis H_0 of a monotone regression function m and it tends in probability to infinity under separate alternatives.

Proposition 3.14. *Under a local alternative with a nonmonotone density $m_n = m + n^{-\frac{s}{2s+1}} r_n$ where m is a monotone function and $(g_n)_{n \geq 1}$ is a sequence of nonmonotone functions of $L^2(w(x) \, dx)$, converging uniformly to a function r as n tends to infinity, then T_n converges weakly to $\int_{\mathcal{I}_{X,h}} \{B_m^*(x) - B_m(x) - r\}^2 w(x) \, dx$.*

It is proved like Proposition 3.9 and the asymptotic biases of the estimators are identical. A bias correction of the regression estimators is performed using an estimator $\widehat{b}_{m,n,h}$ of the bias of $\widehat{m}_{n,h}$, then $n^{\frac{s}{2s+1}}(\widehat{m}_{n,h} - m - \gamma_s \widehat{b}_{m,n,h})$ replaces $n^{\frac{s}{2s+1}}(\widehat{m}_{n,h} - m)$. In the same way, $\widehat{m}_{n,h}^* - m^* - \gamma_s \widehat{b}_{m,n,h}^*$ replaces $\widehat{m}_{n,h}^* - m^*$, the limit of the corrected statistic T_n under H_0 is a weighted integral of a squared centered Gaussian process. Under the alternative of Proposition 3.14, its limit is $\int_{\mathcal{I}_{X,h}} \{B_m^*(x) - B_m(x) - r\}^2 w(x) \, dx$.

3.8 Tests based on observations by intervals

Let X be a real random variable with distribution function F in a real interval $\mathcal{I}_X$, and let $X_1, \ldots, X_n$ be a sample of X. The probability that X belongs to a sub-interval $\mathcal{I}_a(\theta) = [\theta - a, \theta + a]$ of $\mathcal{I}_X$, centered at θ and with length $2a > 0$ is $F(\theta + a) - F(\theta - a)$ and it is estimated by $\widehat{F}_n(\theta + a) - \widehat{F}_n(\theta - a)$. The variations

$$V_{n,a}(\theta) = \nu_n(\theta + a) - \nu_n(\theta - a)$$

of the empirical process on the intervals $\mathcal{I}_a(\theta)$ have the variance

$$\sigma_a^2(\theta) = \{F(\theta + a) - F(\theta - a)\}[1 - \{F(\theta + a) - F(\theta - a)\}].$$

When a tends to zero, $\sigma_a^2(\theta)$ converges to $\Delta F(\theta)\{1 - \Delta F(\theta)\}$ for every θ, where $\Delta F(\theta)$ is the jump of the distribution function F at θ. If F has an uniformly continuous and bounded density f in $\mathcal{I}_X$, the variance $\sigma_a^2(\theta)$ is uniformly bounded and $\lim_{a \to 0} a^{-1} \sigma_a^2(\theta) = 2f(\theta) \leq 2\|f\|_\infty$. The process

$$Y_{n,a_n}(x) = a_n^{-\frac{1}{2}} \{\nu_n(x + a_n) - \nu_n(x - a_n)\}$$

has the mean zero and its variance function does not depend on n but it does not converge in distribution because its higher moments still depend on n and some of them diverge, as proved in Appendix A.5.

The tests about the distribution of the variable X are modified for cumulative observations by intervals. The values of the variable are not observed, but only the numbers $N_i = n \int_{A_i} d\widehat{F}_n(x)$ of observations in sub-intervals A_i of their domain, $i = 1, \ldots, k$. The counting variables $n^{-1}(N_1, \ldots, N_k)$ have a multinomial distribution with probabilities $p_i = \int_{A_i} dF$, for $i = 1, \ldots, k$, and most tests of hypotheses concerning the distribution of the variables $n^{-1} N_i$ have an asymptotic χ^2 distribution. Since the process Y_{n,a_n} diverges, it is not possible to consider tests based on histograms.

3.8.1 *Goodness of fit tests for a density*

Goodness of fit tests of a density to a parametric model $\mathcal{F} = \{F_\theta; \theta \in \Theta\}$, where Θ is a subset of $\mathbb{R}^d$, are based on the vector of the proportions $\widehat{p}_{i,n} = n^{-1} N_i$, for $i = 1, \ldots, k$. Their means $p_i(\theta)$ sum up to one and the vector of the normalized observations

$$W_n(\theta) = [\{p_i(1 - p_i)\}(\theta)]^{-\frac{1}{2}} n^{-1}\{N_i - np_i(\theta)\})_{i=1,\ldots,k}$$

is a vector of k dependent and centered variables $N_i - np_i(\theta)$ normalized by the square root of their variance $v_i(\theta) = n\{p_i(1 - p_i)\}(\theta)$, it converges to a normal vector. Since $\sum_{i=1}^{k} N_i = n$ and the sum of the probabilities is 1, $N_k - np_k(\theta)$ is linearly determined by the $k - 1$ components of $W_n(\theta)$. A goodness of fit test statistic for a single distribution $H_0 : (p_1, \ldots, p_k) = (p_{01}, \ldots, p_{0k})$ such that $p_{0j} = \int_{A_j} dF_0$ is $T_{0n} = \|W_n(\theta_0)\|_2^2$ and its asymptotic distribution is a χ_{k-1}^2, by Cochran's theorem.

The variance of $n^{-\frac{1}{2}}\{N_i - np_i(\theta)\}$ is estimated by the empirical variance

$$\widehat{v}_{i,n} = \widehat{p}_{i,n}\{1 - \widehat{p}_{i,n}\},$$

it is $n^{\frac{1}{2}}$-consistent, for every $i = 1, \ldots, k$. The goodness of fit test statistics

$$S_n = \sum_{i=1}^{k} W_{i,n}^2(\theta_0) = \sum_{i=1}^{k} \frac{\{N_i - np_i(\theta_0)\})^2}{n\{p_i(\theta_0) - p_i^2(\theta_0)\}}$$

and

$$T_{0n} = \sum_{i=1}^{k} \frac{\{N_i - np_i(\theta_0)\}^2}{n\widehat{v}_{i,n}}$$

are asymptotically equivalent as n tends to infinity and they converge weakly to a χ_{k-1}^2 distribution, for every θ_0. An asymptotically free test for the a multinomial distribution with parameter $(p_1, \ldots, p_k)(\theta_0)$ of the observations $n^{-1}(N_1, \ldots, N_k)$ is deduced.

Let K_n be a sequence of local alternatives with probabilities $p_{i,n}$ converging to $p_i(\theta_0)$ and such that $r_{i,n} = n^{\frac{1}{2}}\{p_{i,n} - p_i(\theta_0)\}$ converges to a limit r_i, for $i = 1, \ldots, k$. Under K_n, the statistic T_{0n} converges weakly to a variable $T(r) = \sum_{i=1}^{k}\{X_i + v_i^{-\frac{1}{2}}(\theta_0)r_i\}^2$ where $(X_1, \ldots, X_k)$ is a Gaussian vector asymptotically equivalent to the limit of $(\{n\widehat{v}_{i,n}\}^{-\frac{1}{2}}\{N_i - np_i(\theta_0)\})_{i=1,\ldots,k}$. By Cochran's theorem, the statistic T_{0n} converges weakly, under the alternative to a noncentered χ^2 variable $T(r) = \chi_{k-1}^2 + \mu_k$, with a location parameter $\mu_k(r) = \sum_{i=1}^{k} v_i^{-1}(\theta_0)r_i^2$. Let $c_{k-1,\alpha}$ be the critical value of the test based on the χ_{k-1}^2 distribution, the asymptotic local power of the test based on T_{0n} under K_n is $\inf_r P(T_0 > c_{k-1,\alpha} - \mu_k(r))$, it is strictly larger than the asymptotic level of the test.

A goodness of fit test statistic for the hypothesis H_0 of a parametric model $\mathcal{P}_\Theta = \{p(\theta); \theta \in \Theta\}$ in $C^2(\Theta)$ is

$$T_n = \sum_{i=1}^{k} \frac{\{N_i - np_i(\widehat{\theta}_n)\}^2}{n\widehat{v}_{i,n}},$$

where $\widehat{\theta}_n$ is the maximum likelihood estimator of the unknown parameter θ_0 of $\mathcal{F}$, in the multinomial model with parameter $\{p_i(\theta)\}_{i=1,\ldots,k}$ for the observations by intervals. The estimator of θ_0 maximizes the likelihood

$$L_n(\theta) = \prod_{i=1}^{k-1} p_i^{N_i}(\theta)\left\{1 - \sum_{j=1}^{k-1} p_j(\theta)\right\}^{N_k}.$$

Its logarithm $l_n = \log L_n$ is such that $n^{-1}l_n$ converges uniformly in probability to $l(p) = \sum_{i=1}^{k-1} p_i \log p_i$. Let

$$\Sigma_0 = \sum_{i=1}^{k} p_i^{-1}(\theta_0)p_i^{(1)\otimes 2}(\theta_0). \tag{3.20}$$

Proposition 3.15. *The maximum likelihood estimator of θ_0 has the expansion $n^{\frac{1}{2}}(\widehat{\theta}_n - \theta_0) = n^{\frac{1}{2}}\Sigma^{-1}(\theta_0)l_n^{(1)}(\theta_0) + o_p(1)$ and it converges under H_0 to a centered Gaussian variable Z_0, with variance Σ_0^{-1}.*

Proof. The first two derivatives of $l_n(\theta)$ are

$$l_n^{(1)}(\theta) = \sum_{i=1}^{k} N_i \frac{p_i^{(1)}}{p_i}(\theta), \quad l_n^{(2)}(\theta) = \sum_{i=1}^{k-1} N_i\left(\frac{p_i^{(2)}}{p_i} - \frac{p_i^{(1)\otimes 2}}{p_i^2}\right)(\theta),$$

since the constraint $\sum_{i=1}^{k} p_i = 1$ implies that $\sum_{i=1}^{k} p_i^{(1)} = 0$. It also implies $\sum_{i \neq j} p_i^{(1)} p_j^{(1)} = -\sum_{i=1}^{k} p_i^{(1)2}$. The variable $n^{-\frac{1}{2}} l_n^{(1)}(\theta_0)$ converges weakly to a centered Gaussian variable with variance

$$n^{-1} E_\theta l_n^{(1)2}(\theta) = n \sum_{i=1}^{k} \frac{p_i^{(1)\otimes 2}}{p_i^2}(\theta)\, E_\theta(\widehat{p}_{i,n}^2)$$

$$+ \sum_{j \neq i, j=1}^{k} \frac{p_i^{(1)}}{p_i}(\theta) \frac{p_j^{(1)T}}{p_j}(\theta) n E_\theta\{\widehat{p}_{i,n}\widehat{p}_{j,n}\}$$

$$= \sum_{i=1}^{k} \Big[\frac{p_i^{(1)\otimes 2}}{p_i^2}\{p_i + (n-1)p_i^2\} - n \sum_{i=1}^{k} p_i^{(1)\otimes 2} \Big](\theta)$$

$$= \sum_{i=1}^{k} \frac{p_i^{(1)\otimes 2}}{p_i}(\theta).$$

At θ_0, it equals Σ_0 and $-n^{-1} l_n^{(2)}(\theta)$ converges uniformly in probability to a symmetric matrix function $\Sigma(\theta)$, with value Σ_0 at θ_0. It follows that $n^{-1} l_n(\theta)$ converges uniformly in probability to a concave function $l(\theta)$ with a maximum at θ_0, hence $\widehat{\theta}_n$ converges in probability to θ_0 as n tends to infinity. By a Taylor expansion $l_n^{(1)}(\widehat{\theta}_n) = l_n^{(1)}(\theta_0) + l_n^{(2)}(\widetilde{\theta}_n)(\widehat{\theta}_n - \theta_0)$ where $\widetilde{\theta}_n$ is between $\widehat{\theta}_n$ and θ_0, the expansion of $\widehat{\theta}_n - \theta_0$ follows. □

With an unknown parameter value θ_0, the variances $v_i(\theta_0)$ and Σ_0 are $n^{\frac{1}{2}}$-consistently estimated by $\widehat{v}_{i,n}$ and $\widehat{\Sigma}_n = n^{-1} l_n(\widehat{\theta}_n)$, respectively. Let

$$V_i(\theta_0) = v_i(\theta_0) + \{2p_i(\theta_0) - 1\} p_i^{(1)T}(\theta_0) \Sigma_0^{-1} p_i^{(1)}(\theta_0), \tag{3.21}$$

from Proposition 3.15, it is $n^{\frac{1}{2}}$-consistently estimated by

$$\widehat{V}_{i,n} = \widehat{v}_{i,n} + \{2\widehat{p}_{i,n} - 1\} p_i^{(1)T}(\widehat{\theta}_n) \widehat{\Sigma}_n^{-1} p_i^{(1)}(\widehat{\theta}_n). \tag{3.22}$$

Proposition 3.16. *Under H_0, the goodness of fit test statistic*

$$T_n = \sum_{i=1}^{k} \{n \widehat{V}_{i,n}\}^{-1} \{N_i - n p_i(\widehat{\theta}_n)\}^2, \tag{3.23}$$

for a model $\mathcal{P}_\Theta = \{p(\theta); \theta \in \Theta\}$ in $C^2(\Theta)$, has the asymptotic expansion

$$n^{-1} \sum_{i=1}^{k} V_i^{-1}(\theta_0) [n\{\widehat{p}_{i,n} - p_i(\theta_0)\} + (\widehat{\theta}_n - \theta_0)^t p_i^{(1)}(\theta_0)]^2 + o_p(1)$$

and there exists a set of $k-1$ independent normal variables $(W_1, \ldots, W_{k-1})$ such that T_n converges weakly to a χ_{k-1}^2 variable $\sum_{i=1}^{k-1} W_i^2$.

Proof. From Proposition 3.15, the variable

$$U_{i,n} = n^{\frac{1}{2}}\{n^{-1}N_i - p_i(\widehat{\theta}_n)\} \tag{3.24}$$

has the asymptotic expansion

$$
\begin{aligned}
U_{i,n} &= n^{\frac{1}{2}}\{n^{-1}N_i - p_i(\theta_0)\} - n^{\frac{1}{2}}(\widehat{\theta}_n - \theta_0)^t p_i^{(1)}(\theta_0) + o(1) \\
&= n^{\frac{1}{2}}\{n^{-1}N_i - p_i(\theta_0)\} \\
&\quad - \{p_i^{(1)}(\theta_0)\}^t \Sigma_0^{-1} n^{-\frac{1}{2}} \sum_{j=1}^{k} N_j \frac{p_j^{(1)}}{p_j}(\theta_0) + o_p(1).
\end{aligned}
$$

In this expression, the mean M_i of $\{n^{-1}N_i - p_i(\theta_0)\}\{\sum_{j=1}^{k} N_j (p_j^{-1} p_j^{(1)})(\theta_0)\}$ reduces to

$$M_i = \{n^{-1}EN_i^2 - np_i^2(\theta_0)\}\frac{p_i^{(1)}}{p_i}(\theta_0) = \{1 - p_i(\theta_0)\}p_i^{(1)}(\theta_0),$$

hence $U_{i,n}$ is approximated by the sum of two centered and asymptotically Gaussian variables with covariance $-(1 - p_i)\{p_i^{(1)}(\theta_0)\}^t \Sigma_0^{-1} p_i^{(1)}(\theta_0)$, their respective variances are $v_i(\theta_0)$ and $\{p_i^{(1)}(\theta_0)\}^t \Sigma_0^{-1} p_i^{(1)}(\theta_0)$, under the multinomial distribution indexed by θ_0 for the variable set $(N_1, \ldots, N_k)$. The asymptotic variance of $U_{i,n}$ is then $V_i(\theta_0)$, under the parametrization with θ_0. The variables $U_{i,n}$ are therefore asymptotically Gaussian variables with mean zero and variance $V_i(\theta_0)$. The sum of the main terms in the expansions of variables the $V_i^{-\frac{1}{2}} U_{i,n}$ is zero since $\sum_{i=1}^{k} p_i^{(1)} = 0$. They are dependent and the limiting distribution of the statistic is a consequence of Cochran's theorem. $\qquad\square$

Let $(K_n)_{n\geq 1}$ be a sequence of local alternatives defined by probability functions $p_{i,n}$ in $C^2(\Theta)$, converging uniformly to a function p_i in $\mathcal{F}$ and such that the functions $r_{i,n} = n^{\frac{1}{2}}\{p_{i,n} - p_i\}$ converge uniformly in Θ to a limit r_i of $C^2(\Theta)$, for $i = 1, \ldots, k$. The estimator $\widehat{\theta}_{n,n}$ maximizes $n^{-1}l_{n,n}(\theta) = \sum_{i=1}^{k} \widehat{p}_{i,n} \log\{p_i + n^{-\frac{1}{2}}r_{i,n}\}(\theta)$, for n large enough it also maximizes $n^{-1}l_n(\theta)$, therefore it converges in probability under K_n to a value θ_0 in Θ. The limiting distribution of the normalized estimator converges weakly to a limit depending on Z_0 defined in Proposition 3.15.

Proposition 3.17. *Under K_n, there exists θ_0 in Θ such that $n^{\frac{1}{2}}(\widehat{\theta}_{n,n} - \theta_0)$ converges weakly to $Z_0 + \Sigma^{-1}(\theta_0) \sum_{i=1}^{k} \{p_i^{-1}(p_i r_i^{(1)} - p_i^{(1)} r_i)\}(\theta_0)$.*

Proof.   Under K_n, the derivatives of the log-likelihood are modified as

$$l_{n,n}^{(1)}(\theta) = \sum_{i=1}^{k} N_i \frac{p_i^{(1)} + n^{-\frac{1}{2}} r_{i,n}^{(1)}}{p_{i,n}}(\theta),$$

$$l_{n,n}^{(2)}(\theta) = \sum_{i=1}^{k-1} N_i \left(\frac{p_i^{(2)} + n^{-\frac{1}{2}} r_{i,n}^{(1)}}{p_{i,n}} - \frac{\{p_i^{(1)} + n^{-\frac{1}{2}} r_{i,n}^{(1)}\}^{\otimes 2}}{p_{i,n}^2} \right)(\theta).$$

As n tends to infinity

$$n^{-1}(l_{n,n}^{(1)} - l_n^{(1)})(\theta) = \sum_{i=1}^{k} \widehat{p}_{i,n} \frac{p_i r_{i,n}^{(1)} - p_i^{(1)} r_{i,n}}{p_i p_{i,n}}(\theta)$$

is asymptotically equivalent to $\sum_{i=1}^{k} \widehat{p}_{i,n} \{(p_i r_i^{(1)} - p_i^{(1)} r_i) p_i^{-2}\}(\theta)$, uniformly in Θ. For the second derivative, $\sup_{\Theta} n^{-1} \| l_{n,n}^{(2)} - l_n^{(2)} \|$ converges to zero in probability. The expansion of the estimator of the unknown parameter value in Θ is modified under K_n

$$\widehat{\theta}_{n,n} - \theta_0 = n^{-1} \Sigma_0^{-1} l_{n,n}^{(1)}(\theta_0) + o_p(n^{\frac{1}{2}})$$

$$= \Sigma_0^{-1} \left\{ n^{-1} l_n^{(1)}(\theta_0) + \sum_{i=1}^{k} \widehat{p}_{i,n} \frac{p_i r_i^{(1)} - p_i^{(1)} r_i}{p_i^2}(\theta_0) \right\} + o_p(n^{\frac{1}{2}}). \quad \square$$

The estimator $\widehat{V}_{i,n}$ of the variance $V_i(\theta_0)$ defined by (3.21) is consistent under the alternative K_n. Let

$$E_i = \{m_i^{(1)}(\theta_0)\}^t \Sigma_0^{-1} \sum_{i=1}^{k} \frac{p_i r_i^{(1)} - p_i^{(1)} r_i}{p_i}(\theta_0) \tag{3.25}$$

and let $X_1, \ldots, X_k$ be k independent normal variables such that $\{n \widehat{V}_{i,n}\}^{-\frac{1}{2}} \{N_i - n p_{i,n}(\theta_0)\}$ converges weakly to X_i under K_n, for every $i = 1, \ldots, k$.

Proposition 3.18. *Under K_n, the statistic T_n converges weakly to a non-centered χ^2 variable $T(r) = \chi_{k-1}^2 + \sum_{i=1}^{k} V_i^{-1}(\theta_0) E_i^2$.*

Proof.   Under K_n and from Proposition 3.17, the variable $U_{i,n}$ defined by (3.24), with the estimator $\widehat{\theta}_{n,n}$, has the asymptotic expansion

$$U_{i,n} = n^{\frac{1}{2}} \{\widehat{p}_{i,n} - p_{i,n}(\theta_0)\} - n^{\frac{1}{2}} (\widehat{\theta}_{n,n} - \theta_0)^t p_{i,n}^{(1)}(\theta_0) + o_p(1)$$

$$= n^{\frac{1}{2}} \{\widehat{p}_{i,n} - p_{i,n}(\theta_0)\} - n^{\frac{1}{2}} (\widehat{\theta}_n - \theta_0)^t p_i^{(1)}(\theta_0)$$

$$- \{p_i^{(1)}(\theta_0)\}^t \Sigma_0^{-1} n^{\frac{1}{2}} l_{n,n}^{(1)}(\theta_0) + o_p(1),$$

with the previous expansion of $l_{n,n}^{(1)}$. The variable $U_{i,n}$ is therefore approximated by the sum of two asymptotically Gaussian random variables. It is not centered, its mean is approximated by E_i and its variance converges to $V_i(\theta_0)$ as n tends to infinity. The result is a consequence of Cochran's theorem. $\square$

3.8.2 Goodness of fit tests for a regression function

Let (X, Y) be a set of variables defining a curve $m(x) = E(Y|X = x)$ for x in a subset $\mathcal{I}_X$ of $\mathbb{R}^d$. Goodness of fit hypotheses H_0 for the regression function m in a sub-interval $\mathcal{I}$ of the support of X are tested from observations of the variable X in the sub-intervals of a partition $(A_i)_{i=1,\ldots,k}$, of $\mathcal{I}$. The observations of the regression variable by intervals are those of the indicator of A_i for X

$$\delta_i = 1_{\{X \in A_i\}}.$$

The variable Y is continuously observed in $m(\mathcal{I})$. The parameters of the model are $m_i = E(Y|X \in A_i)$, for $i = 1, \ldots, k$ and their empirical estimators are

$$\widehat{m}_{i,n} = \frac{\sum_{l=1}^n Y_l \delta_{l,i}}{N_i}, \ i = 1, \ldots, k. \tag{3.26}$$

Let $\widehat{\mu}_{i,n} = n^{-1} \sum_{l=1}^n Y_l \delta_{l,i}$ denote its numerator, its denominator is $\widehat{p}_{i,n}$ and their means are $\mu_i = E(Y 1_{\{X \in A_i\}})$ and p_i, respectively.

Proposition 3.19. *Let m be a regression function in $\mathcal{I}$ and let $(A_i)_{i=1,\ldots,k}$ be a partition of $\mathcal{I}$ such that $m(A_i) = m_i$ and all p_i belong to an interval $[c, 1 - c]$, with a strictly positive constant c, the following expansions are satisfied for $i = 1, \ldots, k$, as n tend to infinity*

$$E\widehat{m}_{i,n} - m_i = O(n^{-1}),$$

$$n^{1/2}(\widehat{m}_{i,n} - E\widehat{m}_{i,n}) = n^{1/2} \frac{(\widehat{\mu}_{i,n} - \mu_i) - m_i(\widehat{p}_{i,n} - p_i)}{p_i} + r_{i,n}, \tag{3.27}$$

where $r_{i,n} = o_{L^2}(1)$. Then the variable $\{n^{1/2}(\widehat{m}_{i,n} - m_i)\}_{i=1,\ldots,k}$ converges weakly to a centered Gaussian vector with a diagonal variance matrix, its diagonal terms are

$$v_i = p_i^{-1}[E\{(Y - m_i)^2 | X \in A_i\} - m_i^2(1 - p_i)], \ i = 1, \ldots, k. \tag{3.28}$$

Proof. The order of the bias of $\widehat{m}_{i,n}$ is deduced from the expansion

$$\widehat{m}_{i,n} - m_i = \frac{n^{-1} \sum_{j=1}^n (Y_j \delta_{j,i} - m_i \delta_{j,i})}{p_i}$$

$$- \frac{n^{-1}(\widehat{p}_{i,n} - p_i) \sum_{j=1}^n (Y_j \delta_{j,i} - m_i \delta_{j,i})}{p_i \widehat{p}_{i,n}}.$$

By the consistency of the estimators $\widehat{p}_{i,n}$ and the boundedness condition for the probabilities, for n sufficiently large there exists η in $]0, 1 - c[$ such that p_i^{-1} and $\widehat{p}_{i,n}^{-1}$ are bounded by c^{-1} and $(c - \eta)^{-1}$, respectively. Moreover

$E(Y_j \delta_{j,i}) = m_i E(\delta_{j,i})$ for every $j = 1, \ldots, n$, and by the Cauchy-Schwarz inequality

$$E\left|(\widehat{p}_{i,n} - p_i)\, n^{-1} \sum_{j=1}^{n} (Y_j \delta_{j,i} - m_i \delta_{j,i})\right|$$

$$\leq \left(E\left[\left\{ n^{-1} \sum_{j=1}^{n} (Y_j \delta_{j,i} - m_i \delta_{j,i}) \right\}^2 \right] \right)^{\frac{1}{2}} E\{(\widehat{p}_{i,n} - p_i)^2\}^{\frac{1}{2}},$$

where

$$E\left\{ n^{-1} \sum_{j=1}^{n} (Y_j \delta_{j,i} - m_i \delta_{j,i}) \right\}^2 = n^{-1} \{ E(Y^2 1_{\{X \in A_i\}}) - m_i^2 p_i \}$$

and $E\{(\widehat{p}_{i,n} - p_i)^2\} = O(n^{-1})$, then the result follows. The second expansion is proved like Equation (3.17) for the nonparametric regression estimator. The random vector $\{n^{1/2}(\widehat{m}_{i,n} - m_i)\}_{i=1,\ldots,k}$ has independent components with variances $v_i = p_i^{-2}\{Var\widehat{\mu}_{i,n} + m_i^2 Var\widehat{p}_{i,n} - 2m_i Cov(\widehat{p}_{i,n}, \widehat{\mu}_{i,n})\}$ which yields (3.28). $\square$

Empirical estimators of the variances v_i are defined as

$$\widehat{v}_{i,n} = \frac{1}{\widehat{p}_{i,n}} \left\{ \frac{\sum_{j=1}^{n} (Y_j - \widehat{m}_{i,n})^2 \delta_{j,i}}{N_i} - \widehat{m}_{i,n}^2 (1 - \widehat{p}_{i,n}) \right\}, \quad i = 1, \ldots, k, \quad (3.29)$$

they are $n^{\frac{1}{2}}$-consistent.

A goodness of fit test statistic for the hypothesis $H_0 : m = m_0$, for a known regression curve m_0, is defined as the normalized squared $l_n^2(\mathbb{R}^k)$ distance between $m_{i0} = m_0(A_i)$ and its empirical estimator

$$S_{k,n} = n \sum_{i=1}^{k} \frac{(\widehat{m}_{i,n} - m_{i0})^2}{\widehat{v}_{i,n}}. \qquad (3.30)$$

The statistic $S_{k,n}$ converges weakly under H_0 to a χ_k^2 distribution, for every m_0 and for every k.

Let K_n be a sequence of local alternatives with parameters $m_{i,n}$ converging to m_{i0} and such that $r_{i,n} = n^{\frac{1}{2}}(m_{i,n} - m_{i0})$ converges to a limit r_i, for $i = 1, \ldots, k$. Under K_n, the statistic $S_{k,n}$ converges weakly to a variable $S(r) = \sum_{i=1}^{k} \{X_i + v_{i0}^{-\frac{1}{2}} r_i\}^2$ where $(X_1, \ldots, X_k)$ is a vector of k independent normal variables, asymptotically equivalent to the limit under H_0 of $\{v_{i0}^{-\frac{1}{2}}(\widehat{m}_{i,n} - m_{i0})\}_{i=1,\ldots,k}$. Let c_α be the critical value of the asymptotic test based on the χ_k^2 distribution, the asymptotic local power of the test based on $S_{k,n}$ is $\inf_r P(S(r) > c_\alpha)$.

In a parametric model $\mathcal{M} = \{m; m \in C^2(\Theta)\}$ indexed by a parameter set of $\mathbb{R}^d$, the estimator of the unknown parameter θ_0 which appears in the regression model $m_i(\theta) = E_\theta(Y|X \in A_i)$, for $i = 1, \ldots, k$, can be chosen as maximizing the marginal likelihood of the regression variable X observed by intervals, so that $\widehat{\theta}_n$ satisfies the properties of Proposition 3.15. Let $m_0 = m_{\theta_0}$ be the unknown regression function under H_0, then

$$n^{\frac{1}{2}}\{m_i(\widehat{\theta}_n) - m_{0i}\} = n^{\frac{1}{2}}(\widehat{\theta}_n - \theta_0)^t m_{i0}^{(1)} + o_p(1). \tag{3.31}$$

A test statistic for the hypothesis H_0 that the curve of the variable (X, Y) belongs to $\mathcal{M}$ is

$$S_{k,n} = n \sum_{i=1}^{k} \frac{\{\widehat{m}_{i,n} - m_i(\widehat{\theta}_n)\}^2}{\widehat{V}_{i,n}}. \tag{3.32}$$

The scaling term $\widehat{V}_{i,n}$ in (3.32) is a $n^{\frac{1}{2}}$-consistently estimator of the variance V_{i0} defined like (3.21) from the asymptotic variance (3.28) of the estimator of m_i, under the hypothesis H_0, and from the asymptotic variance of $\widehat{\theta}_n$, where Σ_0 is given by (3.20)

$$V_{i0} = v_{i0} + m_{i0}^{(1)T} \Sigma_0^{-1} m_{i0}^{(1)}. \tag{3.33}$$

From Proposition 3.19, it is estimated by

$$\widehat{V}_{i,n} = \widehat{v}_{i,n} + \{m_i^{(1)}(\widehat{\theta}_n)\}^t \widehat{\Sigma}_n^{-1} m_i^{(1)}(\widehat{\theta}_n), \tag{3.34}$$

where Σ_0 is estimated by $\widehat{\Sigma}_n = \Sigma(\widehat{\theta}_n)$.

Proposition 3.20. *Under the hypothesis H_0 of a model $\mathcal{M}_\Theta$, the statistic $S_{k,n}$ has the asymptotic expansion*

$$n^{-1} \sum_{i=1}^{k} V_{i0}^{-1}\{n(\widehat{m}_{i,n} - m_{i0}) + (\widehat{\theta}_n - \theta_0)^t m_{i0}^{(1)}\}^2 + o_p(1)$$

and it converges weakly to a χ_{k-1}^2 variable.

Proof. From Proposition 3.15, the variable

$$U_{i,n} = n^{\frac{1}{2}}\{m_i(\widehat{\theta}_n) - \widehat{m}_{i,n}\} \tag{3.35}$$

has the asymptotic expansion

$$U_{i,n} = n^{\frac{1}{2}}\{m_i(\widehat{\theta}_n) - m_i(\theta_0)\} - n^{\frac{1}{2}}(\widehat{m}_{i,n} - m_{i0})$$

$$= \{m_{i0}^{(1)}\}^t \Sigma_0^{-1} n^{\frac{1}{2}} \sum_{j=1}^{k} N_j \frac{p_{j0}^{(1)}}{p_{j0}} - n^{\frac{1}{2}}(\widehat{m}_{i,n} - m_{i0}) + o_p(1).$$

The mean of $\{\widehat{\mu}_{i,n} - \mu_{0i}\}\{\sum_{j=1}^{k} N_j(p_j^{-1}p_j^{(1)})(\theta_0)\}$ reduces to $(m_{i0} - \mu_{i0})p_{i0}^{(1)}$ and, using the L^2-approximation 3.27 of the estimator $\widehat{m}_{i,n}$ in Proposition 3.16, the asymptotic covariance of the two terms of the expansion of the variable $U_{i,n}$ is written as

$$C_i = -\{m_{i0}^{(1)}\}^t \Sigma_0^{-1}\{(m_{i0} - \mu_{i0}) - m_{i0}(1 - p_{i0})\}\frac{p_{i0}^{(1)}}{p_{i0}} = 0.$$

The variable $U_{i,n}$ is therefore approximated by the sum of two asymptotically independent Gaussian variables with means zero and variances $v_i(\theta_0)$ defined by (3.28), and $\{m_{i0}^{(1)}\}^t \Sigma_0^{-1} m_{i0}^{(1)}$, respectively, under the multinomial distribution indexed by θ_0 for the variable set $(N_1, \ldots, N_k)$. The variables $U_{i,n}$ are then asymptotically Gaussian variables with means zero and variances V_{i0}, defined by (3.33), under H_0, for every $i = 1, \ldots, k$. Their covariance is $E(U_{i,n}U_{j,n}) = \{m_{i0}^{(1)}\}^t \Sigma_0^{-1} m_{j0}^{(1)} + o(1)$ since

$$E(\widehat{\theta}_n - \theta_0)(\widehat{m}_{i,n} - m_{i0}) = 0,$$
$$E(\widehat{m}_{i,n} - m_{i0})(\widehat{m}_{j,n} - m_{j0}) = 0,$$
$$E(m_i(\widehat{\theta}_n) - m_{i0})(m_j(\widehat{\theta}_n) - m_{j0}) = \{m_{i0}^{(1)}\}^t \Sigma_0^{-1} m_{j0}^{(1)} + o(1),$$

for every $i \neq j$. The arithmetic mean $k^{-1}\sum_{i=1}^{k}\{\widehat{m}_{i,n} - m_i(\widehat{\theta}_n)\}$ converges in probability to zero as n tends to infinity, by the consistency of both estimators, then the proof ends like Proposition 3.16.   □

Let $(K_n)_{n\geq 1}$ be a sequence of local alternatives defined by sequences of functions $m_{i,n}$ in $C^2(\Theta)$, converging uniformly to a function m_i in $\mathcal{F}$ and such that functions $r_{i,n} = n^{\frac{1}{2}}\{p_{i,n} - p_i\}$ and $\rho_{i,n} = n^{\frac{1}{2}}\{m_{i,n} - m_i\}$ converge uniformly in Θ to limits in $C^2(\Theta)$, r_i and ρ_i respectively, for $i = 1, \ldots, k$. The estimator $\widehat{\theta}_{n,n}$ converges in probability under K_n to a value θ_0 in Θ and this limit is also the limit of the estimator $\widehat{\theta}_n$ under H_0, by definition of the local alternatives, then the weak convergence of Proposition 3.17 still applies. Let

$$E_i = \{m_i^{(1)}(\theta_0)\}^t \Sigma_0^{-1} \sum_{i=1}^{k} \frac{p_i r_i^{(1)} - p_i^{(1)} r_i}{p_i}(\theta_0).$$

Proposition 3.21. *Under K_n, the statistic T_n converges weakly to a non-centered χ^2 variable $S(r) = \chi_{k-1}^2 + \sum_{i=1}^{k} V_{0i}^{-1} E_i^2$.*

Proof.   The weak convergence of T_n under K_n is established by the same arguments as that for Proposition 3.18. Now the variable $U_{i,n}$ defined by

(3.35), with the estimators $\widehat{m}_{i,n}$ and $\widehat{\theta}_{n,n}$, has under K_n the asymptotic expansion

$$U_{i,n} = n^{\frac{1}{2}} \{\widehat{m}_{i,n} - m_{i,n}(\theta_0)\} - n^{\frac{1}{2}} (\widehat{\theta}_n - \theta_0)^t m_i^{(1)}(\theta_0)$$
$$- \{m_i^{(1)}(\theta_0)\}^t \Sigma_0^{-1} n^{\frac{1}{2}} l_{n,n}^{(1)}(\theta_0) + o_p(1),$$

with the previous expansion of $l_{n,n}^{(1)}$. The variable $U_{i,n}$ is therefore approximated by a Gaussian random variable. Its mean is approximated by E_i and its variance converges to $V_i(\theta_0)$ as n tends to infinity. □

3.8.3 Tests of symmetry for a density

Let F be the distribution function in $\mathbb{R}$ of a real random variable X with a symmetric density and let $(A_i)_{i=-k,\ldots,k}$ be a symmetric partition of $\mathbb{R}$ in $2k$ sub-intervals A_i and A_{-i}, for $i = 1, \ldots, k$. By symmetry at zero, the probabilities $p_i = \int_{A_i} dF$ and $p_{-i} = \int_{A_{-i}} dF$ are equal. Let N_i be the number of observations of the sample in A_i, for $i = -k, \ldots, -1, 1, \ldots, k$. A test for the hypothesis H_0 of symmetry of the density of X relies on the statistic

$$T_n = \sum_{i=1}^{k} \frac{N_i - N_{-i}}{(2n\widehat{v}_{i,n})^{\frac{1}{2}}}.$$

Proposition 3.22. *Under H_0, the statistic T_n converges weakly to a χ_{k-1}^2 variable T_0. Under the alternative K_n : there exists i in $\{1, \ldots, k\}$ such that $p_{-i,n} = p_{i,n} + n^{-\frac{1}{2}} r_{i,n}$, $\lim_{n\to\infty} p_{i,n} = p_i$ and $\lim_{n\to\infty} r_{i,n} = r_i$ is finite, the statistic T_n converges weakly to $S(r) = T_0 + \sum_{i=1}^{k} (\frac{r_i}{2v_i})^2$.*

Under the hypothesis H_0 of a density with a center of symmetry at an unknown θ, let $\{A_i(\theta)\}_{i\geq 1}$ be a symmetric parametric partition centered at θ, for every real θ. The center of symmetry is estimated by maximum likelihood in a multinomial model with symmetric probabilities around θ, $p_i(\theta) = \int_{A_i} dF(x + \theta) = \int_{\theta+A_i} dF$ and $p_{-i}(\theta) = \int_{\theta-A_i} dF$, for $i = 1, \ldots, k$. Under H_0, the estimator $\widehat{\theta}_n$ of the unknown parameter value θ_0 satisfies the properties of Proposition 3.15. Using then a partition symmetric around $\widehat{\theta}_n$, a test for H_0 relies on the statistic

$$T_n = \sum_{i=1}^{k} \frac{N_i(\widehat{\theta}_n) - N_{-i}(\widehat{\theta}_n)}{(2n\widehat{V}_{i,n})^{\frac{1}{2}}},$$

with the notations $N_i(\theta) = \int_{\theta+A_i} d\widehat{F}_n$ and $N_{-i}(\theta) = \int_{\theta-A_i} d\widehat{F}_n$, for every real θ and $\widehat{V}_{i,n}$ is given by (3.22), for $i = 1, \ldots, k$. Denoting $A_i =]a_i, a_{i+1}]$, the probability $p_i(\theta) = F(\theta + a_{i+1}) - F(\theta + a_i) = \int_{\theta + A_i} dF_\theta$ has the derivative $p_i^{(1)}(\theta) = f(\theta + a_{i+1}) - f(\theta + a_i)$ which is denoted $f(\theta + A_i)$, and according to (3.20), $\Sigma_0 = -2\sum_{i=1}^{k} p_i^{-1}(\theta_0)f^2(\theta_0 + A_i)$. The variances V_i are defined by (3.21).

Proposition 3.23. *Under H_0, the variable T_n converges weakly to a χ^2_{k-1} variable T_0. Under the alternative K_n : there exists i in $\{1, \ldots, k\}$ such that $p_{-i,n} = p_{i,n} + n^{-\frac{1}{2}}r_{i,n}$, $\lim_{n\to\infty} p_{i,n} = p_i$ and $\lim_{n\to\infty} r_{i,n} = r_i$ is finite, the statistic T_n converges weakly to $T(r) = T_0 + \sum_{i=1}^{k} (\frac{r_i}{2V_i})^2$.*

Proof. By the same expansions as in Propositions 3.15 and 3.16, the variable $U_{i,n} = n^{-\frac{1}{2}}\{N_i(\widehat{\theta}_n) - N_{-i}(\widehat{\theta}_n)\} = \nu_n(\widehat{\theta}_n + A_i) - \nu_n(\widehat{\theta}_n - A_i)$ has an asymptotic expansion

$$U_{i,n} = \nu_n(\theta_0 + A_i) - \nu_n(\theta_0 - A_i) + 2f(\theta_0 + A_i)(\widehat{\theta}_n - \theta_0) + o(1),$$

depending on $(p_i^{(1)} - p_{-i}^{(1)})(\theta_0) = 2f(\theta_0 + A_i)$ and on the derivatives of $n^{-1}l_n = n^{-1}\sum_{i=1}^{k}(N_i + N_{-i})$. Under H_0, the independent variables $n^{-\frac{1}{2}}l_{n,i}^{(1)}(\theta_0) = n^{-\frac{1}{2}}(N_i + N_{.})(p_i^{-1}p_{-i}^{(1)})(\theta_0)$ converges weakly to a centered Gaussian variable with variance $2\Sigma_0$ and $n^{-1}l_{n,i}^{(2)}(\theta)$ converges in probability to $2\Sigma(\theta)$, uniformly in Θ.

Under the sequence of the alternative K_n, the limit of $n^{-\frac{1}{2}}l_{n,n}^{(1)}(\theta_0)$ is the limit of $n^{-\frac{1}{2}}\sum_{i=1}^{k} l_{n,i}^{(1)} + r_i^{(1)}(\theta_0) + o_p(1)$ and the limit of $n^{-1}l_{n,n}^{(2)}(\theta)$ is the same as the limit of $n^{-1}l_n^{(2)}(\theta)$ under H_0. With the scaling $(2n\widehat{V}_{i,n})^{-\frac{1}{2}}$ in T_n, the proof ends like for Proposition 3.16.   □

3.8.4   *Tests of a monotone density*

Let X be a real random variable with density f. Let $(A_i)_{i=1,\ldots,k}$ be a partition of $\mathcal{I}$ in k sub-intervals with contant length $a = |A_i|$ and let $p_i = \int_{A_i} f(x)\,dx$. For a n-sample, the vector $(N_i)_{i=1,\ldots,k}$ of the numbers of observations of X in the sub-intervals A_i has a multinomial distribution with probabilities $(p_i)_{i=1,\ldots,k}$.

A test of the hypothesis H_0 of an uniform density in $[0, 1]$ is defined by constant probabilities $(p_i)_{i=1,\ldots,k}$ equal to $p_0 = k^{-1}$ and a test statistic is

$$S_n = \frac{\sqrt{n}k}{k-1} \sum_{i=1}^{k-1} (\widehat{p}_{in} - p_0)^2,$$

with the variance $n^{-1}(k-1)^2k^{-2}$ of the numerator $\sum_{i=1}^{k-1}(\widehat{p}_{in} - \widehat{p}_n)^2$ under the hypothesis H_0. Its limiting distribution is a χ^2_{k-1} variable.

As a tends to zero, the hypothesis of an increasing sequence of probabilities $(p_i)_{i=1,\ldots,k}$ is equivalent to the hypothesis of an increasing density. If n is large enough, the variables $Z_{in} = \widehat{p}_{in} - \widehat{p}_n$ are increasing and their ranks are ordered, hence $Z_{i:n} - Z_{in}$. A rank test can be defined from the sum of squares

$$U_n = n \sum_{i=1}^{k-1} (Z_{i:n} - Z_{in})^2 = n \sum_{i=1}^{k-1} (Z_{i:n} - Z_{R_n,i:n})^2.$$

Under H_0, U_n converges in probability to zero and it is strictly positive under all alternatives.

For n large enough, the numbers $(N_i)_{i=1,\ldots,k}$ are also increasing an we consider a one-sided test with the statistic based on the number of intervals of the partition such that $\widehat{p}_{i+1,n} \geq \widehat{p}_{i,n}$ for every $i = 1,\ldots,k$. Let $Z_{i,n} = 1$ if $\widehat{p}_{i+1,n} \geq \widehat{p}_{i,n}$, which is equivalent to $N_{i+1,n} \geq N_i$, and $Z_{i,n} = 0$ otherwise, $i = 1,\ldots,k$, and let

$$S_{n,k} = \sum_{i=1}^{k-1} (Z_{i,n} - 1)^2. \tag{3.36}$$

Under the hypothesis H_0, the variables $Z_{i,n}$ have the mean $P_0(N_{i+1} \geq N_i)$ which converges to 1 as n tends to infinity, for every $i = 1,\ldots,k$. Their variance tends to zero hence the variable $S_{n,k}$ converges in probability to zero, therefore $S_{n,k}$ converges to zero in probability. Under alternatives, it is strictly positive as n tends to infinity. Bootstrap tests with a normalization by the bootstrap variances can be performed. When $k = k_n$ increases with n, the statistics U_n and $S_{n,k}$ converge to normal variable under the hypothesis H_0, and they diverge as n and k_n tend to infinity. These test can be applied to all curves.

3.9 Exercises

3.9.1. Determine the limiting distribution of the Anderson-Darling type statistics for the tests of symmetry considered in Section 3.3.1 and their asymptotic power against local alternatives of change of location and scale. Hints. The Anderson-Darling test statistic for the hypothesis H_0 of symmetry is defined from the process U_n given in (3.3) as $A_n = \int_{\mathbb{R}} \widehat{\sigma}_{U,n}^{-2} U_n^2 \, d\widehat{F}_n$, with the empirical estimator $\widehat{\sigma}_{U,n}^2$ of the variance σ_U^2 of the variance of U_n under H_0. It converges weakly to $A = \int_{\mathbb{R}} \sigma_U^{-1} U \, dF$. The variable U_n converges to 1 in probability under H_0. Under the alternatives K_n of Proposition 3.3, $\lim_{n \to \infty} E_{\theta_n} U_n^2(x) = 2F(-x) + 4\theta_0^2 f^2(x)$ and

$\lim_{n\to\infty} Var_{\theta_n} U_n(x) = 2F(-x)$, the asymptotic distribution of A_n under K_n and its local power are deduced from these expressions.

3.9.2. Define a test statistic of symmetry for a regression function in a finite interval I_X and determine its local asymptotic properties.

Hints. Under the hypothesis H_0 of symmetry at zero of $m(x)$ in I_X, the integrated regression curve $M(x) = E(Y|X \le x)$ satisfies $M(x) = 1 - M(x)$ and tests are defined as in Section 3.3.

3.9.3. Define a test of linearity of a regression function in a finite interval I_X (see Section 3.6).

Hints. The linear regression $m(x) = a + bx$ is integrated in $I_X = (x_1, x_2)$ as $M(x) = ax + bx^2 + c$ which is estimated by $\widehat{M}_n(x) = \sum_{i=1}^{n} Y_i 1_{X_i \le x}$. The values of $\widehat{M}_n$ at the bounds of I_X and $\bar{X}_n$ determine estimators of the parameters from $\widehat{M}_n$. The estimators are $n^{\frac{1}{2}}$-consistent and asymptotically Gaussian with means zero. As in Section 3.7, a test is deduced from the difference $n^{\frac{1}{2}}(\widehat{m}_n - \widehat{m}_{n,h}) = n^{\frac{1}{2}}(\widehat{m}_n - m_0) + o_p(1)$ and the properties of the test statistics are deduced from the limits of the estimators of the constants. Other tests can be defined like goodness of fit tests for M, from the estimator $\widehat{M}_n$, like the tests defined from the empirical distribution functions.

3.9.4. Define a test of linearity of a regression function with a regression variable observed by intervals (see Section 3.8.2).

Hints. The cumulated mean regression variables in sub-intervals A_j of the interval I_X are $M_Y(A_j) = E(Y|X \in A_i)$ and $M_X(A_j) = E(X|X \in A_i)$, and the linear regression model is $E(Y|X \in A_i) = aE(X|X \in A_i)$, for every $i = 1, \ldots, k$. From the constraint $\sum_{j=1}^{k} M_Y(A_j) = E(Y)$, there exist $k - 1$ free parameters $M_Y(A_j)$ estimated by $\widehat{M}_{Y,n}(A_i)$, and $k - 1$ free parameters $M_X(A_j)$ are estimated by $\widehat{M}_{X,n}(A_i)$. The parameter a is estimated by $\widehat{a}_n = (k-1)^{-1} \sum_{i=1}^{k-1} \{\widehat{M}_{X,n}(A_i)\}^{-1} \widehat{M}_{Y,n}(A_i)$, the variance of $\widehat{a}_n$ is calculated from the cumulated variables. A test statistic is defined as $S_n = V_n^{-\frac{1}{2}} \sum_{i=1}^{k-1} [\{\widehat{M}_{X,n}(A_i)\}^{-1} \widehat{M}_{Y,n}(A_i) - \widehat{a}_n\}^2$, where V_n is the variance of the sum of variables. The statistic S_n converges to an asymptotic χ_{k-1}^2 variable. Its asymptotic local power under alternatives K_n is calculated by an expansion as in Proposition 3.21.

3.9.5. Define a test of $P(X_1 > X_0) = 1$ for the probability of stochastic order of random variables X_1 and X_0 from Equation (1.2).

Chapter 4

Two-sample tests

4.1 Introduction

In parametric models, the likelihood tests are optimal for tests of hypotheses defined by the distribution functions of the variables of two samples against parametric alternatives, they are expressed as rank tests. Such tests are used as omnibus tests with optimality properties in the true model. Many nonparametric tests based on the empirical distribution functions are also rank tests.

In the Kolmogorov-Smirnov and Cramer-von Mises tests of independence of two random variables, the hypothesis is a composite hypothesis of the product of unknown marginal distributions. They are estimated empirically and the limiting distributions of the test statistics depend of the marginal distributions. A transformation of the observations by the empirical distribution yields tests based on the empirical estimator of the dependence function and it is asymptotically free under the hypothesis of independence. Other tests rely on the nonparametric conditional distribution functions of the two-dimensional distribution and asymptotically free statistics are obtained by the same transformation. A classical semiparametric model of dependence of two distribution functions is the model with joint survival function

$$\bar{F}_\alpha(t) = \{\bar{F}_1^{1-\alpha}(t_1) + \bar{F}_2^{1-\alpha}(t_2) - 1\}^{\frac{1}{1-\alpha}}, \ t = (t_1, t_2),$$

with a dependence parameter $\alpha \geq 1$. Its density is

$$f_\alpha(t) = \alpha f_1(t_1)\bar{F}_1^{-\alpha}(t_1)f_2(t_2)\bar{F}_2^{-\alpha}(t_2)\bar{F}^{2\alpha-1}(t).$$

A test for independence is a test of $H_0 : \alpha = 1$ and it relies on the maximum likelihood estimator of the parameter α. Denoting $\Lambda_k = \log \bar{F}_k$ for $k = 1, 2$, the maximum likelihood estimator of α from a n-sample of a bidimensional variable $T = (T_1, T_2)$ is

$$\widehat{\alpha}_n^{-1} = n^{-1} \sum_{i=1}^{n} \{\widehat{\Lambda}_{1,n}(T_{1i}) + \widehat{\Lambda}_{2,n}(T_{2i}) - 2\log \widehat{\bar{F}}_n(T_i)\}, \qquad (4.1)$$

where $\widehat{\Lambda}_{k,n}$ is the estimator of kth marginal cumulative hazard function Λ_k (Chapter 7) and $\widehat{\bar{F}}_n$ is an estimator of $\bar{F}_\alpha$, it may be chosen as a nonparametric estimator or as $\bar{F}_{\alpha_{n,(0)}}$ with a preliminary estimator $\alpha_{n,(0)}$ of the parameter, which yields $\alpha_{n,(1)}$ then this estimator is improved recursively. The question is similar in all semi-parametric models of dependence.

Tests of hypotheses for a bivariate distribution function are often based on the vector of the weighted marginal density functions or distribution functions, like in score tests under independence. The weights depend on the respective sample sizes of the variables under each marginal distributions. Using the empirical variance matrix $\widehat{\Sigma}_n$ of the estimator $\widehat{U}_n$ of this vector leads to a test statistic of the form $\widehat{U}_n^t \widehat{\Sigma}_n \widehat{U}_n$ which has a χ^2 limiting distribution. For a variable with dependent components, such tests are not optimal and using the dependence function of the variable improves the statistics. Some nonparametric tests for two samples have χ^2 limiting distribution such as the tests for observations cumulated by intervals.

Nonparametric tests for the comparison of the distribution functions of two samples, against nonparametric alternatives are based on the empirical distribution functions and they are also expressed as rank tests. The limiting distribution of the test statistics are functions of the standard Wiener process under the hypothesis of equal distribution functions. Test statistics for a homogeneous sample against composite parametric alternatives have a limiting distribution sum of two functions of Gaussian processes, due to the estimation of the unknown parameters of the model, and they are not free distributions. A bootstrap version of the tests is therefore necessary. The asymptotic properties of the bootstrap tests are not detailed for the statistics considered in the following, they are differentiable functionals of the distribution functions hence the bootstrap tests are consistent and their power can be estimated consistently. In a semi-parametric setting, a distribution function belongs to a parametric family and the other one is unspecified. The best estimator of the unknown parameter of the hypothesis is the worst under the alternative.

Tests for the comparison of the regression functions of two samples rely on an integrated difference of their kernel estimators, under semi-parametric models, the tests concern the parameter or the nonparametric function.

Tests concerning the form of a density in $\mathbb{R}^2$ extend the tests for the density of a real variable and their convergence rates differ due to the higher dimension. Finally, tests of independence and tests of comparison of the distribution functions or regression curves of two samples are built for observations by intervals.

4.2 Tests of independence

We first study tests statistics based on the joint and marginal empirical distribution functions of two variables, then tests for the dependence function of the joint variables and finally tests for their conditional distributions.

4.2.1 *Kolmogorov-Smirnov and Cramer-von Mises tests*

On a probability space $(\Omega, \mathcal{F}, P)$, let $X = (X_1, X_2)$ be a two-dimensional real random variable with distribution function F in $\mathbb{R}^2$ and marginal distributions F_1 and F_2. Under the hypothesis H_0 of independence of X_1 and X_2, the joint distribution function factorizes as $F = F_1 F_2$.

Let $X_1, \ldots, X_n$ be a n-sample of the variable X, where $X_i = (X_{1i}, X_{2i})$, and let $\widehat{F}_n(x) = n^{-1} \sum_{i=1}^{n} 1_{\{X_{1i} \leq x_1, X_{2i} \leq x_2\}}$, for $x = (x_1, x_2)$ in $\mathbb{R}^2$, be the joint empirical distribution function of the sample. The marginal empirical distribution functions are $\widehat{F}_{1n}(x_1) = \widehat{F}_n(x_1, \infty)$ and $\widehat{F}_{2n}(x_2) = \widehat{F}_n(\infty, x_2)$. The Kolmogorov-Smirnov test of H_0 is based on the statistic

$$S_n = \sup_{x=(x_1,x_2)\in[0,1]^2} \sqrt{n} |\widehat{F}_n(x) - \widehat{F}_{1n}(x_1)\widehat{F}_{2n}(x_2)| \qquad (4.2)$$

and a Cramer-von Mises test on the statistic

$$T_n = \int_{[0,1]^2} n\{\widehat{F}_n(x) - \widehat{F}_{1n}(x_1)\widehat{F}_{2n}(x_2)\}^2 \, d\widehat{F}_n(x) \qquad (4.3)$$

$$= \sum_{i=1}^{n} \{\widehat{F}_n(X_i) - \widehat{F}_{1n}(X_{1i}))\widehat{F}_{2n}(X_{2i}))\}^2.$$

The joint empirical process is denoted $\nu_n(x) = \sqrt{n}(\widehat{F}_n - F)(x)$ in $\mathbb{R}^2$ and its marginal empirical processes are $\nu_{nk}(x_k) = \sqrt{n}(\widehat{F}_{kn} - F_k)(x_k)$, for all $x = (x_1, x_2)$ in $\mathbb{R}^2$ and $k = 1, 2$. The bivariate survival function of X is

$$\bar{F}(x) = 1 - F_1(x_1) - F_2(x_2) + F(x),$$

and under the independence $\bar{F} = \{1 - F_1\}\{1 - F_2\} = \bar{F}_1 \bar{F}_2$.

Let W, W_1 and W_2 be Wiener processes limits of the empirical processes ν_n, ν_{1n} and ν_{2n}, their covariance functions are

$$E\{W(x)W(y)\} = F(x \wedge y)\{1 - F(x \vee y)\},$$
$$E\{W_k(x_k)W_k(y_k)\} = F_k(x_k \wedge y_k)\bar{F}_k(x_k \vee y_k),$$
$$E\{W(x)W_k(y_k)\} = F(x \wedge y)\bar{F}_k(x_k \vee y_k), \ k = 1, 2.$$

The difference of the empirical processes develops as

$$G_n(x) = \nu_n(x) - n^{-\frac{1}{2}}\nu_{1n}(x_1)\nu_{2n}(x_2),$$
$$G_n(x) = n^{\frac{1}{2}}\{\hat{F}_n(x) - \hat{F}_{1n}(x_1)\hat{F}_{2n}(x_2)\} + F_2(x_2)\nu_{1n}(x_1) + F_1(x_1)\nu_{2n}(x_2)$$
$$- n^{\frac{1}{2}}\{F(x) - F_1(x_1)F_2(x_2)\} \tag{4.4}$$

and it is asymptotically equivalent to $\nu_n(x)$.

Proposition 4.1. *Under H_0, the process $Z_n = n^{\frac{1}{2}}(\hat{F}_n - \hat{F}_{n1}\hat{F}_{n2}) + o_p(1)$ converges weakly to $Z = W - F_2W_1 - F_1W_2$. Under a local alternative with a joint distribution function $F_n = F_{1n}F_{2n} + n^{-\frac{1}{2}}H_n$, where $(H_n)_n$ is a sequence of functions H_n such that $\lim_{n \to \infty} \sup_{x \in \mathbb{R}^2} |H_n(x) - H(x)| = 0$ with a function H of the tangent space to H_0, the process $\nu_n - \nu_{1n}\nu_{2n}$ converges weakly to $H + Z$.*

Proof. By definition, the process G_n converges weakly to W under H_0 and by (4.4), $\sqrt{n}\{\hat{F}_n(x) - \hat{F}_{1n}(x_1)\hat{F}_{2n}(x_2)\} = G_n(x) - F_2(x_2)\nu_{1n}(x_1) - F_1(x_1)\nu_{2n}(x_2) + n^{\frac{1}{2}}\{F(x) - F_1(x_1)F_2(x_2)\}$ converges weakly to the process $Z = W - F_2W_1 - F_1W_2$. Under the local alternative, the empirical processes are centered at F_n and F_{kn} respectively and the process G_n is not centered, it is replaced by

$$G_{nn}(x) = n^{\frac{1}{2}}\{\hat{F}_n(x) - \hat{F}_{1n}(x_1)\hat{F}_{2n}(x_2)\} + F_{2n}(x_2)\{\hat{F}_{1n}(x_1) - F_{1n}(x_1)\}$$
$$+ F_{1n}(x_1)\{\hat{F}_{1n}(x_2) - F_{2n}(x_2)\} - n^{\frac{1}{2}}\{F_n(x) - F_{1n}(x_1)F_{2n}(x_2)\}$$

where $n^{\frac{1}{2}}\{F_n(x) - F_{1n}(x_1)F_{2n}(x_2)\}$ reduces to $H_n(x)$, and the process G_{nn} converges weakly to $W + F_2W_1 + F_1W_2 - H$. It follows that under the local alternative, $\sqrt{n}\{\hat{F}_n(x) - \hat{F}_{1n}(x_1)\hat{F}_{2n}(x_2)\} = G_{nn}(x) - F_{2n}(x_2)\{\hat{F}_{1n}(x_1) - F_{1n}(x_1)\} - F_{2n}(x_2)\{\hat{F}_{1n}(x_1) - F_{1n}(x_1)\} + H_n(x)$ converges weakly to $W - F_2W_1 - F_1W_2 + H$. $\square$

Under H_0, the limiting Wiener process W has the same distribution as W_1W_2, it follows that the statistics S_n and T_n converge weakly to $\sup_{\mathbb{R}^2} |Z|$ and, respectively, $\int_{\mathbb{R}} Z_1 \, dF_1 \int_{\mathbb{R}} Z_2 \, dF_2$ under the hypothesis of independence. These limits depend on the marginal distributions. Under the local alternatives, they converge weakly to $\sup_{\mathbb{R}^2} |Z + H|$ and, respectively, $\int_{\mathbb{R}^2} \{Z + H\}^2 \, dF$. They tend to infinity under every fixed alternative.

For $k = 1, 2$, let $R_{n,ki}$ be the rank of X_{ki} of the kth marginal sample and let $X_{k(1)} < X_{k(2)} < \cdots < X_{k(n)}$ be its order statistics, for every $i = 1, \ldots, n$, $X_{ki} = X_{k(R_{n,ki})}$ and

$$\widehat{F}_{kn}(X_{ki}) = \widehat{F}_{kn}(X_{k(R_{n,ki})}) = n^{-1} R_{n,ki}.$$

The Cramer-von Mises test statistic (4.3) for independence is written as

$$T_n = n^{-\frac{1}{2}} \sum_{i=1}^{n} \left\{ \sum_{j=1}^{n} 1_{\{X_{j1} \leq X_{1i}, X_{j2} \leq X_{2i}\}} - n^{-1} R_{n,1i} R_{n,2i} \right\}^2.$$

The Kolmogorov-Smirnov statistic (4.2) is step-wise constant with jumps at the variables $(X_i)_{i=1,\ldots,n}$, it equals

$$S_n = n^{\frac{1}{2}} \max_{i=1,\ldots,n} |\widehat{F}_n(X_i) - n^{-2} R_{n,1i} R_{n,2i}|.$$

4.2.2   Tests based on the dependence function

The dependence function for the components of the variable X is defined with the inverse of its marginal distribution functions as

$$C(x_1, x_2) = F(F_1^{-1}(x_1), F_2^{-1}(x_2)), \quad (x_1, x_2) \in [0, 1]^2. \qquad (4.5)$$

If the marginals are strictly increasing at $t = (t_1, t_2)$, $F_k^{-1} \circ F_k = id$, $k = 1, 2$, and the joint distribution is uniquely defined by the function C and the marginal distributions as

$$F(t) = P(X_1 \leq t_1, X_2 \leq t_2) = C(F_1(t_1), F_2(t_2)).$$

The variables $F_k(X_k)$ have uniform marginal distributions and their joint distribution is the dependence function C if the distributions F_k are continuous, which implies $F_k \circ F_k^{-1} = id$, for $k = 1, 2$. The hypothesis of independence of the variables X_1 and X_2 is characterized by a unique dependence function defined for every $x = (x_1, x_2)$ of $[0, 1]^2$ by

$$C(x_1, x_2) = \prod_{k=1}^{2} x_k.$$

Let $X_i = (X_{1i}, X_{id})$, $i = 1, \ldots, n$, be a sample of n independent variables having the same distribution as X, their empirical dependence function is defined as the empirical distribution function of the transformed variables $U_{nik} = \widehat{F}_{nk}(X_{ik})$, where $\widehat{F}_{nk}$ is the marginal empirical distribution of the variable X_k

$$\widehat{C}_n(x) = n^{-1} \sum_{i=1}^{n} 1_{\{\widehat{F}_{n1}(X_{1i}) \leq x_1, \widehat{F}_{n2}(X_{i2}) \leq x_2\}}, \quad (x_1, x_2) \in [0, 1]^2. \qquad (4.6)$$

Let $\nu_{nk} = \sqrt{n}(\widehat{F}_{nk} - F_k)$, $k = 1, 2$, be the marginal empirical processes of the sample. The empirical process of the transformed variables is

$$\nu_{n,C}(x) = \sqrt{n}\{\widehat{C}_n(x) - C(x)\}.$$

Lemma 4.1. *If the joint distribution function F is strictly monotone and belongs to $C^1([0,1]^2)$, for every x in $[0,1]^2$*

$$\sqrt{n}\{F(\widehat{F}_{n1}^{-1}(x_1), \widehat{F}_{n2}^{-1}(x_2)) - C(x)\}$$

$$= \sum_{k=1}^{2} \frac{\nu_{nk}}{f_k} \circ F_k^{-1}(x_k) \frac{\partial F}{\partial x_k} \circ \{F_1^{-1}(x_1), F_2^{-1}(x_2)\} + o_p(1).$$

Proof. Let $F_{(k)}^{(1)}$ be the first derivatives of the joint density with respect to the kth coordinate, by the consistency of $\widehat{F}_{nk}$, $k = 1, 2$, and Taylor expansions, we have

$$\sqrt{n}\{F(\widehat{F}_{n1}^{-1}(x_1), \widehat{F}_{n2}^{-1}(x_2)) - F(F_1^{-1}(x_1), F_2^{-1}(x_2))\}$$

$$= \sum_{k=1}^{2} \sqrt{n}\{\widehat{F}_{nk}^{-1}(x_k) - F_k^{-1}(x_k)\} F_{(k)}^{(1)} \circ \{F_1^{-1}(x_1), F_2^{-1}(x_2)\} + o_p(1),$$

$$\sqrt{n}\{\widehat{F}_{nk}^{-1}(x_k) - F_1^{-1}(x_k)\} = \nu_{nk} \circ F_k^{-1}(x_k) \frac{1}{f_k \circ F_k^{-1}(x_k)} + o_p(1). \qquad \square$$

A Kolmogorov-Smirnov type test for the hypothesis H_0 of independence between the variables X_1, X_2 is based on the statistic

$$S_n = \sup_{x \in [0,1]^2} \sqrt{n} |\widehat{C}_n(x) - \prod_{k=1}^{2} x_k| \qquad (4.7)$$

and a Cramer-von Mises type test on the statistic

$$T_n = \int_{[0,1]^2} n\{\widehat{C}_n(x) - \prod_{k=1}^{2} x_k\}^2 \, d\widehat{C}_n(x) = \sum_{i=1}^{n} \{\widehat{C}_n(X_i) - X_{1i}X_{2i}\}^2. \quad (4.8)$$

Theorem 4.1. *Under the hypothesis H_0 of independent variables having a strictly positive density, S_n converges weakly to $S = \sup_{[0,1]^2} |B_2|$ and T_n converges weakly to $T = \int_{[0,1]^2} B_2^2 \, dC$, where B_2 is a Gaussian process with mean zero and covariance $(x_1 \wedge y_1)(x_2 \wedge y_2) - x_1 x_2 y_1 y_2$ at x and y in $[0,1]^2$. Under the alternative, K_n of a dependence function $C_n(x) = x_1 x_2 \{1 + n^{-\frac{1}{2}} H_n(x)\}$ such that H_n converges uniformly in $[0,1]^2$ to a function $H > 0$, the statistics S_n and T_n converge weakly to $S = \sup_{[0,1]^2} |B_2 + H|$ and, respectively, $T = \int_{[0,1]^2} (B_2 + H)^2 \, dC$.*

Proof. The empirical process $\nu_{n,C}$ is written as the sum

$$\nu_{n,C}(x) = \nu_n(\widehat{F}_{n1}^{-1}(x_1), \widehat{F}_{n2}^{-1}(x_2)) + \sqrt{n}\{F(\widehat{F}_{n1}^{-1}(x_1), \widehat{F}_{n2}^{-1}(x_2)) - C(x)\}$$

where ν_n is the empirical process of the bivariate sample. The vector of processes $(\nu_n, \nu_{1n}, \nu_{2n})$ converges weakly to a set of transformed Brownian bridges (B, B_1, B_2) in $[0,1]^2 \times [0,1] \times [0,1]$ such that B_1 and B_2 are independent under H_0 and their covariances at (s,t) are

$$E\{B(t)B(s)\} = F(s_1 \wedge t_1, s_2 \wedge t_2) - F(s)F(t),$$
$$E\{B_k(t)B_k(s)\} = F_k(s_k \wedge t_k) - F_k(s_k)F_k(t_k), \ k = 1, 2,$$
$$E\{B(s)B_1(t_1)\} = F(s_1 \wedge t_1, s_2) - F(s)F_1(t_1),$$
$$E\{B(s)B_2(t_2)\} = F(s_1, s_2 \wedge t_2) - F(s)F_2(t_2),$$

for all s and t. By Lemma 4.1, $\nu_{n,C}$ converges weakly to a standard Brownian bridge

$$B_2(x) = B(F_1^{-1}(x_1), F_2^{-1}(x_2)) + \sum_{k=1}^{2} \frac{\nu_k}{f_k} \circ F_k^{-1}(x_k) \, F_{(k)}^{(1)} \circ \{F_1^{-1}(x_1), F_2^{-1}(x_2)\}$$

and the weak convergence of the statistics S_n and T_n follows. Under H_0, $B = B_1 B_2$ and $F_{(k)}^{(1)} = f_k F_j$, with $j \neq k$, hence

$$B_2(x) = B(F_1^{-1}(x_1), F_2^{-1}(x_2)) + x_2 B_1 \circ F_1^{-1}(x_1) + x_1 B_2 \circ F_2^{-1}(x_2).$$

The last two distributions are free and, under H_0, B_2 has a free distribution with variance and covariance

$$Var B_2(x) = x_1 x_2 (1 - x_1 x_2),$$
$$Cov(B_2(x), B_2(y)) = (x_1 \wedge y_1)(x_2 \wedge y_2) - x_1 x_2 y_1 y_2.$$

The limiting distributions of the test statistics are therefore free distributions under H_0. Under the alternative, $\nu_{n,C} = H_n(F_{1n} \circ \widehat{F}_{1n}^{-1}, F_{2n} \circ \widehat{F}_{2n}^{-1}) + \nu_n(\widehat{F}_{1n}^{-1}, \widehat{F}_{2n}^{-1})$, then the uniform convergence of $\widehat{F}_{kn} - F_{kn}$ to zero in probability, the uniform convergence of $H_n - H$ and $F_{kn} - F_k$ to zero and the weak convergence of the process ν_n to B_2 imply the weak convergence of $\nu_{n,C}$ to $H + B_2$, the result follows. □

The asymptotic threshold of the test for a critical level α is determined by $\lim_{n \to \infty} P_{H_0}(S_n > c_\alpha) = \alpha$ and its power under the local alternative in a family $\mathcal{C}_n$ of distribution functions in $[0,1]^2$, for two dependent variables, is

$$\beta = \lim_{n \to \infty} \inf_{C_n \in \mathcal{C}_n} P_{K_n}(S_n(C_n) > c_\alpha)$$

where $S_n(C)$ is defined by (4.7) for a sample of variables with distribution function C in $\mathcal{C}$.

The critical level and the power of the Cramer-von Mises test statistic at the level α are c'_α such that

$$\alpha = \lim_{n \to \infty} P_{H_0}(T_n > c_\alpha)$$

$$= \lim_{n \to \infty} \inf_{\lambda > 0} E \exp\left\{\lambda \int_{x \in [0,1]^2} v_{n,C}^2 \, d\widehat{C}_n - \lambda c'_\alpha\right\},$$

and its power under the local alternative is

$$\beta(T_n) = \lim_{n \to \infty} \inf_{C \in \mathcal{C}} P_C(T_n(C) > c'_\alpha) = \inf_{C \in \mathcal{C}} P_C\left(\int_{x \in [0,1]^2} (B_2 + H)^2 \, dC > c'_\alpha\right).$$

A test of independence against a positive dependence is defined by the statistic $S_{n+} = \sup_{x \in \mathbb{R}^2}\{\widehat{C}_n(x) - x_1 x_2\}$ and we use $S_{n-} = \sup_{x \in \mathbb{R}^2}\{x_1 x_2 - \widehat{C}_n(x)\}$ for H_0 against an alternative of negative dependence.

Let $X = (X_1, X_2)$ be a two-dimensional variable having a joint density f and marginal distribution functions F_1 and F_2 with densities f_1 and f_2, then

$$f(x) = f_1(x_1) f_2(x_2) \frac{\partial^2 C}{\partial x_1 \partial x_2}(F_1(x_1), F_2(x_2))$$

is denoted $f = f_1 f_2 h \circ (F_1, F_2)$. The log-likelihood ratio test statistic for independence against an alternative of a function h different from 1 is written as

$$l_n = \sum_{i=1}^n \log h \circ (F_1(X_{1i}), F_2(X_{2i})) = \sum_{i=1}^n \log h \circ (U_{1i}, U_{2i}).$$

Under H_0, the log-likelihood is zero if the marginal distribution functions are known. A maximum likelihood test is defined with a family of parametric marginal functions or with the nonparametric estimators of the unknown marginal functions. Its asymptotic behavior is like the parametric maximum likelihood ratio tests in a nonparametric or semi-parametric models where the function C is unspecified under the alternative.

Under local alternatives $h_n = 1 + n^{-\frac{1}{2}}\varphi_n$, with a sequence of functions φ_n in $[0,1]^2$ converging uniformly to a nonzero limit φ, l_n has the expansion

$$l_n = n^{-\frac{1}{2}} \sum_{i=1}^n \varphi_n(X_i) - (2n)^{-1} \sum_{i=1}^n \varphi_n^2(X_i) + o_p(1)$$

and it converges weakly to a Gaussian variable with a finite mean $\mu_\varphi = \frac{1}{2}\sigma_\varphi^2$ and a finite variance $\sigma_\varphi^2 = E\varphi^2(X)$. A critical value corresponding to a nominal level $\alpha = \sup_{F_1, F_2} P_0(|l_n - \mu_0(F_1, F_2)| > \sigma_0(F_1, F_2)c)$ is the $(1-\alpha)$th quantile of the normal distribution. Its asymptotic local power for a class Φ of functions φ is

$$\beta_\Phi(\alpha) = \sup_{F_1, F_2} \inf_{\varphi \in \phi} P_\varphi(|l_n - \mu_\varphi(F_1, F_2)| > \sigma_\varphi(F_1, F_2)c).$$

This test cannot be compared to the tests performed with the statistics (4.7) and (4.8). An alternative to the hypothesis of independence is defined by a parametric class of dependence functions $\phi = \{\varphi_\theta, \theta \in \Theta\}$ indexed by a parameter θ in a bounded open subset of $\mathbb{R}^d$. Under H_0, ratio of the densities is $h_0 = \varphi_{\theta_0} \equiv 1$ and the parameter θ is estimated by

$$\widehat{\theta}_n = \arg\max_{\theta \in \Theta} \sum_{i=1}^{n} \log h_\theta(U_i) = \arg\max_{\theta \in \Theta} l_n(\theta).$$

Proposition 4.2. *Under the assumptions of a class of functions ϕ of $C^2(\Theta)$ such that the integral $I_\theta = \int_{[0,1]^2} \dot{h}_\theta^2 h_\theta^{-1} \, du$ is finite in a neighborhood of θ_0 and nonsingular at θ_0, $n^{\frac{1}{2}}(\widehat{\theta}_n - \theta_0) = -\ddot{l}_{\theta_0}^{-1} n^{-\frac{1}{2}} \dot{l}_{n,\theta_0} + o_p(1)$, it converges weakly to a centered Gaussian variable $I_{\theta_0}^{-1} U_0$, where the variance of U_0 is I_{θ_0}, and the log-likelihood ratio test statistic converges weakly to the χ_d^2 variable $U_0^t I_{\theta_0}^{-1} U_0$ under H_0. Under local alternatives with parameters $\theta_n = \theta_0 + n^{-\frac{1}{2}} \gamma_n$, with γ_n converging to a nonzero limit γ, $n^{\frac{1}{2}}(\widehat{\theta}_n - \theta_n) = -\ddot{l}_{\theta_n}^{-1} n^{-\frac{1}{2}} \dot{l}_{n,\theta_n} + o_p(1)$ where $n^{-\frac{1}{2}} \dot{l}_{n,\theta_n} = n^{-\frac{1}{2}} \dot{l}_{n,\theta_0} + \gamma_n I_{\theta_0} + o_p(1)$ and the log-likelihood ratio test statistic converges to a variable $\{U_0 + \gamma^t I_{\theta_0}\}^t I_{\theta_0}^{-1} \{U_0 + \gamma^t I_{\theta_0}\}$.*

By definition of the parametric maximum likelihood estimator, it satisfies $\dot{l}_{n,\widehat{\theta}_n} = \sum_{i=1}^{n} \dot{h}_{\widehat{\theta}_n}(U_i) h_{\widehat{\theta}_n}^{-1}(U_i) = 0$ and $n^{-1} \sum_{i=1}^{n} \dot{h}_{\theta_0}(U_i) h_{\theta_0}^{-1}(U_i)$ converges in probability $\int_{[0,1]^2} \dot{h}_{\theta_0}(u) \, du_1 \, du_2 = 0$. Under the assumption that I_θ is finite, $\ddot{l}_{n,\theta} = n^{-1} \sum_{i=1}^{n} \{\ddot{h}_\theta(U_i) h_\theta^{-1}(U_i) - \dot{h}_\theta^2(U_i) h_\theta^{-2}(U_i)\}$ converges in probability to I_θ uniformly in a neighborhood of θ_0, then the log-likelihood ratio test statistic has a parametric behavior (Chapter 2).

4.2.3 *Tests based on the conditional distribution*

This section concerns nonparametric tests of independence of real random variables X and Y from a regular nonparametric estimator of the distribution function $F_{X|Y}$ of X conditionally on Y. Let F_{XY} be the joint distribution function a two-dimensional random variable (X, Y) with values in $\mathbb{R}^2$ and let F_Y be the marginal distribution function of Y. They are supposed to be differentiable with respect to the conditioning variable, with respective densities f_Y and $F_{X|Y=y}$ in $C^s(\mathbb{R})$. A test of the hypothesis $H_0 : F_{X|Y} = F_X$ is a test of independence of the variables X and Y against alternatives of dependence. A kernel estimator of $F_{X|Y=y}$ is defined from a sample $(X_i, Y_i)_{i=1,\ldots,n}$ as

$$\widehat{F}_{X|Y,n}(x; y) = \frac{\int K_h(y - s)\widehat{F}_{XY,n}(x, ds)}{\int K_h(y - s)\widehat{F}_{Y,n}(ds)} = \frac{\sum_{i=1}^{n} K_h(y - Y_i)1_{\{X_i \leq x\}}}{\sum_{j=1}^{n} K_h(y - Y_j)},$$

with a kernel density and a bandwidth h_n satisfying Condition 1.1. The bias and the variance of $\widehat{F}_{X|Y,n}$ are $h^s b_{X|Y}$ and $(nh)^{-1}\sigma^2_{X|Y}$, where

$$b_{X|Y}(x; y) = \frac{m_{sK}}{s!} f_Y^{-1}(x)\left\{\frac{\partial^s F_{X|Y}(x; y)}{\partial y^s} - F_{X|Y}(x; y)f_Y^{(s)}(x)\right\},$$

$$\sigma^2_{X|Y}(x) = \kappa_2 f_Y^{-1}(x)\{P(X \leq x|Y = y) - P^2(X \leq x|Y = y)\}.$$

A test of H_0 is built on the empirical processes $\nu_X = n^{\frac{1}{2}}(\widehat{F}_{X,n} - F_X)$ and

$$\nu_{X|Y,n} = n^{\frac{s}{2s+1}}(\widehat{F}_{X|Y,n} - F_{X|Y}), \tag{4.9}$$

from a sample $(X_i, Y_i)_{i=1,\ldots,n}$ and using the L^2-optimal bandwidth for $\widehat{F}_{X|Y,n}$. Since the estimators of the conditional distribution function is biased, a correction improves the test. A Kolmogorov-Smirnov test statistic is

$$T_n^+ = \sup_{x,y} n^{\frac{s}{2s+1}}|\widehat{F}_{X|Y,n}(x; y) - \widehat{F}_{X,n}(x) - \widehat{b}_{F_{X|Y},n}(x; y)|. \tag{4.10}$$

Proposition 4.3. *Under the hypothesis H_0, the statistic T_n^+ is asymptotically equivalent to $\sup_{x,y} |\nu_{X|Y,n}(x; y) - \widehat{b}_{F_{X|Y},n}(x; y)|$ and it converges weakly to the variable $T^+ = \sup_{x,y} |\nu_{X|Y}(x; y)|$. Under the alternative, it tends to infinity.*

Proof. Under the hypothesis H_0, the distribution functions $F_{X|Y}$ and F_X are identical and $T_n^+ = \sup_{x,y} |\nu_{X|Y,n}(x; y) - \widehat{b}_{F_{X|Y},n}(x; y)| + o_p(1)$ since the convergence rates of the processes ν_X and $\nu_{X|Y,n}$ differ and $s > 1$. The weak convergence of the process $\nu_{X|Y,n}$ to a Gaussian process in $\mathbb{R}^2$ implies the weak convergence of the variable T_n^+ to T^+. Under the alternative, $n^{\frac{s}{2s+1}} \sup_{x,y} |F_{X|Y} - F_X|$ diverges whence T_n^+ diverges. □

As the kernel estimator has a lower convergence rate than the empirical distribution function, the statistics behaves as if the marginal distribution function F_X was known. Another test statistic is defined in the same way for the distribution of Y conditionally on X.

Proposition 4.4. *Let $(K_n)_{n \geq 1}$ be a sequence of local alternatives defined by dependent conditional distribution functions $F_n(x; y) = F_X(x) + n^{-\frac{s}{2s+1}} G_n(x; y)$ such that the functions G_n converge uniformly to a bounded function $G(x; y)$ tending to zero in $\mathbb{R}^2$ as x or y tend to infinity. Under Condition 1.1, the statistic T_n^+ converges weakly under $(K_n)_{n \geq 1}$ to $\sup_{x,y} |\nu_{X|Y}(x; y) + G(x; y)|$.*

Proof. Under K_n, $\widehat{F}_{X|Y,n}(x; y)$ is an estimator of $F_X(x) + n^{-\frac{s}{2s+1}} G_n(x; y)$ hence

$$n^{\frac{s}{2s+1}}\{\widehat{F}_{X|Y,n}(x; y) - \widehat{F}_{X,n}(x)\} = \nu_{X|Y,n}(x; y) - n^{\frac{s}{2s+1} - \frac{1}{2}} \nu_{X,n}(x) + G_n(x; y),$$

it is asymptotically equivalent to $\nu_{X|Y,n}(x; y) + \widehat{b}_{F_{X|Y},n}(x; y) + G_n(x; y)$. $\square$

From Proposition 4.4,

$$P_{K_n}(T_n^+ > c) = P_{K_n}(\sup_{x,y} |\nu_{X|Y,n}(x; y) + G_n(x; y)| > c) + o(1)$$

and it converges to $P(\sup_{x,y} |\nu_{X|Y}(x; y) + G(x; y)| > c)$. The supremum is reached as $\nu_{X|Y}(x; y)$ and $G(x; y)$ have the same sign hence $\sup_{x,y} |\nu_{X|Y}(x; y) + G(x; y)|$ is greater than $\sup_{x,y} |\nu_{X|Y}(x; y)|$. It follows that the asymptotic local power of the test is larger than its level α.

4.3 Test of homogeneity

In a population where two independent real samples of a variable X are observed, the question of homogeneity is to determine whether both samples have the same distribution. Let $X_1, \ldots, X_m$ and $X_{m+1}, \ldots, X_{m+n}$ be the samples with respective empirical distribution functions $\widehat{F}_m$ and $\widehat{F}_n$. Under the hypothesis of homogeneity, $\widehat{F}_m$ and $\widehat{F}_n$ converge uniformly in probability to the same distribution function F whereas they converge to distinct limits F_1 and F_2 respectively under the alternative of two different subpopulations. Let $N = n + m$ and let $n = n_N$ and $m = m_N$ tends to infinity with N so that $p = \lim_{N \to \infty} N^{-1} m_N$ belongs to $]0, 1[$. The global empirical distribution function $\widehat{F}_N$ converges uniformly in probability to $F = pF_1 + (1 - p)F_2$ as N tends to infinity.

With two samples, nonparametric tests of homogeneity are defined from the process

$$W_{n,m}(x) = \left(\frac{nm}{n+m}\right)^{\frac{1}{2}}\{\widehat{F}_m(x) - \widehat{F}_n(x)\} \qquad (4.11)$$

$$= \left(\frac{n}{n+m}\right)^{\frac{1}{2}}\nu_m(x) - \left(\frac{m}{n+m}\right)^{\frac{1}{2}}\nu_n(x) + \left(\frac{nm}{n+m}\right)^{\frac{1}{2}}(F_1 - F_2)(x)$$

$$:= W_{0,n,m}(x) + \left(\frac{nm}{n+m}\right)^{\frac{1}{2}}(F_1 - F_2)(x).$$

Proposition 4.5. *Under the hypothesis of homogeneity, the statistic* $\sup_{x\in\mathbb{R}}|W_{n,m}(x)|$ *converges weakly to* $\sup_{x\in\mathbb{R}}|W \circ F(x)|$ *as N tends to infinity, where W is the standard Wiener process. Under fixed alternatives, it diverges.*

Proof. The empirical processes ν_m and ν_n converges weakly to independent Wiener processes $W_1 \circ F_1$ and $W_2 \circ F_1$ respectively. When $F_1 = F_2$, $W_{n,m}$ is asymptotically equivalent to $(1-p)^{\frac{1}{2}}W_1 \circ F - p^{\frac{1}{2}}W_2 \circ F$. The variance of the limit is the variance of the transformed Wiener process $W \circ F$. Under the alternative of distinct sub-distributions F_1 and F_2, $W_{n,m}(x)$ tends to infinity at every value x where $F_1(x)$ differs from $F_2(x)$.   □

The asymptotic critical value of the test of Proposition 4.5 at the nominal level α is a_α such that $P(\sup_{x\in\mathbb{R}}|W \circ F(x)| > a_\alpha) = P(\sup_{x\in[0,1]}|W(x)| > a_\alpha) = \alpha$. By Bienaymé-Chebychev's inequality,

$$P(\sup_{x\in\mathbb{R}}|W(x)| > a_\alpha) \leq a^{-2}E\{(\sup_{x\in\mathbb{R}}|W(x)|)^2\}$$

and for every real x, $E\{W^2(x)\} = x(1 - x) \leq \frac{1}{4}$, which implies $P(\sup_{x\in\mathbb{R}}|W(x)| > a_\alpha) \leq (2a)^{-2}$. With the critical value $a_\alpha = (2\sqrt{\alpha})^{-1}$, $P(\sup_{x\in\mathbb{R}}|W(x)| > a_\alpha) \leq \alpha$.

Proposition 4.6. *Let $(K_{n,m})_{n,m\geq 1}$ be a sequence of local alternatives defined by distribution functions $F_{1m} = F + m^{-\frac{1}{2}}G_m$ and $F_{2n} = F + n^{-\frac{1}{2}}H_n$ such that the functions G_m and H_n converge uniformly to functions G and H, respectively, tending to zero at infinity. Under $(K_n)_{n\geq 1}$, the statistic* $\sup_{\mathbb{R}}|W_{n,m}|$ *converges weakly to* $\sup_{\mathbb{R}}|W + (1-p)^{\frac{1}{2}}G - p^{\frac{1}{2}}H|$.

Proof. Under K_n, $\widehat{F}_m$ is an estimator of $F_{1m} = F + m^{-\frac{1}{2}}G_m$ and $\widehat{F}_n$ is an estimator of $F_{2n} = F + n^{-\frac{1}{2}}H_n$. The approximation of the process $W_{n,m}$ is expressed through the empirical processes $\nu_{m,m} = m^{\frac{1}{2}}(\widehat{F}_m - F_{1m})$ and $\nu_{n,n} = n^{\frac{1}{2}}(\widehat{F}_n - F_{2n})$

$$W_{n,m} = \left(\frac{n}{n+m}\right)^{\frac{1}{2}}\nu_{m,m} - \left(\frac{m}{n+m}\right)^{\frac{1}{2}}\nu_{n,n} + \left(\frac{mn}{n+m}\right)^{\frac{1}{2}}(F_{1m} - F_{2n}).$$

As N tends to infinity, it converges weakly to the same distribution as the process $W \circ F + (1-p)^{\frac{1}{2}} G - p^{\frac{1}{2}} H$. □

The supremum of the limit of the process $W_{n,m}$ is reached as $W \circ F$ and $p^{\frac{1}{2}} H - (1-p)^{\frac{1}{2}} G$ have the same sign, then is greater than the supremum of $W_{0,n,m}$. In that case, the asymptotic local power of the test is larger than its level α.

Other nonparametric statistics of homogeneity are based on the following similar processes

$$W_{2,n} = m^{\frac{1}{2}}(\widehat{F}_m \circ \widehat{F}_n^{-1} - id),$$
$$W_{3,n,m} = n^{\frac{1}{2}}(\widehat{F}_n \circ \widehat{F}_m^{-1} - id).$$

Under the hypothesis of homogeneity

$$W_{2,n} = \nu_m \circ \widehat{F}_n^{-1} + m^{\frac{1}{2}}(F \circ \widehat{F}_n^{-1} - id),$$
$$W_{3,n,m} = \nu_n \circ \widehat{F}_m^{-1} + n^{\frac{1}{2}}(F \circ \widehat{F}_m^{-1} - id)$$

and their limits are obtained from expansions of the quantiles.

Proposition 4.7. *Under the hypothesis of homogeneity, the processes $W_{2,n}$ and $W_{3,n,m}$ converge weakly as N tends to infinity to $(1-p)^{-\frac{1}{2}} W$ and, respectively, $p^{-\frac{1}{2}} W$, where W is the Wiener process. Under the sequence of local alternatives K_n of Proposition 4.6, $W_{2,n}$ converges weakly to $\{1 + p^{\frac{1}{2}}(1-p)^{-\frac{1}{2}}\} W + G \circ F^{-1} - p^{\frac{1}{2}}(1-p)^{-\frac{1}{2}} H \circ F^{-1}$ and $W_{3,n,m}$ converges weakly to $\{1 + p^{-\frac{1}{2}}(1-p)^{\frac{1}{2}}\} W + H \circ F^{-1} - (1-p)^{\frac{1}{2}} p^{-\frac{1}{2}} G \circ F^{-1}$.*

Proof. Arguing as in Lemma 4.1, the process $n^{\frac{1}{2}}(F \circ \widehat{F}_n^{-1} - id)$ is asymptotically equivalent under the hypothesis to $\nu_n \circ F^{-1} + o_p(1)$, therefore

$$W_{2,n} = \nu_m \circ \widehat{F}_n^{-1} + \left(\frac{m}{n}\right)^{\frac{1}{2}} \nu_n \circ F^{-1} + o_p(1)$$

$$= \nu_m \circ F^{-1} + \left(\frac{p}{1-p}\right)^{\frac{1}{2}} \nu_n \circ F^{-1} + o_p(1),$$

$$W_{3,n,m} = \nu_n \circ \widehat{F}_m^{-1} + \left(\frac{n}{m}\right)^{\frac{1}{2}} \nu_m \circ F^{-1} + o_p(1).$$

Their limiting distributions is deduced from the independence of the empirical processes $\nu_m \circ F^{-1}$ and $\nu_n \circ F^{-1}$ which have the same distribution. Under the alternative K_n, the empirical processes are denoted $\nu_{m,m}$ and $\nu_{n,n}$. The process $n^{\frac{1}{2}}\{F_{1m} \circ \widehat{F}_n^{-1} - F_{1m} \circ F_{2n}^{-1}\}$ is approximated by

$$\left(\frac{f_1}{f_2} \nu_{n,n}\right) \circ F_2^{-1}$$

and $m^{\frac{1}{2}}(F_{1m} - F_{2n}) \circ F_{2n}^{-1} = m^{\frac{1}{2}}\{(F_{1m} - F_1) + (F_{1m} - F_{2n})\} \circ F_{2n}^{-1}$ is asymptotically equivalent to $\{G + m^{\frac{1}{2}}(F_{1m} - F_{2n})\} \circ F_{2n}^{-1}$. The statistic $W_{2,n}$ has therefore the expansion

$$W_{2,n} = \nu_{m,m} \circ \widehat{F}_n^{-1} + m^{\frac{1}{2}}\{F_{1m} \circ \widehat{F}_n^{-1} - F_{1m} \circ F_{2n}^{-1}\}$$
$$+ m^{\frac{1}{2}}(F_{1m} - F_{2n}) \circ F_{2n}^{-1} + o_p(1)$$
$$= \nu_{m,m} \circ \widehat{F}_n^{-1} + \left(\frac{m}{n}\right)^{\frac{1}{2}}\left(\frac{f_1}{f_2}\nu_{n,n}\right) \circ F_2^{-1}$$
$$+ \left\{G + m^{\frac{1}{2}}(F_1 - F_2) - \left(\frac{m}{n}\right)^{\frac{1}{2}}H\right\} \circ F_2^{-1} + o_p(1)$$

and its limit is deduced, the expansion of $W_{3,n,m}$ is similar.   □

Under fixed alternatives, the statistic $W_{2,n}$ has the expansion

$$W_{2,n} = \nu_m \circ \widehat{F}_n^{-1} + m^{\frac{1}{2}}\{F_1 \circ \widehat{F}_n^{-1} - F_1 \circ F_2^{-1}\}$$
$$+ m^{\frac{1}{2}}(F_1 - F_2) \circ F_2^{-1} + o_p(1)$$
$$= \nu_m \circ \widehat{F}_n^{-1} + \left(\frac{m}{n}\right)^{\frac{1}{2}}\left(\frac{f_1}{f_2}\nu_n\right) \circ F_2^{-1} + m^{\frac{1}{2}}(F_1 - F_2) \circ F_2^{-1} + o_p(1).$$

It follows that the processes $W_{2,n}$ and $W_{3,n,m}$ diverge under fixed alternatives and the power of the tests converge to 1.

Semi-parametric tests of homogeneity are defined by a family $\{\mathcal{F}\}$ of distribution functions in $\mathbb{R}$, a bounded open parameter subset Θ of $\mathbb{R}^k$ and a regular model $\{\mathcal{M}\} = \{\mathcal{F}_\theta; \mathcal{F} \in \{\mathcal{F}\}, \theta \in \times\}$ such that for every x in $\mathbb{R}$, the map $\theta \mapsto F_\theta$ belongs to $C^2(\Theta)$. The sample $X_1, \ldots, X_n$ is drawn under $F_1 = F$ belonging to $\{\mathcal{F}\}$ and $X_{m+1}, \ldots, X_{m+n}$ under $F_2 = F_{\theta_0}$ belonging to $\{\mathcal{M}\}$. The samples are independent and their distributions are unknown. The hypothesis H_0 of the test is $F = F_{\theta_0}$. The distribution function F is estimated by the empirical distribution function $\widehat{F}_m$ and the unknown parameter is estimated by minimization in Θ of the squared L^2-distance between the empirical estimator of the second sample and its parametric distribution

$$d_n(\theta) = \int_{\mathbb{R}}\{\widehat{F}_n(t) - F_\theta(t)\}^2\,dt.$$

The parametric estimators of θ_0 and F_2 are deduced as

$$\widehat{\theta}_n = \arg\max_{\theta \in \Theta} d_n(\theta),$$
$$\widehat{F}_{n,\theta_0} = F_{\widehat{\theta}_n},$$

they are $n^{\frac{1}{2}}$-consistent, $n^{\frac{1}{2}}(\widehat{\theta}_n - \theta_0)$ and $n^{\frac{1}{2}}(\widehat{F}_{n,\theta_0} - F_{\widehat{\theta}_n})$ converge weakly to centered Gaussian distributions under the hypothesis and all alternatives. A test statistic for the hypothesis $H_0 : F_1 = F_{\theta_0}$ is then defined from the L^2-distance $d(\widehat{F}_n, F_{\widehat{\theta}_m}) = \int_{\mathbb{R}} \{\widehat{F}_n(t) - \widehat{F}_{m,\theta}(t)\}^2 \, dt$ between the empirical estimator F_1 and the parametric estimator of F_{θ_0}. The variable

$$W_{n,m} = \left(\frac{nm}{n+m}\right)^{\frac{1}{2}}(\widehat{F}_m - F_{\widehat{\theta}_n})$$

converges weakly under H_0 to a centered Gaussian variable with a variance depending on F_{θ_0} and the test statistic is modified by a change of variables. Moreover, the distance $d(\widehat{F}_n, F_{\widehat{\theta}_m})$ is not necessarily minimum under H_0 and can be improved. The smaller convergence rate of the parameter estimator is $(m+n)^{-\frac{1}{2}}$ under H_0, it is obtained by minimizing the L^2-distance between the empirical estimator of the whole sample $\widehat{F}_N$ and F_{θ_0}, with $N = n + m$. It is $N^{\frac{1}{2}}$-consistent under H_0 but it is not under fixed alternatives. The value of the test statistic $W_{N,m}$ is therefore expected to be larger than $W_{n,m}$ under the alternatives. Going further in these considerations, the worst estimator of θ_0 and F_{θ_0} under alternatives should be obtained from the first sample. Let us consider the distance

$$d_m(\theta) = \int_{\mathbb{R}} \{\widehat{F}_m(t) - F_\theta(t)\}^2 \, dt$$

and the minimum distance estimators

$$\widehat{\theta}_m = \arg\max_{\theta \in \Theta} d_m(\theta),$$

$$\widehat{F}_{m,\theta_0} = F_{\widehat{\theta}_m}.$$

Under H_0, $m^{\frac{1}{2}}(\widehat{\theta}_m - \theta_0)$ and $m^{\frac{1}{2}}(\widehat{F}_m - F_{\theta_0})$ converge weakly to centered Gaussian distributions, under fixed alternatives $\widehat{\theta}_m$ and $\widehat{F}_m$ diverge like the process

$$W_{n,m} = \left(\frac{nm}{n+m}\right)^{\frac{1}{2}}(\widehat{F}_n - F_{\widehat{\theta}_m}).$$

Under local alternatives $K_m : F_m = F_{\theta_0} + m^{-\frac{1}{2}}\zeta_m$, such that $(\zeta_m)_{m\geq 1}$ is a real sequence converging to $\zeta \neq 0$, the estimators of the distribution function of the first sub-sample satisfies

$$m^{\frac{1}{2}}(\widehat{F}_{m,\theta_0} - F_{\theta_0}) = m^{\frac{1}{2}}(\widehat{F}_{m,\theta_0} - F_m) + \zeta_m,$$

where $m^{\frac{1}{2}}(\widehat{F}_{m,\theta_0} - F_m)$ converges weakly under K_n to the same limit as $m^{\frac{1}{2}}(\widehat{F}_m - F_{\theta_0})$ under H_0. Centered at θ_0, it converge to a Gaussian process with mean ζ, then the process $W_{n,m}$ is bounded in probability under the local alternative K_n.

Proposition 4.8. *Under the assumption of a model of distribution functions in $C^2(\Theta)$ such that the matrix $I_0 = 2 \int_{\mathbb{R}} \dot{F}_{\theta_0}^{\otimes 2}(t) \, dt$ is strictly positive definite, the variable $m^{\frac{1}{2}}(\widehat{\theta}_m - \theta_0)$ converges weakly under H_0 to a centered Gaussian variable*

$$U_0 = \sqrt{2}\left\{ \int_{\mathbb{R}} \dot{F}_{\theta}^{\otimes 2}(t) \, dt \right\}^{-1} \left\{ \int_{\mathbb{R}} W \circ F(t) \, \dot{F}_{\theta_0}(t) \, dt \right\},$$

where W is a Wiener process.

Proof.   The derivatives of d_m are

$$\dot{d}_m(\theta) = -2 \int_{\mathbb{R}} \{\widehat{F}_m(t) - \widehat{F}_{m,\theta}(t)\} \dot{\widehat{F}}_{m,\theta}(t) \, dt,$$

$$\ddot{d}_m(\theta) = -2 \int_{\mathbb{R}} \{\widehat{F}_m(t) - \widehat{F}_{m,\theta}(t)\} \ddot{\widehat{F}}_{m,\theta}(t) \, dt + 2 \int_{\mathbb{R}} \dot{\widehat{F}}_{m,\theta}^{\otimes 2}(t) \, dt,$$

where the estimated distribution function $\widehat{F}_{m,\theta}$ is derived with respect to the parameter θ. The process $\dot{d}_m$ converges uniformly in Θ and in probability to the function $\dot{d}(\theta) = -2 \int_{\mathbb{R}} \{F(t) - F_\theta(t)\} \dot{F}_\theta(t) \, dt$ which is zero at θ_0. The second derivative $\ddot{d}_m(\theta)$ converges uniformly in probability to the matrix $I_\theta = -2 \int_{\mathbb{R}} \{F(t) - F_\theta(t)\} \dot{F}_\theta(t) \, dt + 2 \int_{\mathbb{R}} \dot{F}_\theta^{\otimes 2}(t) \, dt$. Under the hypothesis of homogeneity $I_{\theta_0} = 2 \int_{\mathbb{R}} \dot{F}_\theta^{\otimes 2}(t) \, dt$ is positive definite which implies that the function $l_{m,n}(\theta)$ is convex at θ_0 and the consistency of the minimum of distance estimator of the parameter follows. The first derivative $\dot{d}_m(\theta)$ satisfies

$$m^{\frac{1}{2}} \dot{d}_m(\theta_0) = -2 \int_{\mathbb{R}} \{\nu_m(t) - \nu_{m,\theta_0}(t)\} \dot{\widehat{F}}_{m,\theta_0}(t) \, dt$$

$$= -2 \int_{\mathbb{R}} \{\nu_m(t) - \nu_{m,\theta_0}(t)\} \dot{F}_{\theta_0}(t) \, dt + o_p(1)$$

therefore it converges weakly to the variable

$$-2\left\{ \int_{\mathbb{R}} W_1 \circ F(t) \dot{F}_{\theta_0}(t) \, dt - \int_{\mathbb{R}} W_2 \circ F_{\theta_0}(t) \dot{F}_{\theta_0}(t) \, dt \right\}$$

where W_1 and W_2 are independent Wiener processes. It follows that the variable $m^{\frac{1}{2}} \dot{d}_m(\theta_0)$ converges weakly to a centered Gaussian variable $-2\sqrt{2} \int_{\mathbb{R}} W \circ F_{\theta_0}(t) \, \dot{F}_{\theta_0}(t) \, dt$ under H_0. A first order expansion of $\dot{d}_m(\widehat{\theta}_m)$ and the consistency of the parameter estimator ends the proof like that for the maximum likelihood estimator.   □

Proposition 4.9. *Under the hypothesis of homogeneity and the conditions for* $\mathcal{M}$, *the statistic* $\sup_{x \in \mathbb{R}} |W_{n,m} \circ \widehat{F}_N^{-1}(x)|$ *converges weakly to* $\sup_{x \in \mathbb{R}} |W_0(x)|$ *as* N *tends to infinity, where* $W_0 = W - U_0^t \dot{F}_{\theta_0} \circ F^{-1}$, *with the Wiener process* W *of Proposition 4.8.*

Proof. The process $W_{n,m}$ is expanded as

$$W_{n,m} = \left(\frac{mn}{n+m}\right)^{\frac{1}{2}} \{\widehat{F}_n - \widehat{F}_{m,\theta_0} - (\widehat{\theta}_m - \theta_0)^t \widehat{\dot{F}}_{m,\theta_0}\} + o_p(1)$$

$$= \left(\frac{m}{n+m}\right)^{\frac{1}{2}} \widehat{\nu}_n - \left(\frac{n}{n+m}\right)^{\frac{1}{2}} \{\widehat{\nu}_{m,\theta_0} + (\widehat{\theta}_m - \theta_0)^t \widehat{\dot{F}}_{m,\theta_0}\} + o_p(1)$$

under the hypothesis of a single distribution function $F = F_{\theta_0}$. Then $\lim_{m,n \to} \sup_{x \in \mathbb{R}} |\widehat{F}_n(x) - \widehat{F}_{m,\theta_0}(x)| = 0$ and $\lim_{m,n \to} \sup_{x \in \mathbb{R}} |\widehat{\dot{F}}_{m,\theta_0}(x) - \dot{F}_{\theta_0}(x)| = 0$, and the processes $\widehat{\nu}_m$ and $\widehat{\nu}_{m,\theta_0}$ have the same limiting distribution. By Proposition 4.8, the process $W_{n,m}$ converges weakly to the centered Gaussian process $W \circ F_{\theta_0} - U_0^t \dot{F}_{\theta_0}$, with a Wiener process W. $\square$

The limit of the process $W_{m,n}$ in semi-parametric models differs from the limit of the process in the nonparametric case and it depends on the model defined by $\mathcal{M}$, so the nonparametric tests are of greater interest than tests for specific models. In parametric models of densities, the maximum likelihood estimation of the parameter would also provide an estimator that modifies the asymptotic distribution of $W_{m,n}$. Other test statistics of homogeneity are defined in that case as weighted L^2-distances between the estimated densities of the samples and their distributions are not free, they are specific to the models. The comparison of kernel estimators of the densities of the two samples by L^2-distances provides a nonparametric class of tests for homogeneity, they are again dependent of the densities under the hypothesis and under the alternatives.

4.4 Goodness of fit tests in $\mathbb{R}^2$

Goodness of fit tests for the distribution function F of a bidimensional variable $X = (X_1, X_2)$ with dependent components are performed by testing the goodness of fit of their marginals and of their dependence function C, in two steps. The parametric or nonparametric estimators of the marginals are correlated and an estimator of their covariance is used to build asymptotic χ^2 tests statistics from a n-sample of the variable X.

With a single hypothesis $H_0 : F = F_0$ in $\mathbb{R}^2$, the marginal hypothesis $H_{0m} : F_k = F_{0k}$, for $k = 1, 2$, is performed with the vector of the empirical process under H_{01}, $U_n = n^{\frac{1}{2}}(\widehat{F}_{1n} - F_{01}, \widehat{F}_{2n} - F_{02})^t$. The asymptotic variance of U_n under H_{01} is the matrix function

$$\Sigma = \begin{pmatrix} F_{01}\bar{F}_{01} & F_0 - F_{01}F_{02} \\ F_0 - F_{01}F_{02} & F_{02}\bar{F}_{02} \end{pmatrix}$$

and it is estimated empirically by $\widehat{\Sigma}_n$. For every $x = (x_1, x_2)$, the statistic $U_n(x)^t\widehat{\Sigma}_n(x)U_n(x)$ converges weakly to a free χ_2^2 distribution and this is also the asymptotic distribution of

$$T_n = \sup_{x \in \mathbb{R}^2} U_n(x)^t\widehat{\Sigma}_n(x)U_n(x).$$

The empirical estimator of the dependence funtion C in $\mathbb{R}^2$ is then compared to the function C_0 of the hypothesis H_0 through the empirical process $\nu_n = n^{\frac{1}{2}}(\widehat{C}_n - C_0)$ in $[0,1]^2$. Under the hypothesis H_0, its variance function $\sigma_0^2 = F_0(1 - F_0)$ is estimated by $\widehat{\sigma}_n^2$ empirically and the statistic $\sup_{x \in \mathbb{R}^2}\{\widehat{\sigma}_n^{-1}(x)\nu_n(x)\}^2$ converges weakly to a χ_1^2 distribution. Under local alternatives, the estimators of the distribution functions are not centered and their variance is unchanged, the limiting distributions and their asymptotic local power are deduced.

A semi-parametric goodness of fit test for the parametric hypothesis of a class of bivariate distribution functions $\mathcal{F} = \{F_\theta, \theta \in \Theta\}$ against a parametric alternative $\mathcal{G} \setminus \mathcal{F}$, for a larger class of distribution functions including $\mathcal{F}$, is performed as previously, after estimating the parameters in $\mathcal{F}$ and $\mathcal{G}$. Proposition 3.2 presents the behavior of the free Kolmogorov-Smirnov marginals tests. A similar test is used for the dependence funtion C and the limiting Gaussian distribution is modified in the same way. Asymptotic χ^2 tests are also defined as above. The vector of the differences between the estimated marginal empirical processes under H_{01} is written with the maximum likelihood estimators of the parameters of the marginal distribution functions as

$$\widehat{U}_n = n^{\frac{1}{2}}(\widehat{F}_{1n} - F_{1,\widehat{\theta}_{1n}}, \widehat{F}_{2n} - F_{2,\widehat{\theta}_{2n}})^t$$
$$= U_n - n^{\frac{1}{2}}\{(\widehat{\theta}_{1n} - \theta_{01})^t\dot{F}_{01}, (\widehat{\theta}_{2n} - \theta_{02})^t\dot{F}_{02}\}^t + o_p(1)$$

and its asymptotic variance is

$$V = \Sigma + n(\dot{F}_{01}^t, \dot{F}_{02}^t)Var(\widehat{\theta}_{1n} - \theta_{01}, \widehat{\theta}_{2n} - \theta_{02})(\dot{F}_{01}, \dot{F}_{02})$$
$$- n\dot{F}_{01}^t Cov(\widehat{\theta}_{1n} - \theta_{01}, \widehat{F}_{2n} - F_{02}) - n\dot{F}_{02}^t Cov(\widehat{\theta}_{2n} - \theta_{02}, \widehat{F}_{1n} - F_{01}).$$

Let $I^{-1} = Var_0(\widehat{\theta}_{1n} - \theta_{01}, \widehat{\theta}_{2n} - \theta_{02})$, then

$$Cov\{\widehat{\theta}_{1n} - \theta_{01}, (\widehat{F}_{2n} - F_{02})(x_2)\} = (nI_{11})^{-1}\int_{\mathbb{R}} \frac{\dot{f}_{10}}{f_{10}}(s)\,F(ds, x_2),$$

$$Cov\{\widehat{\theta}_{2n} - \theta_{02}, (\widehat{F}_{1n} - F_{01})(x_1)\} = (nI_{22})^{-1}\int_{\mathbb{R}} \frac{\dot{f}_{20}}{f_{20}}(y)\,F(x_1, dy).$$

Estimating the variance V by plugging yields a matrix $\widehat{V}_n$ and an asymptotic χ_2^2 test for the vector $\widehat{U}_n$ of the marginals. The same arguments apply to the difference of the estimators of the dependence function C, $n^{\frac{1}{2}}(\widehat{C}_n - C_{\widehat{\theta}_n}) = \nu_{n,C} - n^{\frac{1}{2}}(\widehat{\theta}_n - \theta_0)^t \dot{C}_0 + o_p(1)$ and provides an asymptotic χ_1^2 test.

4.5   Tests of symmetry for a bivariate density

A random variable X in $\mathbb{R}^2$ has a symmetric density centered at zero if its distribution F satisfies the hypothesis $H_0 : f(x) = f(-x)$ for every x in $\mathbb{R}^2$. Equivalently, for every x in $\mathbb{R}^2$

$$F(-x) = P(X \in] - \infty, -x]) = P(X \in [x, \infty[)$$
$$= 1 - F_1(x_1^-) - F_2(x_2^-) + F(x^-) := \bar{F}(x^-).$$

Let $X_1, \ldots, X_n$ be a n-sample of X, the right-continuous empirical estimator of $\bar{F}$ is

$$\widehat{\bar{F}}_n(x) = n^{-1} \sum_{i=1}^n 1_{\{X_{1i} > x_1, X_{2i} > x_2\}}.$$

It is also defined as $\widehat{\bar{F}}_n(x) = 1 - \widehat{F}_{1n}(x_1) - \widehat{F}_{2n}(x_2) + \widehat{F}_n(x)$ and the empirical process $\bar{\nu}_n = n^{\frac{1}{2}}(\widehat{\bar{F}}_n - \bar{F})$ related to $\widehat{\bar{F}}_n$ is deduced. It converges weakly to a centered Gaussian process with variance function

$$\bar{\sigma}^2(x) = \bar{F}(x) - \bar{F}^2(x) = \bar{F}(x)\{F_1(x_1^-) + F_2(x_2^-) - F(x^-)\}.$$

For a variable X with independent components, the survival function $\bar{F}$ factorizes as $\bar{F} = \bar{F}_1 \bar{F}_2$ and it is estimated as the product of the estimators of the marginal survival functions.

A Kolmogorov-Smirnov type statistic for the hypothesis of a symmetric density is defined from the process

$$U_n(x) = \sqrt{n}\{\widehat{F}_n(-x) - \widehat{\bar{F}}_n(x^-)\}. \tag{4.12}$$

Let $x_1 > 0$ and let $x_2 > 0$, the covariance of the estimators of $F_1(x_1^-) + F_2(x_2^-) - F(x^-)$ and $F(-x)$ is $F(-x)\bar{F}(x^-)$. Under the hypothesis H_0, $U_n = \nu_n(-x) - \bar{\nu}_n(x^-)$ converges weakly to a centered Gaussian process U with variance function

$$\sigma_U^2(x) = \{F(-x) + \bar{F}(x^-)\}\{1 - F(-x) - \bar{F}(x^-)\}$$
$$= 2F(-x)\{1 - F(-x) - \bar{F}(x^-)\},$$

it is estimated empirically by $\widehat{\sigma}_n^{-1}(x)$.

Under H_0, the normalized statistic $T_n = \sup_{x \in \mathbb{R}^2} \widehat{\sigma}_n^{-1}(x)|U_n(x)|$ converges weakly to $T = \sup_{x \in \mathbb{R}^2} \widehat{\sigma}_n^{-1}(x)|U(x)|$. Under fixed alternatives, $U_n(x) = \nu_n(-x) - \bar{\nu}_n(x^-) + n^{\frac{1}{2}}\{F(-x) - \bar{F}(x^-)\}$ tends to infinity. Its limit under local alternatives are calculated as in Section 3.3.1.

In spherical coordinates, let $\nu_n(A)$ denote the empirical process for the domain A of $\mathbb{R}^2$. We consider the statistic

$$S_n = \sup_{r>0} \sup_{\theta_1 \in [0,2\pi]} \sup_{\theta_2 \in [0,\pi]} |\widehat{F}_n(re^{i\theta_1}, re^{i\theta_2}) - \widehat{F}_n(re^{-i\theta_1}, re^{-i\theta_2})|$$

defined in sections of balls in $\mathbb{R}^2$ with radius $r \leq R_0$, between the angles θ_1 and θ_2 and the opposite section. The value of $F(re^{i\theta_1}, re^{i\theta_2}) - F(re^{-i\theta_1}, re^{-i\theta_2})$ is zero in balls centered at zero and it defines a test for the hypothesis H_0 of symmetric distributions centered at zero, its value is zero in all symmetric sets centered at zero, in particular in balls and ellipses. Under H_0

$$S_n = \sup_{r \leq R_0} \sup_{\theta_1 \in [0,2\pi]} \sup_{\theta_2 \in [0,\pi]} |\nu_n(re^{i\theta_1}, re^{i\theta_2}) - \nu_n(re^{-i\theta_1}, re^{-i\theta_2})|$$

and it converges weakly to the maximum of the difference of the variation of a Wiener process in opposite sections defined by angles θ_1 and θ_2. Consider local alternatives K_n of perturbations of the distribution function of the variable X along the radius r defined by sequences $(r_{1n})_{n \geq 1}$ and $(r_{2n})_{n \geq 1}$ such that $r_{1n} = r + n^{-\frac{1}{2}}\delta_{1n}$ and $r_{2n} = r + n^{-\frac{1}{2}}\delta_{2n}$, where δ_{jn} converges to constants δ_j, for $j = 1, 2$, and $\delta_1 \neq \delta_2$. Then

$$F(r_{1n}e^{i\theta_1}, r_{2n}e^{i\theta_2}) - F(r_{1n}e^{-i\theta_1}, r_{2n}e^{-i\theta_2}) = F(re^{i\theta_1}, re^{i\theta_2})$$
$$- F(re^{-i\theta_1}, re^{-i\theta_2}) + n^{-\frac{1}{2}}\{(\delta_1\dot{F}_1 + \delta_2\dot{F}_2)(re^{i\theta_1}, re^{i\theta_2})$$
$$- (\delta_1\dot{F}_1 + \delta_2\dot{F}_2)(re^{-i\theta_1}, re^{-i\theta_2})\} + o(n^{-\frac{1}{2}}),$$

denoted $F(re^{i\theta_1}, re^{i\theta_2}) - F(re^{-i\theta_1}, re^{-i\theta_2}) + n^{-\frac{1}{2}}R(\theta_1, \theta_2) + o(n^{-\frac{1}{2}})$. The empirical processes under the alternative are defined with the sequences $(r_{1n})_{n \geq 1}$ and $(r_{2n})_{n \geq 1}$ and the limiting distribution of $\nu_n(re^{i\theta_1}, re^{i\theta_2}) - \nu_n(re^{-i\theta_1}, re^{-i\theta_2})$ is modified by the additional term $R(\theta_1, \theta_2)$.

Under the alternative of a perturbation of the center of symmetry as $n^{-\frac{1}{2}}x_n$ such that x_n converges to a non null limit $x = (x_1, x_2)$, the distribution function of the observed variables is $F_n(y) = F(y - x_n)$ and

$$F_n(re^{i\theta_1}, re^{i\theta_2}) - F_n(re^{-i\theta_1}, re^{-i\theta_2}) = F(re^{i\theta_1}, re^{i\theta_2}) - F(re^{-i\theta_1}, re^{-i\theta_2})$$
$$- n^{-\frac{1}{2}}\{(x_1\dot{F}_1 + x_2\dot{F}_2)(re^{i\theta_1}, re^{i\theta_2}) - (x_1\dot{F}_1 + x_2\dot{F}_2)(re^{-i\theta_1}, re^{-i\theta_2})\}$$
$$+ o(n^{-\frac{1}{2}}).$$

Under this sequence of alternatives, the limiting distribution of the process $\nu_{n,n}(r_{1n}e^{i\theta_1}, r_{2n}e^{i\theta_2}) - \nu_{n,n}(r_{1n}e^{-i\theta_1}, r_{2n}e^{-i\theta_2})$ is the same as its limit under H_0, since the additional term $(x_1\dot{F}_1 + x_2\dot{F}_2)(re^{i\theta_1}, re^{i\theta_2}) - (x_1\dot{F}_1 + x_2\dot{F}_2)(re^{-i\theta_1}, re^{-i\theta_2})$ is zero. The test of H_0 under a local change of radius is consistent. Against a local change of center, it is not.

When the center of symmetry θ_0 of X under the hypothesis H_0 is unknown, the variable $X - \theta_0$ has a density symmetric at zero and the parameter θ_0 is estimated under H_0 by the sample mean $\widehat{\theta}_n = \bar{X}_n$. The process U_n is modified as

$$\widehat{U}_n(x) = \sqrt{n}\{\widehat{F}_n(\widehat{\theta}_n - x) - \widehat{\bar{F}}_n(\widehat{\theta}_n + x^-)\}.$$

The mean of the process $\widehat{U}_n$ is expanded as $E_0(\widehat{U}_n(x)|\widehat{\theta}_n) = (\widehat{\theta}_n - \theta_0)\{f(\theta_0 - x) - \bar{f}(\theta_0 + x^-) + o(n^{-\frac{1}{2}}) = o(n^{-\frac{1}{2}})$ and its variance is modified by an additional term due to the variance of the estimator $\widehat{\theta}_n$. Under local alternatives, the mean converges to a nonzero value and the variance to the limiting variance under H_0. Test statistics for the symmetry of the density f_{θ_0} are defined as previously.

4.6 Tests about the form of densities

The density f of a random variable X in $\mathbb{R}^2$ is unimodal if it has a unique global maximum in all directions of the plane, and no local maxima. In the trigonometric parametrization centered at zero, the function defined in $\mathbb{R}^2_+$ by $(r, b) \mapsto f(r\cos\varphi, \frac{r}{b}\sin\varphi)$ has to be unimodal with a mode at zero, uniformly over φ in $]-\pi, \pi[$.

A density f of $C^2(\mathbb{R}^2)$ is estimated from a bivariate sample $X_1, \ldots, X_n$ of X by kernel smoothing with a bivariate kernel

$$K_{h_1,h_2}(u, v) = (h_1 h_2)^{-1} K(h_1^{-1}u, h_2^{-1}v)$$

with a two-dimensional bandwidth $h = (h_1, h_2)^t$ such that h_1 and h_2 tend to zero as n tends to infinity. With a product kernel $K(u, v) = K_1(u)K_2(v)$ such that the functions K_j, $j = 1, 2$, satisfy Condition 1.1

$$\widehat{f}_{n,h}(x) = \int K_{h_1,h_2}(x_1 - u, x_2 - v)\, d\widehat{F}_n(u, v)$$

$$= \frac{1}{n}\sum_{i=1}^{n} K_{h_1,h_2}(x_1 - X_{1i}, x_2 - X_{2i}).$$

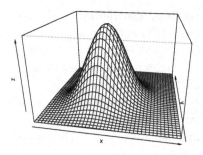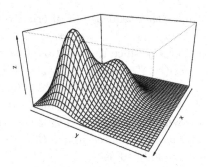

Fig. 4.1 Mixture of two normal densities with mixing probabilities .5 centered at $(0,0)$ and $(1,1)$.

Fig. 4.2 Mixture of two normal densities with mixing probabilities .35 and .65. centered at $(0,0)$ and $(-2,1)$.

The estimator $\widehat{f}_{n,h}$ has a bias and a variance

$$b_h(x) = \frac{1}{2}h^t \ddot{f}(x)h + o(\|h\|^2),$$

$$\sigma_{n,h}^2 = (nh_1h_2)^{-1}f(x) + o((nh_1h_2)^{-1}).$$

With bandwidths $h_1 = h_{1,n}$ and $h_2 = h_{2,n}$ having the same order as n tends to infinity, $h = (h_1^t, h_2^t) = h^t$ and a real symmetric kernel K such that $m_{2K} \neq 0$, its L^2-norm is minimal as h_1 and h_2 are $O(n^{-\frac{1}{6}})$ and $n^{\frac{1}{6}}h_{k,n}$ is supposed to converge to a strictly positive finite limit γ_k, for $k = 1,2$.

The mode M_f of an unimodal density f in $\mathbb{R}^2$ is defined as the location of its unique maximum in all directions and the first derivatives of f satisfy $sign\dot{f}_k(x_k - M_{f,k}) = sign(M_{f,k} - x_k)$, for $k = 1,2$, at $x = (x_1, x_2)^t$ in $\mathbb{R}^2$. The mode is estimated by $\widehat{M}_{n,f} = M_{\widehat{f}_{n,h}}$. The density and its estimators are locally concave in neighbohoods of their local modes and, by the consistency of the kernel estimator, $\widehat{M}_{n,f}$ is therefore consistent. In a neighborhood of the mode, the first derivative of f satisfies $f^{(1)}(M_f) = 0$ and it has a Taylor expansion $f^{(1)}(x) = (x - M_f)^t f^{(2)}(M_f) + o(x - M_f)$. The estimator of the first derivative of the density satisfies $\widehat{f}_{n,h}^{(1)}(\widehat{M}_{f,n,h}) = 0$ and it has an expansion at $\widehat{M}_{f,n,h}$ similar to the expansion for a density in $\mathbb{R}$.

The process $U_{(1),n,h}$ with components $(nh_k^3 h_j)^{1/2}(\widehat{f}_{n,h,k}^{(1)} - f_k^{(1)})$, for $j \neq k$ in $\{1,2\}$, converges weakly in $\mathcal{I}_{X,h}$ to a Gaussian process with mean given in (A.3) and a non degenerate variance (A.4). The convergence rate of the kth component of $(\widehat{M}_{f,n,h} - M_f)$ is the same and by Lemma A.1, their optimal convergence rate is $n^{-\frac{1}{4}}$ as h_1 and h_2 are $O(n^{-\frac{1}{8}})$. Finally, let

$h_1 = \zeta_1 n^{-\frac{1}{8}}$ and $h_2 = \zeta_2 n^{-\frac{1}{8}}$

$$n^{-\frac{1}{4}}(\widehat{M}_{f,n,h} - M_f) = -f^{(2)-1}(M_f)\, n^{-\frac{1}{4}}(\widehat{f}_{n,h}^{(1)} - f^{(1)})(\widehat{M}_{f,n,h}) + o_p(1)$$

$$= -f^{(2)-1}(M_f)\, n^{-\frac{1}{4}}(\widehat{f}_{n,h}^{(1)} - f^{(1)})(M_f) + o_p(1).$$

The weak convergence of the $n^{\frac{1}{4}}(\widehat{f}_{n,h}^{(1)} - f^{(1)})$ implies the weak convergence of $n^{\frac{1}{4}}(\widehat{M}_{f,n,h} - M_f)$ to a centered Gaussian vector with variances proportional to the components of

$$\sigma_{M_f}^2 = f(M_f)\{f^{(2)}(M_f)\}^{-1} \int K^{(1)\otimes 2}(z)\, dz\, \{f^{(2)}(M_f)\}^{-1}.$$

For a density of $C^s(\mathbb{R}^2)$ and under the related conditions for a L^2-optimal bandwidth, the convergence rate of the estimator of the mode of the density is $n^{\frac{s}{2(s+1)}}$.

The derivative $f^{(1)}$ of a monotone density f has components with constant signs. A test of monotony for f can be performed with the sequence of variables $Z_{i,h}(x) = K_h^{(1)}(x - X_i)$, $i = 1, \ldots, n$. For every x in the interior of the interval $\mathcal{I}$, their mean and variance have the next expansions as n tends to infinity (Appendix A.5)

$$EZ_{i,h,k}(x) = \int_{\mathcal{I}_X} K_{h,k}^{(1)}(x - y)\, dF_X(y)$$

$$= f_{X,k}^{(1)}(x) + \frac{1}{2} m_{2K} h_k^2 f_{X,k}^{(3)}(x) + o(\|h\|_2^2),\ k \in \{1,2\}$$

and $Var Z_{i,h,k}(x) = \int \mathcal{I}_X K_h^{(1)2}(x - y)\, dF_X(y)$ is asymptotically equivalent to $(h_k^3 h_j)^{-1} f(x) \int K^{(1)2}(s)\, ds$, with $j \neq k$ in $\{1,2\}$. A test statistic for an increasing density in $\mathcal{I}$ is

$$S_n = \inf_{x \in \mathcal{I}} \widehat{f}_{n,h}^{(1)}(x),$$

it converges to $S = \inf_{x \in \mathcal{I}} f^{(1)}(x)$ which is positive under the hypothesis and strictly negative under the alternative K of a non increasing density. The test based on S_n has then the critical value zero, its risks have the limits

$$\lim_{n \to \infty} P_0(S_n < 0) = 0, \quad \lim_{n \to \infty} P_K(S_n < 0) = 1$$

and the asymptotic power of the test against local alternatives is zero.

The mean and the variance of the process

$$A_{n,h}(t) = \int_{]-\infty, t]} \widehat{f}_{n,h}(x)\, dx$$

have asymptotic expansions similar to (3.9) and (3.10) with the convergence rates of the estimators in $\mathbb{R}^2$

$$
\begin{aligned}
EA_{n,h}(t) - F(t) &= \int_{]-\infty,t]} \left\{ \int_{\mathbb{R}^2} K(u_1)K(u_2)f(x - hu)\,du - f(x) \right\} dx \\
&= \frac{1}{2} \int_{]-\infty,t]} \left[\int_{\mathbb{R}^2} K(u_1)K(u_2)\{h_1^2 f_1^{(2)}(t) + h_2^2 f_2^{(2)}(x) \right. \\
&\quad \left. + o(\|h\|_2^2)\}\,du \right] dx \\
&= \frac{m_{2K}}{2}\{h_1^2 F_1^{(1)}(t) + h_2^2 F_2^{(1)}(t)\} + o(\|h\|_2^2),
\end{aligned}
$$

$$
\begin{aligned}
Var A_{n,h}(t) &= \int_{]-\infty,t]} Var \widehat{f}_{n,h}(x)\,dx \\
&= (nh_1 h_2)^{-1}\kappa_2 F(t) + o((nh_1 h_2)^{-1})
\end{aligned}
$$

and $Cov\{A_{n,h}(t), A_{n,h}(s)\} = Var A_{n,h}(s \wedge t)$. In $C^s(\mathbb{R}^2)$, the expansion of the bias of $A_{n,h}(t)$ is a $O(\|h\|_s^s)$ and its variance has the same expansion as that for a density in $C^2(\mathbb{R}^2)$.

Lemma 4.2. *Under the condition of a kernel such that $m_{Kj} = 0$ for every $j < s$ and $m_{Ks} \neq 0$, the integral estimator $A_{n,h}(t) = \int_{]-\infty,t]} \widehat{f}_{n,h}(x)\,dx$ of a distribution function of $C^s(\mathbb{R}^2)$ converges to a monotone density with the optimal rate $O_p(n^{\frac{s}{2(s+1)}})$, as h_1 and h_2 are $O(n^{-\frac{1}{2(s+1)}})$.*

The isotonic estimator of a monotone decreasing density is defined by monotonization in $\mathbb{R}^2$ of the integrated kernel estimator, as

$$
f^*_{D,n} = \sup_{u_1 \leq x_1} \inf_{v_1 \geq x_1} \sup_{u_2 \leq x_2} \inf_{v_2 \geq x_2} \frac{1}{v_1 - u_1} \frac{1}{v_2 - u_2}\{A_{n,h}(v) - A_{n,h}(u)\}
$$

for a density decreasing in both components. For a density increasing with respect to the first component and decreasing with respect to the second component, it is

$$
f^*_{ID,n} = \inf_{v_1 \geq x_1} \sup_{u_1 \leq x_1} \sup_{u_2 \leq x_2} \inf_{v_2 \geq x_2} \frac{1}{v_1 - u_1} \frac{1}{v_2 - u_2}\{A_{n,h}(v) - A_{n,h}(u)\}
$$

and $f^*_{DI,n}$ defined in a similar way for a density decreasing with respect to the first component and increasing with respect to the second component. Finally

$$
f^*_{I,n} = \inf_{v_1 \geq x_1} \sup_{u_1 \leq x_1} \inf_{v_2 \geq x_2} \sup_{u_2 \leq x_2} \frac{1}{v_1 - u_1} \frac{1}{v_2 - u_2}\{A_{n,h}(v) - A_{n,h}(u)\}
$$

for a density increasing in both components.

Lemma 4.3. *Under the conditions of Lemma 4.2, the estimator $f^*_{n,h}$ of a monotone density f of $C^s(\mathbb{R}^2)$ converges to f with the optimal rate $O_p(n^{\frac{s}{2(s+1)}})$, as h_1 and h_2 are $O(n^{-\frac{1}{2(s+1)}})$.*

This rate is also the optimal convergence rate of the kernel estimator $\widehat{f}_{n,h}$ for a density of $C^s(\mathcal{I}_X)$, $s > 1$. A statistic for a test of a unimodal density f of $C^s(\mathcal{I}_X)$, $s > 1$, is defined as a weighted integrated squared mean

$$T_n = n^{\frac{s}{2(s+1)}} \left\{ \int_{]-\infty,\widehat{M}_{n,f}]\cap\mathcal{I}_{X,h}} (f^*_{n,h} - \widehat{f}_{n,h})^2(x)w(x)\,dx \right.$$

$$\left. + \int_{[\widehat{M}_{n,f},+\infty[\cap\mathcal{I}_{X,h}} (f^*_{n,h} - \widehat{f}_{n,h})^2(x)w(x)\,dx \right\},$$

where $f^*_{n,h}$ is increasing in the first integral and decreasing in the second one, under the conditions of Lemma 4.2. For every x in a compact sub-interval $\mathcal{I}_{X,h_n}$ of the support of the density f, let $B^*_{n,f}(x) = n^{\frac{s}{2(s+1)}}(f^*_{n,h} - f)(x)$ and $B_{f,n} = n^{\frac{s}{s+1}}(\widehat{f}_{n,h} - f)$. Let $b^* = \lim_{n\to\infty}(nh_1h_2)^{\frac{1}{2}}(Ef^*_{n,h} - f)$ and let $b = \lim_{n\to\infty}(nh_1h_2)^{\frac{1}{2}}(E\widehat{f}_{n,h} - f)$ be the asymptotic biases of the estimators of the density f. Under the condition on the bandwidths, the processes $B_{f,n}$ and $B^*_{n,m}$ converge weakly to biased processes $B + b$ and, respectively, $B^* + b^*$ where B and B^* are centered Gaussian processes. Therefore, the statistic T_n converges weakly to the variable $T = \int_{\mathcal{I}_{X,h}}\{B^*(x) - B(x) + b^*(x) - b(x)\}^2 w(x)\,dx$ and it tends in probability to infinity under fixed alternatives. With a bias correction, the statistic is

$$T_n = n^{\frac{s}{2(s+1)}} \left\{ \int_{]-\infty,\widehat{M}_{n,f}]\cap\mathcal{I}_{X,h}} (f^*_{n,h} - \widehat{f}_{n,h} + \widehat{b}_{f,n,h} - \widehat{b}^*_{f,n,h})^2(x)w(x)\,dx \right.$$

$$\left. + \int_{[\widehat{M}_{n,f},+\infty[\cap\mathcal{I}_{X,h}} (f^*_{n,h} - \widehat{f}_{n,h} + \widehat{b}_{f,n,h} - \widehat{b}^*_{f,n,h})^2(x)w(x)\,dx \right\}$$

and the limiting process is centered under the hypothesis H_0.

Proposition 4.10. *Under H_0, the statistic T_n converges weakly to the variable $T = \int_{\mathcal{I}_{X,h}}\{B^*(x) - B(x)\}^2 w(x)\,dx$. Under a local alternative of a nonmonotone density $f_n = f + n^{-\frac{s}{s+1}}g_n$ where f is a monotone function and $(g_n)_{n\geq 1}$ is a sequence of nonmonotone functions of $L^2(w(x)\,dx)$, converging uniformly to a function g as n tends to infinity, the statistic T_n converges weakly to $\int_{\mathcal{I}_{X,h}}\{B^*(x) - B(x) - g(x)\}^2 w(x)\,dx$.*

The proof is the same as the proof of Proposition 3.9. The statistic T_n should be normalized by an estimator of its asymptotic variance. Its asymptotic power under fixed alternatives tends to one and the test is consistent under the local alternatives of Proposition 4.10.

A test for convexity of a density f in an interval $\mathcal{I}$ may be considered as a test of the hypothesis $H_0 : f^{(2)} > 0$ against the alternative H_1 : there exist sub-intervals of $\mathcal{I}$ where $f^{(2)} \leq 0$. An estimator of the second derivative of the density is defined from the second order derivative of kernel K

$$\widehat{f}_{n,h}^{(2)}(x) = h^{-3}n^{-1} \sum_{i=1}^{n} K^{(2)}(h^{-1}(x - X_i)).$$

This estimator is uniformly consistent, its optimal bandwidth is given by Lemma A.2 for $s = 2$. The estimator of the second derivative of a smoother density has the following properties.

Lemma 4.4. *Under the condition of a kernel such that $m_{Kj} = 0$ for every $j < s$ and $m_{Ks} \neq 0$, the estimator $\widehat{f}_{n,h}^{(2)}$ of the second derivative of a monotone density f of $C^s(\mathbb{R}^2)$ converges to $f^{(2)}$ with the optimal rate $n^{\frac{s}{2(s+3)}}$, as h_1 and h_2 are $O(n^{-\frac{1}{2(s+3)}})$.*

Under the same conditions, the empirical process $B_{n,h}^{(2)} = n^{-\frac{s}{2(s+3)}}(\widehat{f}_{n,h}^{(2)} - f^{(2)})$ converges weakly to a biased Gaussian process in $\mathbb{R}^2$. Under the hypothesis of a concave function in $\mathbb{R}^2$, the centered statistic

$$T_{(2),n} = \sup_{x \in \mathcal{I}}\{\widehat{f}_{n,h}^{(2)}(x) - \widehat{b}_{f^{(2)},n,h}(x)\}$$

is negative and $T_{(2),n}$ is strictly positive under the alternative. A test based on $T_{(2),n}$ has the same properties as in $\mathbb{R}$.

Removing the assumption of a degree of differentiability greater than s, the hypothesis of a convex density in an interval $\mathcal{I}$ of $\mathbb{R}^2$ is expressed by

$$f(\alpha x + (1 - \alpha)y) \geq \alpha f(x) + (1 - \alpha)f(y)$$

for every α in $]0,1[$ and for all x and y in $\mathcal{I}$. Under the alternative, there exist α in $]0,1[$, x and y in $\mathbb{R}^2$ where the inequality is reversed. The statistic

$$S_n = \inf_{(x,y) \in \mathcal{I}^2} \inf_{\alpha \in]0,1[} n^{\frac{s}{2(s+1)}}\{\widehat{f}_{n,h}(\alpha x + (1-\alpha)y) - \alpha \widehat{f}_{n,h}(x) - (1-\alpha)\widehat{f}_{n,h}(y)\}$$

is positive under H_0, it is strictly negative under the alternative and it tends to $-\infty$. The critical value for the test based on S_n is therefore zero, and it satisfies the same properties as in $\mathbb{R}$.

4.7 Comparison of two regression curves

The comparison of two regression curves for variables sets (X_1, Y_1) and (X_2, Y_2) relies on tests of the hypothesis

$$H_0 : m_1(x) = E(Y_1|X_1 = x) = E(Y_2|X_2 = x) = m_2(x)$$

for every x in an interval $\mathcal{I}_X$ of the common support of the regression variables X_1 and X_2. The general alternative of two distinct curves on $\mathcal{I}_X$ is K : there exists a sub-interval of $\mathcal{I}_X$ where $E(Y_1|X_1 = x) \neq E(Y_2|X_2 = x)$ for every x of the sub-interval. Let $(X_{ij}, Y_{ij})_{j=1,\dots,n_i, i=1,\dots,I}$ be a sample of independent observations of the variables and let $N = n_1 + n_2$ be the total sample size such that $N^{-1}n_1$ converge to p in $]0, 1[$ as N tends to infinity. Condition 3.1 are assumed for both regression functions m_1 and m_2 and for kernel functions K_1 and K_2.

First, we consider real regression variables. The nonparametric estimators $\widehat{m}_{k,n,h}$ of $m_k(x) = E(Y_k|X_k = x)$, $k = 1, 2$, satisfy the properties presented in Section 3.7. For functions m_1 and m_2 of $C^s(\mathcal{I}_X)$, $s \geq 2$, let h_{k,n_k} be the L^2-optimal bandwidth for $\widehat{m}_{k,n_k,h_k}$ for $k = 1, 2$, and let h_n be the maximum of h_{1,n_1} and h_{2,n_2}. The processes

$$B_{k,n_k,h_k} = n_k^{\frac{s}{2s+1}} (\widehat{m}_{k,n_k,h_k} - m_k) 1_{\{\mathcal{I}_{X,h}\}}, k = 1, 2,$$

converge weakly to Gaussian processes $B_k = \sigma_{m_k} W_1 + \gamma_s b_{m_k}$ where W_1 is a centered Gaussian process on $\mathcal{I}_X$ with variance function 1 and covariance function zero, the constants are defined by the asymptotic bias and variance of the estimators, as in Section 3.7.

The whole sample of regression variables $(X_{1i_1}, X_{2i_2})_{i_1=1,\dots,n_1, i_2=1,\dots,n_2}$ has the density function $f_X = pf_{X,1} + (1-p)f_{X,2}$, where $f_{X,k}$ is the marginal density of X_k, for $k = 1, 2$. Under Conditions 1.1 and 3.1, its kernel estimator is

$$\widehat{f}_{N,h}(x) = \frac{n_1}{N}\widehat{f}_{1,n_1,h}(x) + \frac{n_2}{N}\widehat{f}_{2,n_2,h}(x),$$

with a common bandwidth h for both densities, for every x in $\mathcal{I}_{X,h}$. Let $\mu_k(x) = E(Y_k 1_{\{X_k=x\}})$ and $\mu(x) = p\mu_1(x) + (1-p)\mu_2(x)$, it is estimated by

$$\widehat{\mu}_{N,h}(x) = \frac{1}{Nh}\left\{ \sum_{i=1}^{n_1} Y_{1i} K_h(x - X_{1i}) + \sum_{i=1}^{n_2} Y_{2i} K_h(x - X_{2i}) \right\}.$$

The hypothesis H_0 is equivalent to the equality

$$m_1 = m_2 = m \equiv \frac{p\mu_1 + (1-p)\mu_2}{pf_{X,1} + (1-p)f_{X,2}}. \tag{4.13}$$

The global curve m defined by 4.13 is estimated by $\widehat{m}_N = \widehat{f}_{N,h}^{-1}\widehat{\mu}_{N,h}$ under H_0 and under the alternative. The differences $m - m_1$ and $m - m_2$ are

$$m - m_1 = -(1-p)\frac{\mu_1 f_{X,2} - \mu_2 f_{X,1}}{f_X f_{X,1}},$$

$$m - m_2 = p\frac{\mu_1 f_{X,2} - \mu_2 f_{X,1}}{f_X f_{X,2}}.$$

The hypothesis H_0 of equal curves m_1 and m_2 is therefore also equivalent to the equality of m_1 or m_2 to m. Test statistics can be defined for the different expressions of H_0, we consider only the equality of m_1 and m_2 and they are easily modified for the other forms of the hypothesis H_0.

A statistic for a Kolmogorov-Smirnov test of H_0 relies on the difference of the estimators of the regression functions. The processes B_{1,n_1,h_1} and B_{2,n_2,h_2} have the asymptotic bias $\gamma_s b_{m,k} = \lim_{n\to\infty} h_{n_k}^{-s}(E\widehat{m}_{k,n_k,h} - m_k)$, for $k = 1, 2$. Under H_0, the asymptotic bias of the processes B_{1,n_1,h_1} and B_{2,n_2,h_2} are equal if the bandwidths are equal and if the sub-sample sizes are equal. Let b_m be the common bias under H_0, with equal bandwidths, and let $\widehat{b}_{m,n,h}$ be a consistent estimator of the function b_m calculated using the kernel estimators of the derivatives of the regression curve and of the density of X. For $k = 1, 2$, let γ_k be the bandwidth constant of Condition 1.1.4. Under H_0, the process

$$\left(\frac{n_1 n_2}{N}\right)^{\frac{s}{2s+1}}\{\widehat{m}_{1,n_1,h_1}(x) - \widehat{m}_{2,n_2,h_2})(x)\}$$

is generally noncentered and we consider the statistic

$$T_{1N} = \sup_{x \in \mathcal{I}_{X,h}} \left|\left(\frac{n_1 n_2}{N}\right)^{\frac{s}{2s+1}}\{\widehat{m}_{1,n_1,h_1}(x) - \widehat{m}_{2,n_2,h_2}(x)\}\right.$$

$$\left. -\left\{\left(\frac{n_2}{N}\right)^{\frac{s}{2s+1}}\gamma_1\widehat{b}_{m_1,n_1,h_1}(x) - \left(\frac{n_1}{N}\right)^{\frac{s}{2s+1}}\gamma_2\widehat{b}_{m_2,n_2,h_2}(x)\right\}\right| \quad (4.14)$$

this is the supremum of a centered process with bandwidths converging with the optimal rates.

Proposition 4.11. *Under Conditions 1.1 and 3.1 for each regression curve, the statistic T_{1N} converges under H_0 to the supremum of the centered Gaussian process*

$$\sup_{\mathcal{I}_{X,h}} |(1-p)^{\frac{s}{2s+1}}B_1 - p^{\frac{s}{2s+1}}B_2|.$$

Under the alternative, it tends to infinity.

This is a consequence of the asymptotic behavior of the processes B_{k,n_k,h_k}, $k = 1,2$. Their asymptotic biases under H_0 are equal if the bandwidths, the kernels and the sub-sample sizes are respectively equal for both sub-samples. With unequal sub-sample sizes, the asymptotic biases of B_{1,n_1,h_1} and B_{2,n_2,h_2} differ and the bias correction in (4.14) is only valid under the hypothesis. If the bandwidths converge with faster rates than the optimal rates, the constants γ_1 and γ_2 are zero and the difference of the estimators of the regression curves are centered under the hypothesis.

Proposition 4.12. *Let* $(K_n)_{n\geq 1}$ *be a sequence of local alternatives defined by regression functions* $m_{k,n_k} = m + n_k^{-\frac{s}{2s+1}} r_{k,n_k}$ *such that* $(r_{k,n_k})_{n_k\geq 1}$ *converges uniformly to a function* r_k, $k = 1,2$. *Under Conditions 1.1 and 3.1, the statistic* T_{1N} *converges weakly under* $(K_n)_{n\geq 1}$ *to the supremum of the noncentered Gaussian process*

$$T = \sup_{\mathcal{I}_{X,h}} |(1-p)^{\frac{s}{2s+1}} B_1 - p^{\frac{s}{2s+1}} B_2$$
$$- \{\gamma_1(1-p)^{\frac{s}{2s+1}}(b_{r_1,1} - b_m) - p^{\frac{s}{2s+1}}\gamma_2(b_{r_2,2} - b_m)\}|$$

where $b_{r_k,k}$ *is calculated by replacing* m_k *by* r_k *in the expression of the bias of the regression curves.*

A Cramer-von Mises statistic is an integrated squared difference between the estimators of m_1 and m_2

$$T_{2N} = \left(\frac{n_1 n_2}{N}\right)^{\frac{2s}{2s+1}} \int_{\mathcal{I}_{X,h}} |w_{1N}(x_1)\widehat{m}_{1,n_1,h_1}(x_1) \qquad (4.15)$$
$$- w_{2N}(x_2)\widehat{m}_{2,n_2,h_2})(x_2)|^2 \, d\widehat{F}_{X,N}(x_1,x_2),$$

where the sequences of weighting functions converge uniformly to functions w_k which may be the inverses of estimators of the variances of Y_k conditionally on X_k, $k = 1,2$. With the optimal convergence rates of the bandwidth, a bias correction is performed as in (4.14) using the estimator of the biases under H_0, the corrected statistic is centered under the hypothesis and it diverges under fixed alternatives.

With d-dimensional regression variables, the estimators are unchanged, the convergence rates are modified according to the dimension of the regressors, the rate $n^{\frac{s}{2s+1}}$ of the kernel estimators of the density of X and of the regression function m is replaced by $n^{\frac{s}{2(s+d)-1}}$ and the normalization of the test statistics is modified in accordance.

A regression function in $\mathbb{R}^2$ is defined by a two-dimensional covariate X as the conditional mean $m(x) = E(Y|X = x)$, with x in a subset of $\mathbb{R}^2$. The nonparametric estimator of the density in $\mathbb{R}^2$ is extended to regression estimation and the convergence rates of the estimators are the same. Goodness of fit tests for nonparametric or semi-parametric models of regressions in $\mathbb{R}^2$ are defined by the same statistics as in $\mathbb{R}$ with modified convergence rates and the results are similar.

Tests of symmetry, monotony and unimodality are defined from these estimators like for densities in $\mathbb{R}^2$ and their properties are proved by the same arguments.

4.8   Tests based on observations by intervals

Let $X = (X_1, X_2)$ be a two-dimensional real random variable with distribution function F in $\mathbb{R}^2$ and marginal distributions F_1 and F_2. With observations cumulated by intervals, the values of the variable are not observed, but only the numbers $N_{ij} = n \int_{A_i \times B_j} d\widehat{F}_n$ of observations in sub-intervals $A_i \times B_j$ of a bounded domain $A \times B$ of $\mathbb{R}^2$, for $i = 1, \ldots, k$ and $j = 1, \ldots, l$. The counting variable $(N_{ij})_{i=1,\ldots,k,j=1,\ldots,l}$ has a multinomial distribution with probabilities $p_{ij} = \int_{A_i \times B_j} dF$, for $i = 1, \ldots, k$ and $j = 1, \ldots, l$. In tests of independence, the hypothesis H_0 is the factorization $p_{ij} = p_{1i}p_{2j}$, for $i = 1, \ldots, k$, $j = 1, \ldots, l$.

In tests of homogeneity for two independent sub-samples of a real variable X observed by intervals and with respective distributions functions F_1 and F_2, the number of variables with values in sub-intervals A_i of the kth sub-sample is $N_{ki} = n \int_{A_i} d\widehat{F}_{k,n}$, for $i = 1, \ldots, I$ and $k = 1, 2$. The counting variable $(N_{ki})_{i=1,\ldots,I,k=1,2}$ has a multinomial distribution with probabilities $(p_{ki})_{i=1,\ldots,I,k=1,2}$. The hypothesis H_0 of homogeneity of the distributions of the sub-samples is the hypothesis of $p_{1i} = p_{2i}$, for $i = 1, \ldots, I$.

4.8.1   *Test of independence*

Let $\widehat{p}_{ij,n} = n^{-1}N_{ij}$ be the estimators of the joint probabilities and let $\widehat{p}_{i.,n} = n_1^{-1}N_{1i} = \sum_{j=1}^{n_2} \widehat{p}_{ij,n}$ and $\widehat{p}_{.j,n} = n_2^{-1}N_{2j} = \sum_{i=1}^{n_1} \widehat{p}_{ij,n}$ be the estimators of the marginal probabilities, tests of independence are based on the vector of the differences $\widehat{p}_{ij,n} - \widehat{p}_{i.,n}\widehat{p}_{.j,n}$, $i = 1, \ldots, k$, $j = 1, \ldots, l$. Its components have the means $p_{ij} - p_{1i}p_{2j}$ which equal zero under the hypothesis of independence and they sum up to zero under all alternatives.

Their variances are

$$v_{ij} = \frac{1}{n}[p_{ij}(1-p_{ij})+p_{1i}p_{2j}(1-p_{1i}p_{2j})-2\{n^{-3}E(N_{ij}N_{1i}N_{2j})-p_{ij}p_{1i}p_{2j}\}],$$

where $E(N_{ij}N_{1i}N_{2j}) = p_{ij}^2 + n^{-1}(1-n^{-1})p_{ij}(p_{ij}+p_{1i}+p_{2j}-3)+n^{-2}p_{ij}$
hence

$$\begin{aligned} v_{ij} &= 2p_{ij}(p_{1i}p_{2j}-p_{ij})+n^{-1}\{p_{ij}(1-p_{ij})+p_{1i}p_{2j}(1-p_{1i}p_{2j}) \\ &\quad - 2p_{ij}(p_{ij}+p_{1i}+p_{2j}-3)\}+o(n^{-1}) \\ &= 2p_{ij}(p_{1i}p_{2j}-p_{ij})+n^{-1}\{p_{ij}(7-3p_{ij}-2p_{1i}-2p_{2j}) \\ &\quad + p_{1i}p_{2j}(1-p_{1i}p_{2j})\}+o(n^{-1}). \end{aligned}$$

The covariances of $\widehat{p}_{ij,n}$ and $\widehat{p}_{i'j',n}$ are zero for all $i \neq i' < k$ and $j \neq j' < l$ and the covariance of $\widehat{p}_{ij,n}$ and $\widehat{p}_{k,l,n}$ is $-Var\widehat{p}_{ij,n}$ for all $i < k$ and $j < l$. Under the hypothesis, the variance of $\widehat{p}_{ij,n}$ equals $v_{ij,0} = 2n^{-1}p_{1i}p_{2j}(4-2p_{1i}p_{2j}-p_{1i}-p_{2j})+o(n^{-1})$. The random matrix $n^{\frac{1}{2}}(N_{ij}-N_{1i}N_{2j})_{i=1,\dots,k,j=1,\dots,l}$ converges weakly to a centered Gaussian matrix with dependent elements having covariances defined by the variances $v_{ij,0}$. Its components are linearly dependent since $\sum_{i=1,\dots,k}(N_{ij}-N_{1i}N_{2j})=0$ and $\sum_{j=1,\dots,l}(N_{ij}-N_{1i}N_{2j})=0$. The variances are estimated empirically under H_0 by $\widehat{v}_{ij,0} = O(n^{-1})$ and the matrix of the normalized observations

$$W_n = \left\{\widehat{v}_{ij,0}^{-\frac{1}{2}}\Big(\frac{N_{ij}}{n}-\frac{N_{1i}}{n}\frac{N_{2j}}{n}\Big)\right\}_{i=1,\dots,k,j=1,\dots,l}$$

converges weakly to a $k \times l$-dimensional matrix of linearly dependent normal variables.

Proposition 4.13. *The test statistic defined as*

$$T_n = \sum_{i=1}^{k-1}\sum_{j=1}^{l-1}\widehat{v}_{ij,0}^{-1}\Big(\frac{N_{ij}}{n}-\frac{N_{1i}}{n}\frac{N_{2j}}{n}\Big)^2$$

converges weakly to a $\chi^2_{(k-1)(l-1)}$ variable under H_0. Under fixed alternatives, T_n diverges.

Under an alternative K, there exist integers $i_K \leq k$ and $j_K \leq l$ such that $p_{i_K j_K}$ differs from $p_{1i_K}p_{2j_K}$. The variances v_{ij} are equivalent to $2p_{ij}(p_{1i}p_{2j}-p_{ij})$ which differs from zero at (i_K,j_K) and the statistic T_n diverges since its normalization does not have the order $n^{-\frac{1}{2}}$. The test based on the statistic T_n is therefore consistent with an asymptotic power 1 for every fixed alternatives.

Let K_n be a sequence of local alternatives with probabilities $p_{ij,n}$ converging to $p_{ij} = p_{1i}p_{2j}$ and such that $r_{ij,n} = n^{-\frac{1}{2}}(p_{1i,n}p_{2j,n} - p_{ij,n})$ converges to a limit r_{ij}, for all $i = 1, \ldots, k-1$ and $j = 1, \ldots, l-1$. Under K_n, the variances v_{ij} are $O(n^{-1})$ and they are approximated by

$$v_{ij,K} = n^{-1}\{p_{ij}(7 - 3p_{ij} - 2p_{1i} - 2p_{2j} + 2r_{ij,n}) + p_{1i}p_{2j}(1 - p_{1i}p_{2j})\}.$$

Let $(Z_{ij})_{i=1,\ldots,k-1,j=1,\ldots,l-1}$ be a matrix of centered Gaussian variables with variances 1. The limiting distribution of the statistic under K_n is obtained by rescaling the variables of T_n.

Proposition 4.14. *The statistic T_n converges weakly under K_n to a variable $T(r) = \sum_{i=1}^{k-1} \sum_{j=1}^{l-1} v_{ij,K} v_{ij,0}^{-1} (Z_{ij} + r_{ij})^2.$*

4.8.2 Test of homogeneity

Tests of homogeneity for two independent sub-samples of respective sizes n_1 and n_2 of a variable X observed by intervals are based on the vector of the differences $n_1^{-1}N_{1i} - n_2^{-1}N_{2i}$, $i = 1, \ldots, k$. Under the hypothesis H_0, its components have the means $p_{1i} - p_{2i}$ which equal zero and the variances

$$v_i = \frac{1}{n_1}p_{1i}(1 - p_{1i}) + \frac{1}{n_2}p_{2i}(1 - p_{2i})$$

have the same expression under H_0 and under all alternatives. Let $\widehat{v}_{i,n}$ be the empirical estimator of v_i, the vector of the normalized observations

$$W_n = \{\widehat{v}_{i,n}^{-\frac{1}{2}}(n_1^{-1}N_{1i} - n_2^{-1}N_{2i})\}_{i=1,\ldots,k-1}$$

converges weakly to a k-dimensional vector of independent normal variables $(Z_i)_{i=1,\ldots,k-1}$.

Proposition 4.15. *The test statistic defined as*

$$T_n = \sum_{i=1}^{k-1} \widehat{v}_{i,n}^{-1}(n_1^{-1}N_{1i} - n_2^{-1}N_{2i})^2$$

converges weakly to a χ_{k-1}^2 variable under H_0. Under fixed alternatives, T_n diverges.

Under fixed alternatives, at least one component of the vector $(p_{1i} - p_{2i})_{i=1,\ldots,k-1}$ is nonzero and the mean of the vector W_n diverges. The test based on the statistic T_n is therefore consistent and its asymptotic power is 1 for every fixed alternatives.

Let K_n be a sequence of local alternatives with probabilities $p_{1i,n}$ and $p_{2i,n}$ converging to p_{1i} and, respectively, p_{2i} and such that the differences $r_{ji,n} = (n_1 n_2)^{\frac{1}{2}} n^{-\frac{1}{2}} (p_{ji,n} - p_i)$ converge to limits r_{ji} different from zero, for all $i = 1, \ldots, k-1$ and $j = 1, 2$. Under K_n, the means of the variables $n_1^{-1} N_{1i} - n_2^{-1} N_{2i}$ are $p_{1i,n} - p_{2i,n} = (n_1 n_2)^{-\frac{1}{2}} n^{\frac{1}{2}} (r_{1i} - r_{2i})$ and they converge to a nonnull value. Their variances are asymptotically equivalent to their variances under H_0 and they are proportional to $n(n_1 n_2)^{-1}$.

Proposition 4.16. *The statistic T_n converges weakly under K_n to a variable $T(r) = \sum_{i=1,\ldots,k-1} (Z_i + (r_{1i} - r_{2i})\{p_i(1 - p_i)\}^{-\frac{1}{2}})^2$.*

4.8.3 *Comparison of two regression curves*

Let (X_1, Y_1) and (X_2, Y_2) be two independent sets of variables defined by the curves $m_j(x) = E(Y_j | X_j = x)$ for x in a subset $\mathcal{I}_X$ of $\mathbb{R}^d$. A test of the hypothesis H_0 of identical regression curves m_1 and m_2 in a sub-interval $\mathcal{I}$ of the support of X_1 and X_2 is performed from observations of the variables X_1 and X_2 cumulated in the sub-intervals of a partition $(A_i)_{i=1,\ldots,k}$ of $\mathcal{I}$. The observations of the regression variables X_j by intervals are those of the indicator of X_j in the set A_i, $\delta_{i,j} = 1_{\{X_j \in A_i\}}$, and the variables Y_j are continuously observed in $m(\mathcal{I})$.

The parameters of the model are $m_{i,j} = E(Y_j | X_j \in A_i)$, for $i = 1, \ldots, k$ and $j = 1, 2$. Their empirical estimators calculated from two independent samples of sizes n_j are defined by (3.26)

$$\widehat{m}_{i,j,n_j} = \frac{\sum_{l=1}^{n_j} Y_{l,j} \delta_{l,i}}{N_{i,j}}, \ i = 1, \ldots, k, \ j = 1, 2.$$

Let $\widehat{\mu}_{i,j,n} = n_j^{-1} \sum_{l=1}^{n_j} Y_{l,j} \delta_{l,i}$ denote its numerator, its denominator is $\widehat{p}_{i,j,n_j} = n_j^{-1} N_{i,j}$ and their means are $\mu_{i,j} = E(Y_j 1_{\{X_j \in A_i\}})$ and $p_{i,j}$, respectively. The asymptotic behavior of the estimators $\widehat{m}_{i,j,n_j}$ is given by Proposition 3.19. The variable $\{n_j^{1/2}(\widehat{m}_{i,j,n_j} - m_{i,j})\}_{i=1,\ldots,k}$ converges weakly to a centered Gaussian vector having a diagonal variance matrix with components $v_{i,j}$ given by (3.28)

$$v_{i,j} = p_{i,j}^{-1}[E\{(Y_j - m_{i,j})^2 | X_j \in A_i\} - m_{i,j}^2(1 - p_{i,j})], \ i = 1, \ldots, k.$$

Empirical estimators $\widehat{v}_{i,j,n_j}$ of the variances $v_{i,j}$ are defined by (3.29) they are $n_j^{\frac{1}{2}}$-consistent, for $j = 1, 2$.

A goodness of fit test statistic for the hypothesis $H_0 : m_1(A_i) = m_2(A_i)$ for every $i = 1, \ldots, k$ is defined as the normalized squared $l^2(\mathbb{R}^k)$-distance between the vectors with components $m_1(A_i)$ $m_2(A_i)$. It is empirically estimated by

$$S_{n_1,n_2} = \sum_{i=1}^{k} \frac{(\widehat{m}_{i,1,n} - \widehat{m}_{i,2,n})^2}{n_1^{-1}\widehat{v}_{i,1,n_1} + n_2^{-1}\widehat{v}_{i,2,n_2}}.$$

The mean of $\widehat{m}_{i,1,n} - \widehat{m}_{i,2,n}$ is zero for every $i = 1, \ldots, k$ under H_0 and there exists at least an interger $i_K \leq k$ such that they differ for $i = i_K$ under an alternative K.

Proposition 4.17. *The statistic S_{n_1,n_2} converges weakly under H_0 to a χ_k^2 distribution as n_1 and n_2 tend to infinity. Under fixed alternatives, it diverges.*

By definition, the vector of components the variables $\{n_1^{-1}\widehat{v}_{i,1,n_1} + n_2^{-1}\widehat{v}_{i,2,n_2}\}^{-\frac{1}{2}}(\widehat{m}_{i,1,n} - \widehat{m}_{i,2,n})^2$, $i = 1, \ldots, k$, converges under H_0 to a vector $(X_1, \ldots, X_k)$ of k independent normal variables. Under alternatives, at least a component has a nonzero mean and it tends to infinity by the normalization.

Let K_n be a sequence of local alternatives with conditional mean values $m_{i,1,n_1}$ for Y_1 and $m_{i,2,n_2}$ for Y_2, such that $m_{i,1,n_1}$ and $m_{i,2,n_2}$ converge to the same limit m_i and $r_{i,j,n_j} = (n_1 n_2)^{\frac{1}{2}} n^{-\frac{1}{2}}(m_{i,j,n_j} - m_i)$ converges to a limit $r_{i,j}$, for all $i = 1, \ldots, k$ and $j = 1, 2$. The variances $v_{ji,n}$ is such that all $n_j^{-1} v_{ji,n}$ converge to the same limit v_i as n tends to infinity.

Proposition 4.18. *Under K_n, the statistic S_{n_1,n_2} converges weakly to a noncentered χ_k^2 variable $S(r) = \sum_{i=1}^{k}\{X_{i,k} + v_i^{-\frac{1}{2}}(r_{i,1} - r_{i,2})\}^2$.*

4.9   Exercises

4.9.1. Determine the limiting distribution of the Cramer-von Mises type statistic (4.3) for the independence in a semi-parametric bivariate model with distribution function F such that $\bar{F}_\theta(x) = \bar{F}_1(x_1)\bar{F}_2(x_2)\psi_\theta(x)$, defined for all $x = (x_1, x_2)$ in $\mathbb{R}^2$ by a parametric function ψ_θ of $C^2(\Theta)$, where Θ is a bounded open subset of R^d.

Hints. The parameter θ can be estimated by maximum of likelihood or by minimization of the L^2-distance between the empirical estimator $\widehat{F}_n$ and F_θ where the marginal distributions are replaced by their nonparametric estimators, as in Section 4.3. The asymptotic variance of the estimator of θ is calculated from the derivatives of ψ_θ. Proposition 4.8 is modified according to the semi-parametric model.

4.9.2. Express the log-likelihood statistic of independence in the bivariate exponential model with the Marshall-Olkin distribution function F such that $\bar{F}(x) = e^{-\theta_1 x_1} e^{-\theta_2 x_2} e^{-\theta_3 (x_1 \vee x_2)}$, for every $x = (x_1, x_2)$ in $\mathbb{R}_+^2$, with parameters $\theta_k > 0$, $k = 1, 2, 3$.

Hints. The parameters are estimated by maximum likelihood, noting that the marginal distributions are exponential with respective parameters θ_1 and θ_2, only the third parameter can be estimated from the bivariate distribution. The asymptotic properties of the estimators are standard. The independence is equivalent to $\theta_3 = +\infty$ and a test is deduced from the asymptotic properties of the estimator of this parameter. A test can also be deduced from the previous exercise.

4.9.3. Define the log-likelihood test statistic of independence in the parametric exponential and power archimedian models (1.4) defined with the functions

$$p_\lambda(y) = e^{\lambda y}, \lambda > 0, y < 0,$$
$$p_\theta(y) = y^\theta, \theta > 0, y \in [0, 1].$$

From an expansion of the statistic in terms of the marginal empirical processes and of the parameter estimator, determine their asymptotic distribution, their asymptotic local power and their relative efficiency.

Hints. Follow the same proofs.

4.9.4. Determine the asymptotic distributions of the process $W_{m,n}$ defined by (4.11) in parametric models for both samples and in a model where the distribution function of the first sample is known and the second one is parametric. Compare the power of the test defined from this statistic to the power of the test defined by (4.11).

Hints. This is similar to a goodness of fit statistic for one sample, except that the parameter is estimated by minimization of $d_n(\theta)$ as in the semi-parametric test of Proposition 4.9.

4.9.5. Determine the asymptotic distribution of a statistic of homogeneity defined as the L^2-distance of the kernel estimator of a density and a parametric density with a parameter estimated by maximum likelihood, when the alternative is determined by a parametric model.

Hints. The asymptotic distribution of the maximum likelihood estimators of the parameters are given in Section 2.2, they apply to the semi-parametric process $W_{n,m}$ defined as in Proposition 4.9.

4.9.6. Generalize the goodness of fit tests for a parametric family of Section 3.8.1 to the density function of a two-dimensional variable X observed cumulatively on rectangles.

Hints. Let $(A_{lj})_{l=1,\cdots,k_1, j=1,\cdots,k_2}$ be a partition of the observation domain in rectangles. For a n-sample, the observations are the discrete counting variables N_{lj} of the number of items in A_{lj}, they have a multinomial distribution with parameters p_{lj}. Under the independence H_0, $p_{lj} = p_{lj}(\theta)$ and the parameters are estimated by $\widehat{p}_{lj,n} = n^{-1}N_{lj}$, $\widehat{\theta}_n = \arg\min_{\theta \in \Theta} d_n(\theta)$, where

$$d_n(\theta) = \sum_{l=1,\cdots,k_1} \sum_{j=1,\cdots,k_2} \frac{\{\widehat{p}_{lj,n} - p_{lj}(\theta)\}^2}{p_{lj}(\theta)\{1 - p_{lj}(\theta)\}},$$

its asymptotic properties are determined as a minimum distance estimator. The test statistic is $d_n(\widehat{\theta}_n)$ and by a Taylor expansion of $p_{lj}(\widehat{\theta}_n)$ at the parameter value, it has an asymptotic $\chi^2_{(k_1-1)(k_2-1)}$ distribution under H_0. Under a local hypothesis, its asymptotic mean modifies this distribution and the power of the test is deduced.

4.9.7. Generalize the goodness of fit tests of Section 3.8.2, to a regression function for a bidimensional response variable $Y = (Y_1, Y_2)$ observed by intervals.

Hints. Let $(A_{lj})_{l=1,\cdots,k_1, j=1,\cdots,k_2}$ be a partition of the observation domain of X in rectangles and let $(B_{lj})_{l=1,\cdots,k_1, j=1,\cdots,k_2}$ be a partition of the observation domain of Y. A parametric model is defined as $m_{lj}(x; \theta) = P_\theta(Y \in B_{lj} | X \in A_{lj})$, with θ in an open convex subset Θ of $\mathbb{R}^d$. The parameter value under H_0 is estimated by minimum of the distance d_n defined for (X, Y). The asymptotic properties of the estimator of the parameter and of $d_n(\widehat{\theta}_n)$ are deduced. Tests statistics are defined by a L^2-distance of the matrix functions $\widehat{m}(x; \theta)$ and $m(x; \widehat{\theta}_n)$.

Chapter 5

Multi-dimensional tests

5.1 Introduction

The main tests of the previous section are generalized to k sub-populations and statistics have the form of $l^2(\mathbb{R}^k)$-distances, generally their limit for large samples are χ^2 distributions. The goodness of fit test for k populations and the comparison of k populations by χ^2 tests were introduced by Pearson (1900) and Cochran (1947), they are parametric tests for cumulated observation and they generalize the Student or the χ_1^2 tests for the comparison of two populations. This chapter presents nonparametric tests statistics for the hypotheses defined in the previous chapter to k sub-populations. The degree of freedom of the limiting distribution is generally smaller than the dimension of the observed variables due to the linear relationship of their components. Cochran's theorem applies and the degrees of freedom are directly established in each case.

In tests of independence or equality of the distributions of sub-samples, the test for k components can be used recursively in a decreasing algorithm to determine which components of the observed variable are dependent or have equal distribution functions. In tests of homogeneity, when a test for k sub-samples reject the null hypothesis, the sub-samples which differ from the kth can be determined step by step in a decreasing sequence of tests of sub-hypotheses. Let H_{0j} be the hypothesis similar to the H_0 for the sub-samples indexed by $\{1, \ldots, k\} \setminus \{j\}$, for $j = 1, \ldots, k$, then $k(k-2)$ tests of the hypotheses H_{0j} are performed with $(k-2)$ statistics of comparison of the samples indexed by $\{1, \ldots, k\} \setminus \{j\}$ to a reference sample which is not necessarily the kth sample. This algorithm is repeated for all tests performed with m sub-samples, m in $\{2, \ldots, k-1\}$.

Under sampling by intervals, the tests statistics for the integral of the distribution functions in an interval or the mean of a random variable conditionally on a regression function in an interval have asymptotic χ^2 distributions.

5.2   Tests of independence

Let $X = (X_1, \ldots, X_k)$ be a k-dimensional variable with distribution function $F(x) = P(X_1 \leq x_1, \ldots, X_k \leq x_k)$ in $\mathbb{R}^k$ and marginal distributions F_j, $j = 1, \ldots, k$. Under hypothesis H_0 of independence of components of X, the joint distribution function factorizes as $F = \prod_{j=1,\ldots,k} F_j$. The joint and marginal distribution functions are estimated empirically from a n-sample. Tests of H_0 based on the empirical distribution functions $\widehat{F}_n$ and the marginal estimators $\widehat{F}_{j,n}$ are the Kolmogorov-Smirnov test with statistic

$$S_n = \sup_{x=(x_1,\ldots,x_k)\in\mathbb{R}^k} \sqrt{n}\left|\widehat{F}_n(x) - \prod_{j=1}^{k} \widehat{F}_{j,n}(x_j)\right| \tag{5.1}$$

the Cramer-von Mises test defined by the statistic

$$T_n = \int_{\mathbb{R}^k} n\left\{\widehat{F}_n(x) - \prod_{j=1}^{k} \widehat{F}_{j,n}(x_j)\right\}^2 d\widehat{F}_n(x). \tag{5.2}$$

and their transform by the marginal quantiles which defines the dependence function of the k components of X. Let $\nu_n = \sqrt{n}(\widehat{F}_n - F)$ be the joint empirical process in $\mathbb{R}^k$ and let $\nu_{j,n}$ be the marginal empirical processes, their limits are the Wiener processes W and W_j, $j = 1, \ldots, k$. Under the independence, the empirical process is

$$\nu_n = n^{\frac{1}{2}}\left(\prod_{j=1}^{k} \widehat{F}_{n_j} - \prod_{j=1}^{k} F_j\right).$$

By the expansion

$$\prod_{j=1}^{k} A_j - \prod_{j=1}^{k} a_j = \prod_{j=1}^{k}(A_j - a_j) - \sum_{j=1}^{k} a_j^{-1}(A_j - a_j)\prod_{l=1}^{k} a_l + o(1)$$

for all series $(A_j)_{j=1,\ldots,k}$ and $(a_j)_{j=1,\ldots,k}$ such that $A_j - a_j = o(1)$, it follows that

$$\prod_{j=1}^{k} \nu_{j,n}(x_j) = n^{\frac{k-1}{2}}\left\{\nu_n(x) + \sum_{j=1}^{k} \nu_{j,n}(x_j)\prod_{m\neq j,m=1}^{k} F_m(x_m)\right\} + o_p(1)$$

Under H_0 and all alternatives, the difference of the empirical processes

$$G_n(x) = \nu_n(x) - n^{-\frac{k-1}{2}} \prod_{j=1}^{k} \nu_{j,n}(x_j)$$

has the expansion

$$G_n(x) = n^{\frac{1}{2}} \Big\{ \widehat{F}_n(x) - \prod_{j=1}^{k} \widehat{F}_{j,n}(x_j) + \sum_{j=1}^{k} (\widehat{F}_{j,n} - F_j)(x_j) \prod_{l \neq k, l=1}^{k} F_l(x_l)$$

$$+ \prod_{j=1}^{k} F_j(x_j) - F(x) \Big\} + o_p(1). \tag{5.3}$$

Proposition 5.1. *Under H_0, the process $Z_n = n^{\frac{1}{2}}(\widehat{F}_n - \prod_{j=1,\ldots,k} \widehat{F}_{j,n})$ converges weakly to $Z = W - \sum_{j=1}^{k} W_j \prod_{l \neq k, l=1}^{k} F_l = W - F \sum_{j=1}^{k} F_j^{-1} W_j$. Under a local alternative with a joint distribution function $F_n = \prod_{j=1}^{k} F_{j,n} + n^{-\frac{1}{2}} H_n$, where $(H_n)_n$ is a sequence of strictly positive functions H_n converging uniformly to a function $H > 0$ in $\mathbb{R}^k$, the process Z_n converges weakly to $H + Z$.*

Proof. The limiting distribution of Z_n under H_0 is a straightforward consequence of the expansion (5.3) and of the asymptotic equivalence of the processes G_n and ν_n. Under the local alternative, the empirical processes are centered at F_n and $F_{j,n}$ respectively and the above expansion is rewritten as

$$G_{nn}(x) = n^{\frac{1}{2}} \Big\{ \widehat{F}_n(x) - \prod_{j=1}^{k} \widehat{F}_{j,n}(x_j) + \prod_{j=1}^{k} F_{j,n}(x_j) - F_n(x)$$

$$+ \sum_{j=1}^{k} (\widehat{F}_{j,n} - F_{j,n})(x_j) \prod_{l \neq k, l=1}^{k} F_{l,n}(x_l) \Big\} + o_p(1)$$

where the mean of $G_{nn}(x)$ is $n^{\frac{1}{2}} \{ \prod_{j=1}^{k} F_{j,n}(x_j) - F_n(x) \} = -H_n(x)$ and the process G_{nn} converges weakly to $Z - H$. It follows that under the local alternative

$$\sqrt{n} \Big\{ \widehat{F}_n(x) - \prod_{j=1}^{k} \widehat{F}_{j,n}(x_j) \Big\} = G_{nn}(x) + H_n(x)$$

$$- \sqrt{n} \sum_{j=1}^{k} (\widehat{F}_{j,n} - F_{j,n})(x_j) \prod_{l \neq k, l=1}^{k} F_{l,n}(x_l)$$

converges weakly to $W - \sum_{j=1}^{k} W_j \prod_{l \neq k, l=1}^{k} F_l + H$. $\qquad \square$

The statistics are still expressed with the product of the ranks of the variables in each sub-sample. Under H_0, the statistic S_n converges weakly to $S = \sup_{\mathbb{R}^k} |W| = \sup_{[0,1]^k} |Z \circ (F_1^{-1}, \ldots, F_k^{-1})|$ and T_n converges weakly to $T = \int_{\mathbb{R}^k} Z^2 \prod_{j=1}^{k} dF_j = \int_{[0,1]^k} Z^2 \circ (F_1^{-1}, \ldots, F_k^{-1}) \prod_{j=1}^{k} dx_j$. They are not free distributions since the limiting process of Proposition 5.1 is transformed as

$$Z \circ (F_1^{-1}, \ldots, F_k^{-1})(x) = W \circ (F_1^{-1}, \ldots, F_k^{-1})(x)$$
$$- F \circ (F_1^{-1}, \ldots, F_k^{-1}) \sum_{j=1}^{k} \frac{1}{x_j} W_j \circ F_j^{-1}(x_j).$$

Under the local alternative, they converge to $S_K = \sup_{\mathbb{R}^k} |W + H|$ and, respectively, $T_K = \int_{\mathbb{R}^k} (W + H)^2 \prod_{j=1}^{k} dF_j$. Under a fixed alternative, $n^{\frac{1}{2}}(\prod_{j=1}^{k} F_j - F)$ tends to infinity on an interval of $\mathbb{R}^k$ and both statistics diverge, the asymptotic power of the tests is therefore one.

The dependence function of the components of X in $\mathbb{R}^k$ is

$$C(x_1, \ldots, x_k) = F(F_1^{-1}(x_1), \ldots, F_k^{-1}(x_k)), \quad (x_1, \ldots, x_k) \in [0, 1]^k. \quad (5.4)$$

If the marginals are strictly increasing at $t = (t_1, \ldots, t_k)$, the joint distribution is expressed as

$$F(t) = P(X_1 \leq t_1, \ldots, X_k \leq t_k) = C(F_1(t_1), \ldots, F_k(t_k)).$$

The variables $F_k(X_k)$ have uniform marginal distributions and their joint distribution is the dependence function C if the distributions F_k are continuous, which implies $F_k \circ F_k^{-1} = id$, for $j = 1, \ldots, k$. The independence of the variables $X_1, \ldots, X_k$ is characterized by $C(x_1, \ldots, x_k) = \prod_{j=1}^{k} x_j$ for every x in $[0, 1]^k$.

Let $X_i = (X_{i1}, \ldots, X_{ik})$, $i = 1, \ldots, n$, be a sample of n independent variables having the same continuous distribution as X, their empirical dependence function is defined from the transformed variables $U_{ij,n} = \widehat{F}_{j,n}(X_{ij})$ and $\widehat{G}_{j,n}(x) = n^{-1} \sum_{i=1}^{n} 1_{\{\widehat{F}_{j,n}(X_{ij}) \leq x_j\}}$ is the empirical distribution of the marginal uniform variables, their joint empirical distribution is

$$\widehat{C}_n(x) = n^{-1} \sum_{i=1}^{n} 1_{\{\widehat{F}_{1,n}(X_{i1}) \leq x_1, \ldots, \widehat{F}_{k,n}(X_{ik}) \leq x_k\}}$$
$$= n^{-1} \sum_{i=1}^{n} 1_{\{U_{i,n} \leq x\}},$$

for all $(x_1, \ldots, x_k)$ in $[0, 1]^k$. The empirical process

$$\nu_{n,C}(x) = \sqrt{n}\{\widehat{C}_n(x) - C(x)\}$$

has the mean

$$E\nu_{n,C}(x) = \sqrt{n}\{P(\widehat{F}_{1,n}(X_{i1}) \leq x_1, \ldots, \widehat{F}_{k,n}(X_{ik}) \leq x_k) - C(x)\}$$
$$= \sqrt{n}\{C(F_1 \circ \widehat{F}_{1,n}(x_1), \ldots, F_k \circ \widehat{F}_{k,n}(x_k)) - C(x)\},$$

its marginals are denoted $\nu_{j,n,C}$. Lemma 4.1 extends to $[0,1]^k$.

Lemma 5.1. *For every x in $[0,1]^k$*

$$\sqrt{n}\{F(\widehat{F}_{1,n}^{-1}(x_1), \ldots, \widehat{F}_{k,n}^{-1}(x_k)) - C(x)\}$$

$$= \sum_{j=1}^{k} \frac{\nu_{k,n}}{f_k} \circ F_k^{-1}(x_k) \frac{\partial F}{\partial x_k} \circ \{F_1^{-1}(x_1), \ldots, F_j^{-1}(x_j)\}\{1 + o(1)\}.$$

It follows that $\lim_{n\to\infty} E\nu_{n,C}(x) = 0$ for every x in $[0,1]^k$. Under the null hypothesis of independence between the variables $X_1, \ldots, X_k$, the marginal distributions are known and the statistics are simpler than the statistics defined from the process G_n. A Kolmogorov-Smirnov type test for the hypothesis H_0 is based on the goodness of fit statistic

$$S_n = \sup_{x \in [0,1]^k} \sqrt{n} \left| \widehat{C}_n(x) - \prod_{j=1}^{k} x_j \right| \tag{5.5}$$

and a Cramer-von Mises test goodness of fit statistic independence of the components is

$$T_n = \int_{[0,1]^k} n \left\{ \widehat{C}_n(x) - \prod_{j=1}^{k} x_j \right\}^2 \prod_{j=1}^{k} dx_j. \tag{5.6}$$

By the weak convergence under the null hypothesis of the empirical process $\nu_{n,C}$ to a Brownian bridge B_k with covariance function $\prod_{j=1}^{k}(x_j \wedge y_j - x_j y_j)$, the statistics S_n and T_n converge weakly to $S = \sup_{[0,1]^k} |B_k|$ and, respectively, $T = \int_{[0,1]^k} B_k^2 \, dC$, they are free distributions.

Proposition 5.2. *Under the local alternative of a dependence function $C_n(x) = \prod_{j=1}^{k} x_j \{1 + n^{-\frac{1}{2}} H_n(x)\}$, where $(H_n)_n$ is a sequence of strictly positive functions H_n converging uniformly to a function $H > 0$ in $\mathbb{R}^k$, the process $\nu_{C,n}(x) - \prod_{j=1}^{k} x_j$ converges weakly to $H + B_k$.*

The local asymptotic distributions of the statistics is deduced. The asymptotic threshold of the test for a critical level α is determined by the asymptotically free distribution of their limit

$$\lim_{n\to\infty} P_{H_0}(S_n > c_\alpha) = P(\sup_{\mathbb{R}^k} |B_k| > c_\alpha) = \alpha$$

and its power under a sequence of alternatives in a family $\mathcal{C}_n$ of distribution functions in $[0,1]^k$ with dependent components is

$$\beta = \lim_{n\to\infty} \inf_{C_n\in\mathcal{C}_n} P_C(S_n(C_n) > c_\alpha) = \inf_H P(\sup_{\mathbb{R}^k} |H + B_k| > c_\alpha)$$

where $S_n(C)$ is defined like (5.5) for a sample of variables with a known distribution function C in $\mathcal{C}$. The statistic T_n has similar properties.

5.3   Test of homogeneity of k sub-samples

A n-sample is drawn in a population of k independent sub-populations with distinct distributions, the observations consist in k sub-samples of size n_j, $j = 1,\ldots,k$, such that $n = \sum_{j=1}^{k} n_j$ and $\lim_{n\to\infty} n^{-1}n_j = \lambda_j$ belongs to $]0,1[$. In the jth sub-population, the sample $(X_{ij})_{i=1,\ldots,n_j}$ consists in independent and identically distributed random variables, each with distribution function F_j. The empirical distribution functions are denoted $\widehat{F}_{j,n}$ in the jth sub-population and, in the global population, $\widehat{F}_n = \sum_{j=1}^{k} n^{-1}n_j\widehat{F}_{j,n}$ which converges uniformly in probability to $F = \sum_{j=1}^{k} \lambda_j F_j$. Nonparametric tests of the hypothesis $H_0 : F_1 = \ldots = F_k$ are defined from the empirical processes of the sub-populations. For every real x, let W_n be the process with components

$$W_{j,n}(x) = \left(\frac{n_j n_k}{n}\right)^{\frac{1}{2}} \{\widehat{F}_{j,n}(x) - \widehat{F}_{k,n}(x)\} \tag{5.7}$$

$$= \left(\frac{n_k}{n}\right)^{\frac{1}{2}} \nu_{j,n}(x) - \left(\frac{n_j}{n}\right)^{\frac{1}{2}} \nu_{k,n}(x)$$

$$+ \left(\frac{n_j n_k}{n}\right)^{\frac{1}{2}} \{F_j(x) - F_k(x)\}, \ j = 1,\ldots,k-1,$$

where the first sum is denoted $W_{0,j,n}$ and the second one is zero under H_0. Let $(W_1,\ldots,W_k)$ be a vector of independent Wiener processes and let

$$W_{j,\lambda} = \lambda_k^{\frac{1}{2}} W_j \circ F_j - \lambda_j^{\frac{1}{2}} W_k \circ F_k, \ j = 1,\ldots,k-1.$$

The variance of $W_{j,\lambda}$ is $v_{W,j}(x) = \lambda_k F_j(x)\{1 - F_j(x)\} + \lambda_j F_k(x)\{1 - F_k(x)\}$ at x, and the empirical estimator of $v_{W,j}$ is denoted $\widehat{v}_{W,j,n}$. A test statistic for the hypothesis of homogeneity is defined as

$$T_n = \sum_{j=1,\ldots,k-1} \int_{\mathbb{R}} W_{j,n}^2(x) \, d\widehat{F}_n(x).$$

Proposition 5.3. *Under H_0, the statistic T_n converges weakly, as n tends to infinity, to the χ_{k-1}^2 variable $T_0 = \sum_{j=1,\ldots,k-1} \int_{\mathbb{R}} W_{j,\lambda}^2(x) \, dF(x)$, where* *the processes $W_{j,\lambda}$ are defined with identical marginal distribution functions. It diverges under fixed alternatives.*

Proof. The empirical processes $\nu_{j,n}$ converge weakly to independent Wiener processes $W_j \circ F_j$, for $j = 1, \ldots, k-1$, and the statistic W_n is asymptotically equivalent to W_λ. Under all alternatives, there exists at least a sub-distribution F_j distinct of F_k and $W_n(x)$ tends to infinity at x where $F_j(x)$ differs from $F_k(x)$. With the normalization $n_j n_k n^{-1}$, a component of the second sum in the expression of W_n diverges. $\square$

The variable T has the mean $ET = \sum_{j=1,\ldots,k-1} \int_{\mathbb{R}} v_{W,j} \, dF$, this is not a standard χ^2 variable and a bootstrap test must be performed in order to determine its critical value at a nominal level α.

Proposition 5.4. *Let $(K_n)_{n \geq 1}$ be a sequence of local alternatives defined by distribution functions $F_{j,n} = F + n_j^{-\frac{1}{2}} H_{j,n}$ such that the functions $H_{j,n}$ converge uniformly to functions H_j vanishing at infinity, respectively for $j = 1, \ldots, k$. Under $(K_n)_{n \geq 1}$, the statistic T_n converges weakly to*
$$\sum_{j=1,\ldots,k-1} \int_{\mathbb{R}} v_{W,j}^{-1} \{\lambda_k^{\frac{1}{2}}(W_j \circ F + H_j) - \lambda_j^{\frac{1}{2}}(W_k \circ F + H_k)\}^2 \, dF.$$

Proof. Under K_n, $\widehat{F}_{j,n}$ is an estimator of $F_{j,n}$. The approximation of the process $W_{j,n}$ is expressed through the empirical processes $\nu_{j,n,n} = n^{\frac{1}{2}}(\widehat{F}_{j,n} - F_{j,n})$ and
$$W_{j,n} = \left(\frac{n_k}{n}\right)^{\frac{1}{2}} \nu_{j,n} - \left(\frac{n_j}{n}\right)^{\frac{1}{2}} \nu_{k,n} + \left(\frac{n_j n_k}{n+m}\right)^{\frac{1}{2}} (F_{j,n} - F_{k,n}).$$

As n tends to infinity, it converges weakly to the process $\lambda_k^{\frac{1}{2}}(W_j \circ F + H_j) - \lambda_j^{\frac{1}{2}}(W_k \circ F + H_k)$, with independent Wiener processes W_j. $\square$

The variance matrix Σ_W of the vector $(W_{j,n})_{j=1,\ldots,k-1}$ has diagonal terms $v_{W,j}$ and the extra-diagonal terms of the matrix equal $(\lambda_j \lambda_{j'})^{\frac{1}{2}} F_k(1 - F_k)$. Let $\widehat{\Sigma}_{W,n}$ be the empirical estimator of the matrix Σ_W and let W_n be the vector with components $(W_{j,n})_{j=1\ldots,k}$. A normalized statistic of H_0 is defined as
$$T_n = \int_{x \in \mathbb{R}^k} (W_n^t \widehat{\Sigma}_{W,n}^{-1} W_n)(x) \, d\widehat{F}_n(x).$$

Under H_0, it converges weakly to a χ_{k-1}^2 variable and its limit under the local alternatives of Proposition 5.4 is a sum of noncentered normal variables with respective means $\lambda_k^{\frac{1}{2}} H_j - \lambda_j^{\frac{1}{2}} H_k$.

Other nonparametric statistics of homogeneity are based on the following similar processes with a change of variables

$$W_{2,n} = \sum_{j=1}^{k-1} n_k \int_{[0,1]} \{\widehat{F}_{k,n} \circ \widehat{F}_{j,n}^{-1}(x) - x\}^2 \, dx = \sum_{j=1}^{k-1} W_{2,n,j},$$

$$W_{3,n} = \sum_{j=1}^{k-1} n_k \int_{[0,1]} \{\widehat{F}_{j,n} \circ \widehat{F}_{k,n}^{-1}(x) - x\}^2 \, dx = \sum_{j=1}^{k-1} W_{3,n,j}.$$

Under the hypothesis of homogeneity

$$W_{2,n,j} = \int_{[0,1]} [\nu_{k,n} \circ \widehat{F}_{j,n}^{-1}(x) + n_k^{\frac{1}{2}} \{F \circ \widehat{F}_{j,n}^{-1}(x) - x\}]^2 \, dx, \qquad (5.8)$$

$$W_{3,n,j} = \int_{[0,1]} [\nu_{j,n} \circ \widehat{F}_{k,n}^{-1}(x) + n_j^{\frac{1}{2}} \{F \circ \widehat{F}_{k,n}^{-1}(x) - x\}]^2 \, dx$$

and their limits are obtained from expansions of the quantiles.

Proposition 5.5. *Under the hypothesis of homogeneity, the processes $W_{2,n}$ and $W_{3,n}$ converge weakly, as n tends to infinity, to the variables*

$$W_{20} = \sum_{j=1}^{k-1} \int_{[0,1]} \left\{ W_k(x) + \left(\frac{\lambda_k}{\lambda_j}\right)^{\frac{1}{2}} W_j(x) \right\}^2 \, dx,$$

$$W_{30} = \sum_{j=1}^{k-1} \int_{[0,1]} \left\{ W_j(x) + \left(\frac{\lambda_j}{\lambda_k}\right)^{\frac{1}{2}} W_k(x) \right\}^2 \, dx,$$

where $W = (W_1, \ldots, W_k)$ is Wiener process with independent components. Under the sequence of local alternatives K_n of Proposition 5.4, they converge weakly to

$$W_2 = \sum_{j=1}^{k-1} \int_0^1 \left((W_k \circ F_k + H_k) \circ F_j^{-1} - \left(\frac{\lambda_j}{\lambda_k}\right)^{\frac{1}{2}} (W_j + H_j \circ F_j^{-1}) \right.$$

$$\left. + \left[f_j^{-1} W_k \circ F_k \left\{ h_k - \left(\frac{\lambda_j}{\lambda_k}\right)^{\frac{1}{2}} h_j \right\} \right] \circ F_j^{-1} \right)^2 \, dx$$

$$= W_{20} + \sum_{j=1}^{k-1} \int_{\mathbb{R}} \left[(H_k - H_j) + W_k \left\{ h_k - \left(\frac{\lambda_j}{\lambda_k}\right)^{\frac{1}{2}} h_j \right\} \right]^2 \, dF,$$

$$W_3 = \sum_{j=1}^{k-1} \int_0^1 \left((W_j \circ F_j + H_j) \circ F_k^{-1} - \left(\frac{\lambda_k}{\lambda_j}\right)^{\frac{1}{2}} (W_k + H_k \circ F_k^{-1}) \right.$$

$$\left. + \left[f_k^{-1} W_j \circ F_j \left\{ h_j - \left(\frac{\lambda_k}{\lambda_j}\right)^{\frac{1}{2}} h_k \right\} \right] \circ F_k^{-1} \right)^2 \, dx$$

$$= W_{30} + \sum_{j=1}^{k-1} \int_{\mathbb{R}} \left[(H_j - H_k) + W_k \left\{ h_j - \left(\frac{\lambda_k}{\lambda_j}\right)^{\frac{1}{2}} h_k \right\} \right]^2 \, dF.$$

Proof. Arguing as in Lemma 4.1, the process $n_j^{\frac{1}{2}}(\varphi \circ \widehat{F}_{j,n}^{-1} - \varphi \circ F_j^{-1})$ is asymptotically equivalent under the hypothesis to $\{\varphi^{(1)}(f_j)^{-1}\nu_{j,n}\} \circ F_j^{-1}$, therefore

$$
A_{j,n} = n_k^{\frac{1}{2}}(\widehat{F}_{k,n} \circ \widehat{F}_{j,n}^{-1} - id)
$$

$$
= \nu_{k,n} \circ \widehat{F}_{j,n}^{-1} + \left(\frac{n_k}{n_j}\right)^{\frac{1}{2}}\nu_{j,n} \circ F^{-1} + o_p(1)
$$

$$
= \nu_{k,n} \circ F^{-1} + \left(\frac{\lambda_k}{\lambda_j}\right)^{\frac{1}{2}}\nu_{j,n} \circ F^{-1} + o_p(1),
$$

$$
B_{j,n} = n_j^{\frac{1}{2}}(\widehat{F}_{j,n} \circ \widehat{F}_{k,n}^{-1} - id)
$$

$$
= \nu_{j,n} \circ \widehat{F}_{k,n}^{-1} + \left(\frac{n_j}{n_k}\right)^{\frac{1}{2}}\nu_{k,n} \circ F^{-1} + o_p(1).
$$

Their limiting distributions is deduced from the independence of the empirical processes $\nu_{k,n} \circ F^{-1}$ and $\nu_{j,n} \circ F^{-1}$ which have the same distribution. Under the alternative K_n, the empirical processes are denoted $\nu_{n,n,j}$, for $j = 1, \ldots, k$. The process $n_j^{\frac{1}{2}}(\widehat{F}_{k,n} - F_{k,n}) \circ F_{j,n}^{-1}$ is approximated by

$$
\left(\frac{f_k}{f_j}\nu_{n,k,n}\right) \circ F_k^{-1}
$$

and $n_k^{\frac{1}{2}}(F_{j,n} - F_{k,n}) \circ \widehat{F}_{k,n}^{-1} = n_k^{\frac{1}{2}}\{(F_{j,n} - F_j) + (F_k - F_{k,n}) + (F_j - F_k)\} \circ \widehat{F}_{k,n}^{-1}$ is asymptotically equivalent to $\{H_k + (n_j^{-1}n_k)^{\frac{1}{2}}H_j\} \circ F_k^{-1} + \{f_k^{-1}\nu_{n,n,k}(h_k + (n_j^{-1}n_k)^{\frac{1}{2}}h_j)\} \circ F_k^{-1}$. The variable $W_{2,n,j}$ has therefore the expansion

$$
W_{2,n,j} = n_k \int_0^1 \{(\widehat{F}_{k,n} - \widehat{F}_{j,n}) \circ \widehat{F}_{j,n}^{-1}(x)\}^2 \, dx
$$

$$
= \int_0^1 \left((\nu_{n,n,k} + H_{k,n}) \circ F_{j,n}^{-1} - \left\{\left(\frac{\lambda_j}{\lambda_k}\right)^{\frac{1}{2}}(\nu_{n,n,j} + H_{j,n})\right\} \circ F_{j,n}^{-1} \right.
$$

$$
\left. + \left[f_j^{-1}\nu_{n,n,k}\left\{h_k - \left(\frac{\lambda_j}{\lambda_k}\right)^{\frac{1}{2}}h_j\right\}\right] \circ F_{j,n}^{-1}\right)^2 dx + o_p(1)
$$

and its limit is deduced, since the marginal distribution functions of the alternative converge to those of the hypothesis, the expansion of $W_{3,n}$ is similar. $\qquad \square$

Under fixed alternatives, a term $n_k(F_k - F_j) \circ F_j^{-1}$ appears in the expansion of the statistic $W_{2,n}$ and it tends to infinity, therefore the processes $W_{2,n}$ and $W_{3,n}$ diverge under fixed alternatives and the power of the tests converge to 1.

Let $\{\mathcal{M}\} = \{\mathcal{F}_\theta; \mathcal{F} \in \{\mathcal{F}\}, \theta \in \Theta\}$ be a regular class of distribution functions indexed by a bounded open parameter set Θ. Let us consider the hypothesis H_0 of equality of k distribution functions when one of them, say the kth, belongs to the parametric family $\{\mathcal{M}\}$. Semi-parametric tests of H_0 are built by minimizing the L^2-distance between the estimator of the distribution function with the unknown parameter value θ_0 of the kth sub-sample and the empirical distribution function of the first $k-1$ sub-samples.

Let $m_n = \sum_{j=1}^{k-1} n_j$, the estimator $\widehat{\theta}_n$ of θ_0 is defined as in Section 4.3 for the first $k-1$ sub-samples with total size $m_n = n - n_k$ and it satisfies Proposition (4.8) under the hypothesis H_0. The test statistic relies on the difference between the empirical estimator of the kth sub-sample and its parametric estimator

$$W_{n,k} = \left\{ \frac{n_k(n - n_k)}{n} \right\}^{\frac{1}{2}} (\widehat{F}_{k,n} - F_{\widehat{\theta}_n}).$$

Under the hypothesis H_0, the statistic $\sup_{x \in \mathbb{R}} |W_{n,k} \circ \widehat{F}_n^{-1}(x)|$ satisfies Proposition (4.9). Under fixed alternatives, the estimators $\widehat{\theta}_m$ and $\widehat{F}_m$ and the process $W_{n,k}$ diverge.

Under local alternatives $K_n : n_j^{\frac{1}{2}}(F_{j,n} - F_{\theta_0}) = \zeta_{j,n}$, such that $(\zeta_{j,n})_{n\geq 1}$ is a real sequence converging to $\zeta_j \neq 0$, for every $j = 1, \ldots, k-1$, the mean $F_n = (n - n_k)^{-1} \sum_{j=1}^{k-1} n_j F_{j,n}$ is distribution function of the first $k-1$ sub-samples and it tends to F_{θ_0} as n tends to infinity.

Proposition 5.6. *Under the local alternatives* K_n, *the statistic* $\sup_{x \in \mathbb{R}} |W_{n,k} \circ \widehat{F}_n^{-1}(x)|$ *converges weakly to the supremum of a noncentered process, as n tends to infinity.*

Proof. The global empirical estimator $\widehat{F}_m$ of the first $k-1$ sub-samples is such that $(n - n_k)^{\frac{1}{2}}(\widehat{F}_m - F_n)$ converges weakly under K_n to a centered Gaussian process $W \circ F_{\theta_0}$, where W is the Wiener process. Its discrepancy to the centering under the hypothesis is such that

$$\mu_{k,n} = (n - n_k)^{-1}(F_n - F_{\theta_0})$$

$$= (n - n_k)^{-1} \sum_{j=1}^{k-1} n_j(F_{j,n} - F_{\theta_0}) = (n - n_k)^{-1} \sum_{j=1}^{k-1} n_j^{\frac{1}{2}} \zeta_{j,n}$$

and $(n - n_k)^{\frac{1}{2}} \mu_{k,n}$ converges to a nonzero limit $\sum_{j=1}^{k-1} q_j^{\frac{1}{2}} \zeta_j$ with the weights $q_j = \lambda_j (1 - \lambda_k)^{-1}$, as n tends to infinity. It follows that the minimum distance estimator is such that $m^{\frac{1}{2}}(\widehat{F}_{m,\theta_0} - F_{\theta_0})$ converge weakly under K_n to a noncentered Gaussian process which implies the result. $\qquad \square$

The local asymptotic power of the test defined as in Section 4.3 is therefore equal to one. The models of the k sub-populations are not necessarily the same and a test of equal distribution functions belonging to different models extends this test.

5.4 Test of homogeneity of k rescaled distributions

Let us consider the function $\varphi = F_1 \circ F_0^{-1}$ for the control of a sample with a distribution function F_1 as compared to a reference sample with distribution function F_0. The estimator of the distribution function F_l of a variable X_l is estimated from a sub-sample of size n_l of X_l, for $l = 0, 1$, and the limiting distribution of

$$\widehat{\varphi}_n - \varphi = \widehat{F}_{1,n_1} \circ \widehat{F}_{0,n_0}^{-1} - F_1 \circ F_0^{-1}$$

is studied as in Section 5.3 for the processes $W_{2,n}$ and $W_{3,m}$.

Let $\lambda_l = \lim_{n_0+n_1 \to \infty}(n_0 + n_1)^{-1}n_l$ be in $]0,1[$ for $l = 0, 1$ and let $a_n = (n_0 + n_1)^{-1}n_0, n_1 \sim \lambda_0\lambda_1(n_0 + n_1)$. The weak convergence of the process $\{n_0 n_1(n_0 + n_1)^{-1}\}^{\frac{1}{2}}(\widehat{\varphi}_n - \varphi) = a_n^{\frac{1}{2}}(\widehat{\varphi}_n - \varphi)$ is obtained by the same arguments as Lemma 4.1

$$a_n^{\frac{1}{2}}(\widehat{\varphi}_n - \varphi) = \left(\frac{n_0}{n_0 + n_1}\right)^{\frac{1}{2}} \nu_{1,n_1} \circ F_0^{-1}$$

$$+ \left(\frac{n_1}{n_0 + n_1}\right)^{\frac{1}{2}} \left(\frac{f_1}{f_0}\nu_{0,n_0}\right) \circ F_0^{-1} + o_p(1) \qquad (5.9)$$

$$= \lambda_0^{\frac{1}{2}} \nu_{1,n_1} \circ F_0^{-1} + \lambda_1^{\frac{1}{2}} \left(\frac{f_1}{f_0}\nu_{0,n_0}\right) \circ F_0^{-1} + o_p(1).$$

Lemma 5.2. *The process* $a_n^{\frac{1}{2}}(\widehat{\varphi}_n - \varphi)$ *converges weakly to a centered Gaussian process* $W_\varphi = \lambda_0^{\frac{1}{2}} W_1 \circ F_1 \circ F_0^{-1} + \lambda_1^{\frac{1}{2}} \frac{f_1}{f_0} \circ F_0^{-1} W_0$, *where* W_0 *and* W_1 *are independent Wiener processes.*

Test statistics for the hypothesis $\varphi = id$ are defined in Section 5.3 by the processes $W_{2,n}$ and $W_{3,m}$. For the comparison of k curves, k samples of independent variables X_{0j} and X_{1j}, $j = 1, \ldots, k$, are observed where the variables X_{0j} are reference variables and the variables X_{1j} are new measurements. For example, the altitude and the temperature modify the adjustment of devices and the norms are different according to the place where materials are produced. The distribution functions of the variables X_{0j} and X_{1j} are denoted F_{0j} and F_{1j} and the functions $\varphi_j = F_{1j} \circ F_{0j}^{-1}$ are compared.

The hypothesis of the test is $H_0 : \varphi_1 = \ldots = \varphi_k$. Under an alternative K, there exist j and $l \leq k$ such that φ_j differs from φ_l. The distribution functions are estimated from sub-samples of $n_{\rho,j}$ observations of the variables $X_{\rho,j}$ by $\widehat{F}_{n,\rho,j}$, for $\rho = 0, 1$ and $j = 1, \ldots, k$, and the curves φ_j is estimated by $\widehat{\varphi}_{j,n} = \widehat{F}_{n,1,j} \circ \widehat{F}_{n,0,j}^{-1}$. Let $n_0 = \sum_{j=1}^{k} n_{0j}$ and $n_1 = \sum_{j=1}^{k} n_{1j}$, the total sampling size is $n = n_0 + n_1$ and the sub-samples sizes are supposed to have the same order and such that $\lim_{n \to \infty} n^{-1} n_{\rho,j}$ belongs to $]0, 1[$, for all $\rho = 0, 1$ and $j = 1, \ldots, k$. The above notations are extendend to k functions. Let $\lambda_{\rho,j} = \lim_{n_0 + n_1 \to \infty} (n_{0j} + n_{1j})^{-1} n_{\rho,j}$ in $]0, 1[$ and $a_{j,n} = (n_{0j} + n_{1j})^{-1} n_{0j} n_{1j} \sim \lambda_{0j} \lambda_{1j} (n_{0j} + n_{1j})$, for $j = 1, \ldots, k$. Finally, let $b_{j,n} = (a_{j,n} + a_{k,n})^{-1} a_{j,n}$ and $b_j = \lim_{n_0 + n_1 \to \infty} b_{j,n}$ be in $]0, 1[$, and let $u_{j,n} = a_{j,n} a_{k,n} (a_{j,n} + a_{k,n})^{-1}$, for every $j = 1, \ldots, k - 1$. A test statistic is defined like the process W_n in Section 5.8 for the homogeneity of k sub-samples. Let W_n be the process with components

$$W_{j,n}(x) = u_{j,n}^{\frac{1}{2}} \{ \widehat{\varphi}_{j,n}(x) - \widehat{\varphi}_{k,n}(x) \}, \tag{5.10}$$

for every $j = 1, \ldots, k - 1$. By the expansion (5.9)

$$W_{j,n}(x) = (b_{k,n} a_{j,n})^{\frac{1}{2}} (\widehat{\varphi}_{j,n} - \varphi_j)(x) - (b_{j,n} a_{k,n})^{\frac{1}{2}} (\widehat{\varphi}_{k,n} - \varphi_k)(x)$$
$$+ u_{j,n}^{\frac{1}{2}} \{ \varphi_j(x) - \varphi_k(x) \} + o_p(1).$$

From Lemma 5.2 and under H_0, the process $W_{j,n}$ converges weakly to $W_j = b_k^{\frac{1}{2}} W_{\varphi,j} - b_j^{\frac{1}{2}} W_{\varphi,k}$, for $j = 1, \ldots, k - 1$.

Local alternatives K_n are defined by sequences of functions $\varphi_{j,n}$ converging uniformly to φ_j and such that there exist non null functions r_j satisfying $\lim_{n \to \infty} \sup_{\mathbb{R}} |u_{k,n}^{\frac{1}{2}} \{ \varphi_{k,n} - \varphi_{j,n} \} - r_j| = 0$, for $j = 1, \ldots, k - 1$. Under K_n, the processes $\widehat{\varphi}_{j,n}$ are centered at $\varphi_{j,n}$ and the processes $W_{j,n}$ are asymptotically equivalent to $W_j = b_k^{\frac{1}{2}} W_{\varphi,j} - b_j^{\frac{1}{2}} W_{\varphi,k} - u_{k,n}^{\frac{1}{2}} \{ \varphi_{k,n} - \varphi_{j,n} \}$, for $j = 1, \ldots, k - 1$.

Proposition 5.7. *Under H_0, the statistic*

$$T_n = \sum_{j=1,\ldots,k-1} \int_{\mathbb{R}} W_{j,n}^2(x) \, dx$$

converges weakly to the variable $T_0 = \sum_{j=1,\ldots,k-1} \int_{\mathbb{R}} W_j^2(x) \, dx$, as n tends to infinity. Under fixed alternatives, it diverges and under K_n, it converges weakly to $T = \sum_{j=1,\ldots,k-1} \int_{\mathbb{R}} (W_j - r_j)^2(x) \, dx$.

The variables $W_{j,n}$ are not independent, the asymptotic covariance of $W_{j,n}$ and $W_{j',n}$ is $(b_j b_{j'})^{\frac{1}{2}} v_k$ and the variance $v_j(x)$ of $Var W_j(x)$ is

$$v_j(x) = \lambda_{0j} \{ F_{1j}(1 - F_{1j}) \} \circ F_{0j}^{-1}(x) + \lambda_{1j} \left[\left(\frac{f_{1j}}{f_{0j}} \right)^2 \circ F_{0j}^{-1}(x) \{ x(1 - x) \} \right].$$

Let Σ be the covariance matrix of the random vector $W = (W_j)_{j=1,\ldots,k-1}$ and let $\widehat{\Sigma}_n$ be its empirical estimator. The statistic

$$T_n = \sum_{j=1,\ldots,k-1} \int_{\mathbb{R}} (W_{j,n}^t \widehat{\Sigma}_n^{-1} W_{j,n})(x)\, dx$$

converges weakly to the χ_{k-1}^2 variable under H_0 and it converges to a noncentered χ_{k-1}^2 variable $T = \sum_{j=1,\ldots,k-1} \int_{\mathbb{R}} \{(W-r)^t \Sigma^{-1}(W-r)\}(x)\, dx$ under K_n.

5.5 Test of homogeneity of several variables of $\mathbb{R}^k$

Let $X_j = (X_{j1},\ldots,X_{jk})$, $j = 1,\ldots,m$, be m independent random variables in $\mathbb{R}^k$ with respective distribution functions F_j and with marginal distribution functions F_{jl}, for all $j = 1,\ldots,m$ and $l = 1,\ldots,k$. The problem is to test the hypothesis $H_0 : F_1 = F_2 = \ldots = F_m$ from m independent samples of respective sizes n_j, $X_{j,1},\ldots,X_{j,n_j}$ of the variables X_j, $j = 1,\ldots,m$. The components of the variables X_j are supposed to be dependent and the sample sizes have the same convergence rate, $\lim_{n \to \infty} n^{-1} n_j = \lambda_j$, where $n = \sum_{i=1,\ldots,n} n_j$. In a linear model with dependent normal error variables ε_{jl}, $X_{jl} = \mu_j + \sigma_j \varepsilon_{jl}$, $l = 1,\ldots,n_j$, $j = 1,\ldots,m$, the mean vectors and variance matrices are estimated independently from each component. A test of H_0 is a test of equality of the means of the variables and of their variance matrices. The approach is similar for all distribution of the error, the hypothesis H_0 is the equality of the means of the variables and of the dependence functions of their components.

With unspecified distribution functions, the hypothesis H_0 is equivalent to the equality of all marginal distribution functions and of the m dependence functions of the random vectors. The dependence function C_j of the components of X_j is estimated by $\widehat{C}_{j,n_j}$ defined by (5.5) and the first $m-1$ estimators $\widehat{C}_{j,n_j}$ are compared by $\widehat{C}_{m,n_m}$, like the distribution functions in the previous section. The processes $n^{\frac{1}{2}}(\widehat{C}_{j,n_j} - C_j)$ converge weakly to independent and centered Gaussian processes $W_{j,C}$ in $\mathbb{R}^k$. The test statistic

$$T_n = \sum_{j=1,\ldots,m-1} \int_{x \in \mathbb{R}^k} \frac{n_j n_m}{n_j + n_m} \{\widehat{C}_{j,n} - \widehat{C}_{m,n}\}^2(x)\, dx$$

converges weakly under H_0 to the variable

$$T = \sum_{j=1,\ldots,m-1} \int_{x \in \mathbb{R}^k} \left\{ \left(\frac{\lambda_m}{\lambda_j + \lambda_m} \right)^{\frac{1}{2}} W_{j,C}(x) - \left(\frac{\lambda_m}{\lambda_j + \lambda_m} \right)^{\frac{1}{2}} W_{m,C}(x) \right\}^2 dx.$$

Let $a_j = \{\lambda_j(\lambda_j + \lambda_m)\}^{\frac{1}{2}}$, the mean of T is the sum of the integrated variances of the processes $a_m W_{j,C} - a_j W_{m,C}$, that is $\int_{x \in \mathbb{R}^k} \{a_m^2 Var W_{j,C} + a_j^2 Var W_{m,C}\}^2(x)\,dx$, where $Var W_{j,C}(x)$ depends on the distribution function of the variable X_j, for every $j = 1, \ldots, m - 1$. The variance of T depends on higher moments of the processes $a_m W_{j,C} - a_j W_{m,C}$. They can be estimated by bootstrap and a bootstrap statistic is deduced as $T_n^* \widehat{Var}_n^{* - \frac{1}{2}}$.

Under local alternatives defined by sequences of dependence functions $C_{j,n} = C + n_j^{-\frac{1}{2}} H_{j,n}$ such that $\lim_{n \to \infty} \sup_{x \in \mathbb{R}^k} |H_{j,n} - H_j| = 0$ for every $j = 1, \ldots, m$, and $H_{j_0} \neq H_m$ for at least one variable X_{j_0}, the process $\{n_{j_0} n_m (n_{j_0} + n_m)^{-1}\}^{\frac{1}{2}} (\widehat{C}_{j_0,n} - \widehat{C}_{m,n})$ converges weakly to $a_m W_{j_0,C} - a_j W_{m,C} + a_m H_{j_0} - a_{j_0} H_m$ and the statistic T_n converges to the sum of the L^2 norms of the limit of such processes, for all j_0 such that $H_{j_0} \neq H_m$. Tests of equality of the marginal distribution functions are considered in the next section.

5.6   Test of equality of marginal distributions

In Section 5.2, k independent samples with respective distribution functions F_j, $j = 1, \ldots, k$, are observed. Considering dependent samples with margins F_j, the hypothesis $H_0 : F_1 = \cdots = F_k$ is considered in a model with an unknown dependence function C for the k observed variables. Their joint distribution is $F(x) = C(F_1(x_1), \ldots, F_k(x_k))$ in $\mathbb{R}^k$.

The marginal empirical distribution functions $\widehat{F}_{j,n_j}$ obtained from a n_j sample of the variable X_j with the distribution F_j are dependent, their covariances $c_{jl,n} = cov\{\widehat{F}_{j,n_j}(x_j), \widehat{F}_{ln_l}(x_l)\}$ are

$$c_{jl,n}(x_j, x_l) = \frac{1}{n_j \vee n_l} P(X_j \leq x_j, X_l \leq x_l) - \frac{1}{n_j \wedge n_l} F_j(x_j) F_l(x_l).$$

The covariance matrix $V_{n,k}$ of the vector $(\widehat{F}_{j,n_j})_{j=1,\ldots,k}$ is estimated empirically and a test statistic for H_0 is built from the empirical estimators of the differences $(F_j - F_k)(x)$, for all $j = 1, \ldots, k$ and x in $\mathbb{R}$. Their variance under H_0 is $v_{jl,n}(x) = (n_j + n_k)(n_j n_k)^{-1} F(x)\{1 - F(x)\} - 2c_{jl,n}(x)$. The statistics of Section 5.3 have to be modified by a normalization with the empirical estimator $\widehat{\Sigma}_{n,k}$ of their variance matrix $\Sigma_{n,k}$.

A normalized statistic of H_0 is defined as

$$T_n = \int_{x \in \mathbb{R}^k} (W_n^t \widehat{\Sigma}_{W,n}^{-1} W_n)(x)\,d\widehat{F}_n(x).$$

Under H_0, it converges weakly to a χ^2_{k-1} variable since the marginals of the process $W_n^t \widehat{\Sigma}_{W,n}^{-1} W_n$ have asymptotic χ^2_{k-1} distributions, and their limits under the local alternatives of Proposition 5.4 are sums of noncentered normal variables with respective means $\lambda_k^{\frac{1}{2}} H_j - \lambda_j^{\frac{1}{2}} H_k$.

The limits of the statistics $W_{2,n}$ and $W_{3,n}$ defined from a transform by the quantiles are also modified. The processes $A_{j,n} = n_k^{\frac{1}{2}} (\widehat{F}_{k,n} \circ \widehat{F}_{j,n}^{-1} - id)$ defining $W_{2,n,j}$ in (5.8) are dependent, their asymptotic variances and covariances are calculated from their approximation

$$A_{j,n} = \nu_{k,n} \circ F_j^{-1} + \left(\frac{\lambda_k}{\lambda_j}\right)^{\frac{1}{2}} \left(\frac{f_k}{f_j} \nu_{j,n}\right) \circ F_j^{-1} + o_p(1).$$

The covariance of $\nu_{k,n} \circ F_j^{-1}(x_j)$ and $\nu_{k,n} \circ F_l^{-1}(x_l)$ is

$$C_{jl,k}(x_j, x_l) = F_k \circ (F_j^{-1}(x_j) \wedge F_l^{-1}(x_l)) - F_k \circ F_j^{-1}(x_j) F_k \circ F_l^{-1}(x_l),$$

with the above notation for the covariance of $\widehat{F}_{j,n_j}(x_j)$ and $\widehat{F}_{l,n_l}(x_l)$, this implies

$$\begin{aligned}
Var A_{j,n}(x) = & \{F_k(1 - F_k)\} \circ F_j^{-1}(x) \\
& + \frac{\lambda_k}{\lambda_j}\left(\frac{f_k}{f_j}\right)^2 \circ F_j^{-1}(x) x_j(1 - x_j) + o(1),
\end{aligned}$$

$$\begin{aligned}
Cov\{A_{j,n}(x_j), A_{l,n}(x_l)\} = & C_{jl,k}(x_j, x_l) \\
& + \frac{\lambda_k}{(\lambda_j \lambda_l)^{\frac{1}{2}}} \frac{f_k}{f_j} \circ F_j^{-1}(x_j) \frac{f_k}{f_l} \circ F_l^{-1}(x_l) (n_j n_l)^{\frac{1}{2}} c_{jl,n} \circ (F_j^{-1}(x_j), F_l^{-1}(x_l)) \\
& + \left(\frac{\lambda_k}{\lambda_j}\right)^{\frac{1}{2}} \frac{f_k}{f_j} \circ F_j^{-1}(x_j) (n_j n_k)^{\frac{1}{2}} c_{jk,n} \circ (F_j^{-1}(x_j), F_l^{-1}(x_l)) \\
& + \left(\frac{\lambda_k}{\lambda_l}\right)^{\frac{1}{2}} \frac{f_k}{f_l} \circ F_l^{-1}(x_l) (n_l n_k)^{\frac{1}{2}} c_{lk,n} \circ (F_j^{-1}(x_j), F_l^{-1}(x_l)) + o(1).
\end{aligned}$$

The statistic $W_{2,n}$ is replaced by $\int_{[0,1]^k} (A_n^t \widehat{V}_n^{-1} A_n)(x)\, dx$ and it converges weakly under H_0 to a χ^2_{k-1} variable. Under local alternatives, the normal variables of the limit under H_0 are not centered and their means have the limits given in Proposition 5.5.

5.7 Test of exchangeable components for a random variable

A variable X in $\mathbb{R}^k$ has exchangeable components if X and $X_{\sigma_k(1)}, \ldots, X_{\sigma_k(k)}$ have the same distributions, for every permutation σ_k of $\{1, \ldots, k\}$, where $\sigma_k(j)$ denotes the jth component of $\sigma_k(\{1, \ldots, k\})$.

Let $\mathcal{P}_k$ be the set of all permutations of $\{1, \ldots, k\}$ and let $X_1, \ldots, X_n$ be a n-sample of the variable X. A test statistic for the hypothesis of exchangeability of the components of X relies on the supremum over all permutations of the difference of their distribution functions mean components of the n observations

$$S_{k,n} = \sup_{x \in \mathbb{R}^k} \sup_{\sigma_k \in \mathcal{P}_k} n^{\frac{1}{2}} |\widehat{F}_{n,\sigma_k}(x) - \widehat{F}_n(x)|.$$

All components of the mean vector of the sample have the same distribution under H_0 but not under alternatives. If the components of the variable X are independent, the test of H_0 is also a test of homogeneity of the distributions of k independent variables and the statistics of Section 5.3 can be used. With dependent components, the statistics for dependent components defined in Section 5.6 apply.

5.8   Tests in single-index models

In the semi-parametric regression model

$$m(X) = E(Y \mid X) = g \circ \varphi_\theta(X), \tag{5.11}$$

the real variable Y is the response to a d-dimensional vector X of explanatory variables, φ_θ is a parametric function from $\mathbb{R}^d$ onto $\mathbb{R}$, with unknown parameter vector θ, with true value θ_0, and g is an unknown function. The parameter set Θ is supposed to be an open and bounded subset of $\mathbb{R}^d$ containing θ_0. In order to reduce the effect of the dimension d in the nonparametric estimation of the functions m and $\sigma^2(x) = Var(Y|X = x)$, several authors have studied the single-index model with a transformed linear function $\varphi_\theta(x)$ and a constant variance. The model was extended by Pons (2011) to allow a functional linear model for the conditional mean and functional variance

$$m(X) = g(\theta^T X), \tag{5.12}$$
$$Var(Y|X) = \sigma^2(X).$$

With a known function g, the estimator of the parameter maximizing the likelihood or a quasi-likelihood in models defined by (5.11) or (5.12) is $n^{1/2}$-consistent and asymptotically Gaussian. Estimators for a polynomial approximation of (5.12) have similar properties. The best rate for θ_0 is nonparametric due to the kernel estimator of the curve g and it is parametric only in parametric models for the function g.

Let $(X_i, Y_i)_{i=1,...,n}$ be a sample of $(d+1)$-dimensional variable of the model with values in a metric space $(\mathcal{X}^d, \mathcal{Y})^{\otimes n}$ endowed with the L^2 metric with respect to the Lebesgue measure. A regular nonparametric estimator of the conditional mean function $m(x) = E(Y \mid X = x)$ can be defined for every continuous function g in model (5.12). Let K be a symmetric continuous density, let $h = h_n$ be a bandwidth converging to zero as n tends to infinity. Assuming that the parameter θ is known, kernel estimators of the functions m and g are defined at x in $\mathcal{X}$ and, respectively, $z = \theta^T x$ in $\mathcal{Z} = \{\theta^T x; \theta \in \Theta, x \in \mathcal{X}\}$ by

$$\widehat{m}_{n,h}(x) = \frac{\sum_{i=1}^n Y_i K_h(x - X_i)}{\sum_{i=1}^n K_h(x - X_i)},$$

$$\widehat{g}_{n,h}(z; \theta) = \frac{\sum_{i=1}^n Y_i K_h(z - \theta^T X_i)}{\sum_{i=1}^n K_h(z - \theta^T X_i)},$$

at fixed θ. The estimator of g is defined as the ratio of mean estimators

$$\mu_{Z_\theta, n, h}(z; \theta) = n^{-1} \sum_{i=1}^n Y_i K_h(z - \theta^T X_i)$$

and the density estimator at $z = \theta^T x$

$$\widehat{f}_{Z_\theta, n, h}(x) = \int K_h(x - s) \, d\widehat{F}_{Z_\theta, n}(s) = n^{-1} \sum_{i=1}^n K_h(z - \theta^T X_i)$$

with the empirical distribution $\widehat{F}_{Z_\theta, n}$ of the real variable $Z_\theta = \theta^T X$. The parameter θ is then estimated by minimizing a mean squared error of estimation, which is a goodness-of-fit criterium for the model. The global goodness-of-fit error and the estimator of θ minimizing this error are

$$\widehat{V}_{n,h}(\theta) = n^{-1} \sum_{i=1}^n \{Y_i - \widehat{g}_{n,h}(\theta^T X_i; \theta)\}^2, \tag{5.13}$$

$$\widehat{\theta}_{n,h} = \arg\min_{\theta \in \Theta} \widehat{V}_{n,h}(\theta), \tag{5.14}$$

and $\widehat{m}_{n,h}(x) = \widehat{g}_{n,h}(\widehat{\theta}_{n,h}^T x; \widehat{\theta}_{n,h})$, with the estimator (5.14).

The estimators of θ in the single-index model (5.11) are defined similarly at $z = \varphi_\theta(x)$, replacing the scalar product $Z_\theta = \theta^T X$ by $Z_\theta = \varphi_\theta(X)$ in $\mathcal{Z} = \{\varphi_\theta(x); \theta \in \Theta, x \in \mathcal{X}\}$. The estimator $\widehat{\theta}_n$ minimizes an empirical distance $\widehat{V}_{n,h}$ which converges uniformly to a function V of $C^2(\Theta)$ under the assumption of functions m in $C^2(\mathcal{Z})$, and such that the limiting function is minimal and locally convex at the true parameter value θ_0. The empirical distance $\widehat{V}_{n,h}$ belongs to $C^2(\Theta)$ and the first derivative satisfy $\widehat{V}_{n,h}^{(1)}(\widehat{\theta}_n) = 0$

and $V^{(1)}(\theta_0) = 0$. The derivatives of the conditional mean function m and g are estimated by the derivatives of the estimators $\widehat{m}_{n,h}$ and $\widehat{g}_{n,h}$. Under the classical conditions for the kernel function K and for a function g in $C^2(\mathcal{X})$, the convergence rate of $\widehat{m}_{n,h}$ is $(nh^d)^{1/2}$ and the convergence rate of its derivative is $(nh^{d+2})^{1/2}$, which depends on the dimension d of the regression variable. In model (5.12), the convergence rates for the estimator of g and its derivative, at fixed θ, are respectively $(nh)^{1/2}$ and $(nh^3)^{1/2}$, so the estimators in the semi-parametric model converge faster than those of the nonparametric model for m.

Expansions of the bias and the variance of the estimators for the parameter θ and the function $g(\theta^T x)$ were established in Pons (2011), with their weak convergence. The pointwise and uniform convergence of the kernel estimators of the function g in $\mathcal{I}_{X,h}$ are established under the conditions of Section 3.7. Let Σ_{V,θ_0} be the asymptotic variance of $\widehat{V}_{n,h}^{(1)}(\theta_0^T x; \theta_0)$.

Proposition 5.8. *Under Conditions 1.1 and 3.1 and with the optimal bandwidth $h_{V,n} = O(n^{-1/(2s+3)})$, $s \geq 2$, the variable $n^{s/(2s+3)}(\widehat{\theta}_{n,h} - \theta_0) - \{V_{n,h}^{(2)}(\theta_0)\}^{-1}V_{n,h}^{(1)}(\theta_0)$ converges weakly to a centered Gaussian variable with variance $v_0 = V_{\theta_0}^{(2)-1}\Sigma_{V,\theta_0}V_{\theta_0}^{(2)-1}$. The mean squared error of the estimator $\widehat{m}_{n,h}$ is $O(n^{-s/(2s+3)})$ and it converges weakly to Gaussian processes with the optimal rate $n^{s/(2s+3)}$.*

Let V_θ be the limit in probability of $\widehat{V}_{n,h}(\theta)$ as n tends to infinity. Expanding the first derivative $\widehat{V}_{n,h}^{(1)}(\widehat{\theta}_{n,h})$, for $\widehat{\theta}_{n,h}$ in a neighborhood of θ_0 implies

$$\widehat{\theta}_{n,h} - \theta_0 = -\{\widehat{V}_{n,h}^{(2)}(\theta_0)\}^{-1}\widehat{V}_{n,h}^{(1)}(\theta_0) + o_P(\widehat{\theta}_{n,h} - \theta_0) \qquad (5.15)$$

and Proposition 5.8 is a consequence of the consistency of the estimator $V_{n,h}^{(2)}(\theta_0)$ and of the weak convergence of $n^{s/(2s+3)}V_{n,h}^{(1)}(\theta_0)$ to a biased Gaussian process.

In a model with an unknown variance function σ^2, the error criteria and the estimator of g are normalized by an estimator of σ^{-1}. The variance $\sigma(x)^2 = E[\{(Y - m(X))\}^2 | X = x]$ is assumed to be continuous and it is estimated by smoothing the empirical error using the kernel K with a bandwidth δ_n converging to zero as n tends to infinity

$$\widehat{\sigma}_{n,h,\delta}^2(x) = \frac{\sum_{i=1}^n \{Y_i - \widehat{m}_{n,h}(X_i)\}^2 K_\delta(x - X_i)}{\sum_{i=1}^n K_\delta(x - X_i)}. \qquad (5.16)$$

The estimator is denoted $\widehat{\sigma}_{n,h,\delta}^2(x) = \widehat{f}_{X,n,\delta}^{-1}(x)\widehat{S}_{n,h,\delta}(x)$, with the weighted mean squared error

$$\widehat{S}_{n,h,\delta}(x) = n^{-1} \sum_{i=1}^{n} \{Y_i - \widehat{m}_{n,h}(X_i)\}^2 K_\delta(x - X_i)$$

$$= \int \{y - \widehat{m}_{n,h}(s)\}^2 K_\delta(x - s) \, d\widehat{F}_{X,Y,n}(s, y).$$

Let $\beta_{n,h,\delta}$ be the bias of $\widehat{\sigma}^2_{n,h,\delta}$.

Proposition 5.9. *Assume that for a density f_X and a function μ of $C^r(\mathcal{I}_X)$ and for a variance σ^2 of $C^k(\mathcal{I}_X)$, with k and $r \geq 2$, the bandwidth sequences $(\delta_n)_n$ and $(h_n)_n$ satisfy*

$$\delta_n = O(n^{-1/(2k+1)}), \quad h_n = O(n^{-1/\{2s+1\}}),$$

as n tends to infinity. Under the conditions of Proposition 5.8, the process $(n\delta)^{1/2}(\widehat{\sigma}^2_{n,h,\delta} - \sigma^2) - \gamma_k \beta_{n,h,\delta}$ converges weakly to a Gaussian process with mean and covariances zero and with variance function $\kappa_2 \sigma_4(x)$.

Under the above conditions

$$E\{Y - \widehat{m}_{nh}(x)\}^2 = \sigma^2(x) + O(h^{4s}) + O((nh)^{-1}),$$

hence the bias of the estimator $\widehat{S}_{n,h,\delta}(x)$ is written as an expression of the bias $b^2_{m,n,h}(x)$ and the variance $\sigma^2_{m,n,h}(x)$ of the estimator of the regression function

$$\beta_{n,h,\delta}(x) = b^2_{m,n,h}(x)f_X(x) + \sigma^2_{m,n,h}(x)f_X(x) + \frac{\delta^{2k}}{(k!)^2}(\sigma^2(x)f_X(x))^{(2)}$$

$$+ o(\delta^{2k} + h^{2s} + (nh)^{-1})$$

and its variance is expanded as $(n\delta)^{-1}\kappa_2 \sigma_4(x) + o(n^{-1})$, depending on the fourth moment $\sigma_4(x) = E[\{(Y - m(X))\}^4 | X = x]$.

With a single-index model for the conditional variance, let η be a parameter in an open and bounded subset of $\mathbb{R}^d$ and let σ^2 be a semi-parametric variance function $\sigma^2(\eta^T x) = E[\{(Y - m(X))\}^2 | \eta^T X = \eta^T x]$. It is estimated at fixed η by $\widehat{\sigma}^2_{n,h,\delta}(\eta^T x)$ like in (5.16). The global error (5.13) is modified as

$$\widehat{V}_{n,h}(\eta, \theta) = n^{-1} \sum_{i=1}^{n} \widehat{\sigma}^{-1}_{n,h}(\eta^T X_i)\{Y_i - \widehat{g}_{n,h}(\theta^T X_i; \theta)\}^2 \qquad (5.17)$$

and the estimator (5.14) is replaced by the random vector $(\widehat{\eta}^T_{n,h}, \widehat{\theta}^T_{n,h})^T$ which minimizes $\widehat{V}_{n,h}$. Semiparametric estimators of the regression function m and the variance function $\sigma^2(\eta^T x)$ are deduced for the estimators of parameters η and θ, and the real functions g and σ^2. From the previous convergences, all estimators are uniformly consistent. For functions f_X, g and σ^2 in C^s, the previous expansions imply that the estimator obtained by minimizing (5.17) have a bias $O(h^s)$ and their variance has the same order as the estimator (5.14) with bandwidths $\delta_n = h_n$. The weak convergence of the estimators is similar to the results of Proposition 5.8.

In the nonparametric regression (5.11) with a change of variables, the linear expression $\theta^T X$ is replaced by a transformation of X using a parametric family of functions defined in R^d, $\phi = \{\varphi_\theta\}_{\theta \in \Theta}$ subset of the class C^2. The semi-parametric regression model is

$$Y = g \circ \varphi_\theta(X) + \sigma^2(X)\varepsilon. \tag{5.18}$$

The error is still $V(\theta, \sigma) = E\sigma^{-1}(X)\{Y - m(X)\}^2$ with $m = g \circ \varphi_\theta$, at fixed θ and σ, and the parameter θ is estimated by minimizing

$$\widehat{V}_{n,h}(\theta) = n^{-1} \sum_{i=1}^{n} \widehat{\sigma}_{n,h}^{-1}(X_i)\{Y_i - \widehat{g}_{n,h} \circ \varphi_\theta(X_i)\}^2. \tag{5.19}$$

The variance σ^2 is estimated by (5.16) and its estimator satisfies the properties of Proposition 5.9 as in model (5.12). The results of Proposition 5.8 apply to the estimators of θ_0 and m.

In a class of models (5.11), let us consider a test of significative effect of the components of the variable X. The hypothesis is written as $H_0 : \theta$ belongs to Θ_0, an open subset of the parameter set Θ. The hypothesis may be a linear hypothesis and $\Theta_0 = \{\theta \in \Theta; A\theta = 0\}$ where A is a $d_0 \times d$-matrix and $d_0 < d$, or any other constraint for components of θ. The parameter θ of the model is estimated by $\widehat{\theta}_{n,h,0}$ under the hypothesis and by $\widehat{\theta}_{n,h}$ in Θ.

A test statistic for H_0 against the alternative $K : \theta$ belongs to $\Theta \setminus \Theta_0$ is the difference of the mean squared errors under K and H_0

$$T_{n,h} = n^{2s/(2s+3)}\{\widehat{V}_{n,h}(\widehat{\theta}_{n,h}) - \widehat{V}_{n,h,0}(\widehat{\theta}_{n,h,0})\},$$

with the process $\widehat{V}_{n,h,0}$ defined in Θ_0.

Proposition 5.10. *Under the conditions of Proposition 5.8, the statistic $T_{n,h}$ converges weakly under H_0 to a noncentered $\chi^2_{d-d_0}$ variable.*

Proof. The proof is similar to the proof of the converges weak convergence of the log-likelihood ratio test statistic to a $\chi^2_{d-d_0}$ variable when the hypothesis concerns the parameter of a parametric density function. It uses again the approximation of $\widehat{\theta}_{n,h} - \theta_0$ by (5.15) with $\widehat{V}_{n,h}$ defined by (5.13) or (5.19) according to the model. Under H_0

$$\widehat{\theta}_{n,h,0} - \theta_0 = -2\{\widehat{V}_{n,h,0}^{(2)}(\theta_0)\}^{-1}\widehat{V}_{n,h,0}^{(1)}(\theta_0)\{1 + o_p(1)\},$$

and the estimator under the alternative satisfies the same expansion. The difference $\widehat{\theta}_{n,h} - \widehat{\theta}_{n,h,0}$ is therefore a $O_p((nh^3)^{-1/2})$. A first order expansion of $\widehat{V}_{n,h}(\widehat{\theta}_{n,h})$ and $\widehat{V}_{n,h,0}(\widehat{\theta}_{n,h,0})$, as n tends to infinity, yields

$$\widehat{V}_{n,h}(\widehat{\theta}_{n,h}) - \widehat{V}_{n,h,0}(\widehat{\theta}_{n,h,0}) = \widehat{V}_{n,h}^{(1)T}(\theta_0)\{\widehat{V}_{n,h}^{(2)}(\theta_0)\}^{-1}\widehat{V}_{n,h}^{(1)}(\theta_0)$$
$$- \widehat{V}_{n,h,0}^{(1)T}(\theta_0)\{\widehat{V}_{n,h,0}^{(2)}(\theta_0)\}^{-1}\widehat{V}_{n,h,0}^{(1)}(\theta_0)$$
$$+ o_p((nh_n^3)^{-1})$$

and the asymptotic $\chi^2_{d-d_0}$ distribution of the statistic $T_{n,h}$ follows by inverting explicitly $\widehat{V}_{n,h}^{(2)}(\theta_0)$. Since the bias of $\widehat{V}_{n,h}^{(1)T}(\theta_0)$ and $\widehat{V}_{n,h}^{(1)T}(\theta_0)$ do not vanish, the limiting $\chi^2_{d-d_0}(\mu_T)$ variable with a noncentrality term. $\square$

Under the alternative K, $T_{n,h}$ tends to infinity and the power of the test with level α tends to one for every α in $[0,1]$. Its asymptotic power against a local alternative K_n : there exists θ_0 in Θ_0 and a real sequence $(\delta_n)_n$ converging to $\delta > 0$ such that $\theta_n = \theta_0(1 + n^{-s/(2s+3)}\delta_n)$ is obtained from expansions of $\widehat{V}_{n,h}(\widehat{\theta}_{n,h})$ and $\widehat{V}_{n,h,0}(\widehat{\theta}_{n,h,0})$, as $\widehat{\theta}_{n,h} - \theta_n$ and $\widehat{\theta}_{n,h,0} - \theta_n$ tends to zero. Under K_n, $\widehat{V}_{n,h}(\widehat{\theta}_{n,h}) - \widehat{V}_{n,h,0}(\widehat{\theta}_{n,h,0})$ develops as

$$\{\widehat{V}_{n,h}^{(1)}(\theta_0) + \theta_n - \theta_0\}^T\{\widehat{V}_{n,h}^{(2)}(\theta_0)\}^{-1}\{\widehat{V}_{n,h}^{(1)}(\theta_0) + \theta_n - \theta_0\}$$
$$- \{\widehat{V}_{n,h,0}^{(1)}(\theta_0) + \theta_n - \theta_0\}^T\{\widehat{V}_{n,h,0}^{(2)}(\theta_0)\}^{-1}\{\widehat{V}_{n,h,0}^{(1)}(\theta_0) + \theta_n - \theta_0\}$$
$$+ o_p((nh_n^3)^{-1}).$$

It follows that $T_{n,h}$ converges weakly under K_n to a noncentered $\chi^2_{d-d_0}$ distribution with a noncentrality term depending on the limit δ of δ_n

$$\lim_{n\to\infty}(nh_n^3)[\{V_{n,h}^{(1)}(\theta_0) + \theta_n - \theta_0\}^T\{V^{(2)}(\theta_0)\}^{-1}\{V_{n,h}^{(1)}(\theta_0) + \theta_n - \theta_0\}$$
$$- \{V_{n,h,0}^{(1)}(\theta_0) + \theta_n - \theta_0\}^T\{V^{(2)}(\theta_0)\}^{-1}\{V_{n,h,0}^{(1)}(\theta_0) + \theta_n - \theta_0\}].$$

Tests of mispecification of the transformed linear model (5.12) as it is defined by a family of nonlinear semi-parametric transformations of the form (5.12) can be defined from a comparison of the estimators $\widehat{m}_{n,h}$ of the function m in a model defined by (5.11) and $\widehat{m}_{n,h,lin}$ in the linear model (5.12). A Kolmogorov-Smirnov test statistic for the hypothesis H_0 of the linear model against the alternative K of model (5.11) is

$$S_{n,h} = \sup_{x\in\mathcal{I}_{X,h}} n^{s/(2s+3)}|\widehat{m}_{n,h}(x) - \widehat{m}_{n,h,lin}(x)|.$$

Proposition 5.11. *Under Conditions 1.(1)-1.(3) and with the bandwidth* $h_n = O(n^{-1/(2s+3)})$, $s \geq 2$, *the statistic* $S_{n,h}$ *converges weakly under* H_0 *to the supremum of the difference of two Gaussian processes. Under the alternative, it tends to infinity.*

Proof. Let $\widehat{\theta}_{n,h}$ be the estimator of the parameter of the model under the alternative and $\widehat{\theta}_{n,h,0}$ be its estimator under H_0 and let θ_0 be the parameter value under H_0. Under H_0, the parameter value minimizing the mean squared error V of the alternative model is denoted θ_1. The process $U_{n,h} = n^{s/(2s+3)}(\widehat{m}_{n,h} - \widehat{m}_{n,h,lin})$ has the expansion

$$\begin{aligned}
U_{n,h}(x) &= n^{s/(2s+3)}(\widehat{\theta}_{n,h} - \theta_1)^T \widehat{g}_{n,h}^{(1)} \circ \varphi_{\theta_0}(x)\, \varphi_{\theta_0}^{(1)}(x) \\
&\quad - n^{s/(2s+3)}(\widehat{\theta}_{n,h,0} - \theta_0)^T x \widehat{g}_{n,h}^{(1)}(\theta_1^T x) + o_p(1) \\
&= n^{s/(2s+3)}(\widehat{\theta}_{n,h} - \theta_1)^T g^{(1)} \circ \varphi_{\theta_0}(x)\, \varphi_{\theta_0}^{(1)}(x) \\
&\quad - n^{s/(2s+3)}(\widehat{\theta}_{n,h,0} - \theta_0)^T x g^{(1)}(\theta_1^T x) + o_p(1)
\end{aligned}$$

and it converges weakly to the difference of Gaussian processes. □

A comparison of two regression curves is performed as in Section 4.7 using the weak convergence of the estimator of the semi-parametric regression function m of Proposition 5.8. The goodness of fit test of a semi-parametric model against a nonparametric alternative relies on the comparison of the nonparametric estimator of the regression function m and its semi-parametric estimator. Since their convergence rates differ, this statistic is asymptotically equivalent to a comparison of the semi-parametric estimator to the true unknown regression function.

5.9 Comparison of k curves

Consider k regression curves $m_j(x) = E(Y_j | X_j = x)$ defined for variables sets $(X_j, Y_j)_{j=1,\ldots,k}$, they are compared by the means of tests of the hypothesis $H_0 : m_1(x) = \cdots = m_k(x)$ for every x in an interval $\mathcal{I}_X$ of the common support of the regression variables X_j, for every $j = 1, \ldots, k$. The general alternative of distinct curves on $\mathcal{I}_X$ is K : there exist j_1 and j_2 in $\{1, \ldots, k\}$ and a sub-interval of $\mathcal{I}_X$ where $m_{j_1} \neq m_{j_2}$. Let $(X_{ij}, Y_{ij})_{l=1,\ldots,n_i, i=1,\ldots,n}$ be a sample of independent observations of the variables in k homogeneous sub-samples of size n_j and let $n = \sum_{j=1,\ldots,k} n_j$ be the total sample size. The sub-sample sizes are supposed to be $O(n)$ and such that $\lambda_j = \lim_{n \to \infty} n^{-1} n_j$ belong to $]0, 1[$. Condition 3.1 are assumed for the regression functions m_j, the kernel functions K_j and their bandwidth h_j, $j = 1, \ldots, k$. With real regression variables, the tests rely on the nonparametric estimators $\widehat{m}_{j,n,h}$ of $m_j = E(Y_j|X_j = x)$, $j = 1, \ldots, k$, satisfying the asymptotic properties of Section 3.7. For m_j in $C^s(\mathcal{I}_X)$, $s \geq 2$, let h_{j,n_j} be the L^2-optimal bandwidth for $\widehat{m}_{j,n_j,h_j}$. The processes

$$B_{j,n_j,h_j} = n_j^{\frac{s}{2s+1}} (\widehat{m}_{j,n_j,h_j} - m_j) 1_{\{\mathcal{I}_{X,h}\}}, j = 1, \ldots, k$$

converge weakly to Gaussian processes $B_j = \sigma_{m_j} W_1 + \gamma_s b_{m_j}$ where W_1 is a centered Gaussian process on $\mathcal{I}_X$ with variance function 1 and çovariance function zero, the constants are defined by the asymptotic bias and variance of the estimators.

The whole sample of regression variables has the mean density function $f_X = \sum_{j=1,\ldots,k} \lambda_j f_j$, where f_j is the marginal density of X_j, $j = 1, \ldots, k$. Its kernel estimator is the empirical mean of the kernel estimators of the densities of the sub-samples of X

$$\widehat{f}_{n,h}(x) = \sum_{j=1,\ldots,k} \frac{n_j}{n} \widehat{f}_{j,n_j,h}$$

with a common bandwidth h for both densities, for every x in $\mathcal{I}_{X,h}$. Let $\mu_j(x) = E(Y_j 1_{\{X_j=x\}})$ and $\mu(x) = \sum_{j=1,\ldots,k} \lambda_j \mu_j(x)$, $\mu_j(x)$ is estimated by $\widehat{\mu}_{j,n,h_j}(x) = (n_j h_j)^{-1} \sum_{i=1}^{n_j} Y_{ij} K_h(x - X_{ij})$ and $\mu(x)$ it is estimated by

$$\widehat{\mu}_{n,h}(x) = \sum_{j=1,\ldots,k} \frac{n_j}{n} \widehat{\mu}_{j,n_j,h}.$$

The notations are those of Section 4.7. A statistic for a Kolmogorov-Smirnov test for H_0 relies on the differences of the regression functions of the kth sub-sample and the other $k - 1$ sub-samples.

Let us consider the estimator $\widehat{b}_{m,n,h}$ of the bias constant b_m of the regression estimators under H_0, defined in Section 4.7 and the statistic

$$S_n = \sum_{j=1,\ldots,k-1} \sup_{x \in \mathcal{I}_{X,h}} \left| \left(\frac{n_j n_k}{n_j + n_k} \right)^{\frac{s}{2s+1}} (\widehat{m}_{j,n_j,h_j} - \widehat{m}_{k,n_k,h_k})(x) \right.$$

$$\left. - \left\{ \gamma_j \left(\frac{n_k}{n_j + n_k} \right)^{\frac{s}{2s+1}} - \gamma_k \left(\frac{n_j}{n_j + n_k} \right)^{\frac{s}{2s+1}} \right\} \widehat{b}_{m,n,h}(x) \right| \quad (5.20)$$

with bandwidths converging with the optimal rates.

Proposition 5.12. *Under Conditions 1.1 and 3.1 for each regression curve, the statistic S_n converges under H_0 to the supremum of the centered Gaussian process*

$$\sum_{j=1,\ldots,k-1} \sup_{\mathcal{I}_{X,h}} \left| \left(\frac{\lambda_k}{\lambda_j + \lambda_k} \right)^{\frac{s}{2s+1}} B_j - \left(\frac{\lambda_j}{\lambda_j + \lambda_k} \right)^{\frac{s}{2s+1}} B_k \right|.$$

Under the alternative, it tends to infinity.

This is a consequence of the asymptotic behavior of the processes B_{j,n_j,h_j}, $j = 1, \ldots, k$. Their asymptotic biases under H_0 are equal if the bandwidths are equal and if the sub-sample sizes are equal. With unequal sub-sample sizes, the asymptotic biases of B_{1,n_1,h_1} and B_{2,n_2,h_2} differ and the bias correction in (5.20) is only valid under this condition.

Proposition 5.13. *Let* $(K_n)_{n \geq 1}$ *be local alternatives defined by regression functions* $m_{j,n_j} = m + n_j^{-\frac{s}{2s+1}} r_{j,n_j}$ *such that* $(r_{j,n_j})_{n_j \geq 1}$ *converges uniformly to a function* r_j, $j = 1, \ldots, k$. *Under Conditions 1.1 and 3.1, the statistic* S_n *converges weakly under* $(K_n)_{n \geq 1}$ *to the supremum of the noncentered Gaussian process*

$$S_k = \sum_{j=1,\ldots,k-1} \sup_{\mathcal{I}_{X,h}} \left| \left(\frac{\lambda_k}{\lambda_j + \lambda_k} \right)^{\frac{s}{2s+1}} B_j - \left(\frac{\lambda_j}{\lambda_j + \lambda_k} \right)^{\frac{s}{2s+1}} B_k \right.$$

$$\left. - \gamma_j \left(\frac{\lambda_k}{\lambda_j + \lambda_k} \right)^{\frac{s}{2s+1}} b_{r_j,j} + \gamma_k \left(\frac{\lambda_j}{\lambda_j + \lambda_k} \right)^{\frac{s}{2s+1}} b_{r_k,k} \right|$$

where $b_{r_j,j}$ *is calculated by replacing* m_j *by* r_j *in the expression of the bias of the regression curves.*

A Cramer-von Mises statistic is the sum of the integrated squared differences between the estimators of m_j and m_k, for $j = 1, \ldots, k-1$

$$T_n = \sum_{j=1,\ldots,k-1} \left(\frac{n_j n_k}{n_j + n_k} \right)^{\frac{2s}{2s+1}} \int_{\mathcal{I}_{X,h}} |w_{j,n}(x) \widehat{m}_{j,n_j,h_j}(x) \quad (5.21)$$

$$- w_{kn}(x) \widehat{m}_{k,n_k,h_k})(x)|^2 \, d\widehat{F}_{X,n}(x),$$

where the sequences of weighting functions converge uniformly to functions w_j which may be the inverses of estimators of the variances of Y_j conditionally on X_j, $j = 1, \ldots, k$. With the optimal convergence rates of the bandwidths, a bias correction is performed as in (5.20) using the estimator of the biases under H_0, the corrected statistic is centered under the hypothesis and it diverges under fixed alternatives. With d-dimensional regression variables, the estimators are unchanged, the convergence rates are modified according to the dimension of the regressors, they are replaced by $n^{\frac{s}{2(s+d)-1}}$ and the normalization of the test statistics is modified by this rate.

5.10 Tests in proportional odds models

Consider a regression model with a discrete response variable Y corresponding to a categorization of an unobserved continuous real variable Z in a partition $(I_k)_{k \leq K}$ of its range, with the probabilities $\Pr(Z \in I_k) = \Pr(Y = k)$.

With a regression variable X and intervals $I_k = (a_{k-1}, a_k)$, the cumulated conditional probabilities are

$$\pi_k(X) = \Pr(Y \leq k \mid X) = \Pr(Z \leq a_k \mid X),$$

and $E\pi_K(X) = 1$. In Pons (2011), the proportional odds model is defined through the logistic model

$$\pi_k(X) = p(a_k - m(X)),$$

with the logistic probability $p(y) = \exp(y)/\{1 - \exp(y)\}$ and a regression function m, the nonparametric function m was estimated by kernel smoothing from the empirical odds ratios. This model is equivalent to

$$\pi_k(X)\{1 - \pi_k(X)\}^{-1} = \exp\{a_k - m(X)\}$$

for every function π_k such that $0 < \pi_k(x) < 1$ for every x in $\mathcal{I}_X$ and for $1 \leq k < K$. This implies that the odds-ratio for the observations (X_i, Y_i) and (X_j, Y_j) with Y_i and Y_j in the same class does not depend on the class

$$\frac{\pi_k(X_i)\{1 - \pi_k(X_j)\}}{\{1 - \pi_k(X_i)\}\pi_k(X_j)} = \exp\{m(X_j) - m(X_i)\},$$

for every $k = 1, \ldots, K$, this is the proportional odds model.

The probability functions $\pi_k(x)$ are estimated by the proportions $\widehat{\pi}_{k,n}(x)$ of observations of the variable Y in class k, conditionally on the regressor value x. Let

$$U_{ik} = \log \frac{\widehat{\pi}_{k,n}(X_i)}{1 - \widehat{\pi}_{k,n}(X_i)}, \ i = 1, \ldots, n,$$

calculated from the observations $(X_i, Y_i)_{i=1,\ldots,n}$ such that $Y_i = k$. In the logistic model, the value $x_k = \pi_k^{-1}(\frac{1}{2})$ such that $\pi_k(X)\{1 - \pi_k(X)\}^{-1} = 1$ is estimated by $\widehat{x}_k = \widehat{\pi}_{k,n}^{-1}(\frac{1}{2})$ such that $\widehat{\pi}_{k,n}(\widehat{x}_k)\{1 - \widehat{\pi}_{k,n}(\widehat{x}_k)\}^{-1} = 1$ hence $U_{ik} = 0$. The constant a_k is then estimated by

$$\widehat{a}_{n,h,k} = \widehat{m}_{n,h}(\widehat{x}_k).$$

The variations of the regression function m between two values x and y are estimated by

$$\widehat{m}_{n,h}(x) - \widehat{m}_{n,h}(y) = K^{-1} \sum_{k=1}^{K} \left\{ \frac{\sum_{i=1}^{n} U_{ik} K_h(X_i - x)}{\sum_{i=1}^{n} K_h(X_i - x)} - \frac{\sum_{i=1}^{n} U_{ik} K_h(X_i - y)}{\sum_{i=1}^{n} K_h(X_i - y)} \right\}.$$

This estimator yields an estimator for the derivative of the regression function, $\widehat{m}_{n,h}^{(1)}(x) = \lim_{|x-y| \to 0}(x-y)^{-1}\{\widehat{m}_{n,h}(x) - \widehat{m}_{n,h}(y)\}$, it is the derivative of the kernel estimator of a regression curve with responses U_{ik} to the regression variable X_i. In practice, a choice of $x - y = 2h$ with a kernel function K defined in $[-1, 1]$ yields a difference of two values of the kernel estimators. Integrating the mean derivative from an initial value where it is known provides a nonparametric estimator of the regression function m.

A test of equality of the conditional probabilities π_k and π_l is equivalent to a test of the hypothesis $H_0 : a_k = a_l$ since the odds ratio is a constant

$$\frac{\pi_k(X)\{1 - \pi_l(X)\}}{\{1 - \pi_k(X)\}\pi_l(X)} = \exp\{a_k - a_l\},$$

for k and l in $\{1, \dots, K\}$. Let

$$\beta_k(x) = \frac{\pi_k(x)}{1 - \pi_k(x)}, \quad \beta_l(x) = \frac{\pi_l(x)}{1 - \pi_l(x)},$$

under H_0 and at $x_k = x_l$

$$\frac{\pi_k(x_k)\{1 - \pi_l(x_l)\}}{\{1 - \pi_k(x_k)\}\pi_l(x_l)} = \exp\{m(x_k) - m(x_l)\} = 1.$$

The hypothesis H_0 is equivalent to $\beta_k(x) = \beta_l(x)$ for every x in the support of X and we consider the difference of the estimators or

$$U_{n,kl} = n^{\frac{1}{2}} \sup_x |\widehat{\beta}_{n,h,k}(x) - \widehat{\beta}_{n,h,l}(x)|.$$

Under H_0, the variable $U_{n,kl}$ converges weakly to the supremum of a Gaussian process, as in the previous section. A bias correction must again be performed to get a centered limiting process.

5.11   Tests for observations by intervals

The tests built in Chapter 4.8 are generalized to k sub-samples of observations by intervals, with the same notations. Let $A_{i_1,\dots,i_k}$, $i_j = 1, \dots, l_j$, $j = 1, \dots, k$ denote the i_jth interval of observation for the jth component of a random variable X which is observed in $A_i = A_{i_1} \times \cdots \times A_{i_k}$, for $i = (i_1, \cdots, i_k)$ such that $i_j = 1, \dots, l_j$ for $j = 1, \dots, k$. The observations cumulated by intervals are the random integer vectors

$$N_i = n \int_{A_i} d\widehat{F}_n, \; i = (i_1, \dots, i_k), \tag{5.22}$$

for all $i_j = 1, \dots, l_j$, $j = 1, \dots, k$ defined from the cumulated empirical distribution function of X. The array of counting variables $n^{-1}(N_{i_1,\dots,i_k})_{i_j=1,\dots,l_j, j=1,\dots,k i=1,\dots,k}$ has a multinomial distribution with a $l_1 \times \cdots \times l_k$ dimensional array of probabilities $p_i = p_{i_1,\dots,i_k} = \int_{A_i} dF$.

5.11.1 *Test of independence*

The hypothesis H_0 of independence is the factorization of the probabilities p_i as $p_i = EN_i = \prod_{j=1,\ldots,k} EN_{j,i_j} = \prod_{j=1,\ldots,k} p_{j,i_j}$, where N_{j,i_j} is the jth marginal of N_i, from (5.22), for all sub-intervals $i_j = 1, \ldots, l_j$ where the jth component of the variable X is observed, for $j = 1, \ldots, l$. Tests of independence rely on the vector of the differences $n^{-1}(N_i - \prod_{j=1,\ldots,k} N_{j,i_j})$, for $j = 1, \ldots, k$. The probabilities of their multinomial distribution are zero under the hypothesis of independence and they sum up to zero under all alternatives. The arrays v_{in} of their variances are calculated under the multinomial distribution, they have the form $n^{-1}v_{i,0}$ under H_0. The random array $n^{\frac{1}{2}}(N_i - \prod_{j=1,\ldots,k} N_{j,i_j})_{i_j=1,\ldots,l_j-1,j=1,\ldots,k}$ converges weakly to a centered Gaussian array of elements with variances $v_{i,0}$ under H_0. Their covariances are zero under H_0 since their components are linearly independent. Under H_0, let $\widehat{v}_{i,0} = O(n^{-1})$ be the empirical estimator of $n^{-1}v_{i,0}$ under H_0, the array W_n of the independent components of the variables $N_i - \prod_{j=1,\ldots,k} N_{j,i_j}$ normalized by their variances converges weakly to an array of linearly dependent normal variables. Let $m_k = \prod_{j=1,\ldots,k}(l_k - 1)$.

Proposition 5.14. *The test statistic defined as*

$$T_n = \sum_{i_1=1,\ldots,l_1-1} \cdots \sum_{i_k=1,\ldots,l_k-1} \widehat{v}_{i_1,\ldots,i_k,0}^{-1} \left(N_{i_1,\ldots,i_k} - \prod_{j=1,\ldots,k} N_{j,i_j}\right)^2$$

converges weakly to a $\chi^2_{m_k}$ variable under H_0. Under fixed alternatives, T_n diverges.

The proof is similar to the proof of Proposition 4.13. The test based on the statistic T_n is therefore consistent with an asymptotic power 1 for every fixed alternatives.

Under fixed alternatives, the variances v_i are $O(1)$ and the statistic diverges. Let K_n be a sequence of local alternatives with probabilities $p_{i,n}$ converging to $p_i = \prod_{j=1,\ldots,k} p_{j,i_j}$ and such that $r_{i,n} = n^{-\frac{1}{2}}(\prod_{j=1,\ldots,k} p_{j,i_j} - p_{i,n})$ converges to a limit r_i, for all $i = (i_1, \ldots, i_k)$ such that i_j belongs to $\{1, \ldots, l_j - 1\}$, for $j = 1, \ldots, k$. Under K_n, the variances $v_{i,n}$ are $O(n^{-1})$ and $nv_{i,n}$ converges to a non null and finite limit $v_{i,K}$ differing from the limit $v_{i,0}$ under H_0. Let $(Z_{i_1,\ldots,i_k})_{i_j=1,\ldots,l_j-1,j=1,\ldots,k}$ be an array of independent and centered Gaussian variables with variance 1. The limiting distribution of the statistic under K_n is obtained by rescaling the variables of T_n.

Proposition 5.15. *The statistic T_n converges weakly under K_n to a variable $T(r) = \sum_{i=1}^{k} \sum_{j=1}^{l} v_{i,K} v_{i,0}^{-1}(Z_i + r_i)^2$.*

5.11.2   *Test of homogeneity*

In tests of homogeneity for k independent sub-samples of a real variable X, the cumulative observations of the occurence of the kth variable in the same real intervals A_i are denoted $N_{ji} = n_j \int_{A_i} d\widehat{F}_{j,n_j}$, for $i = 1, \ldots, l$ and $j = 1, \ldots, k$, with the respective empirical distributions functions $\widehat{F}_{j,n_j}$ of F_j obtained from a sub-sample of size n_j, for every $j = 1, \ldots, k$. The matrix of counting variables $n_j^{-1}(N_{ji})_{i=1,\ldots,l,j=1,\ldots,k}$ has a multinomial distribution with probabilities $(p_{ji})_{i=1,\ldots,l,j=1,\ldots,k}$.

The hypothesis H_0 of homogeneity of the distributions of the sub-samples is the hypothesis of $p_{ji} = p_{ki}$, for all $i = 1, \ldots, l$ and $j = 1, \ldots, k-1$. Test statistics of homogeneity for k independent sub-samples of respective sizes n_j, $j = 1, \ldots, k$ of a variable X observed by intervals rely on the vectors of the differences $(n_j^{-1}N_{ji} - n_k^{-1}N_{ki})_{i=1,\ldots,l-1}$, for every $j = 1, \ldots, k-1$. Under the hypothesis H_0, their components have the means $p_{ji} - p_{ki}$ which equal zero and the variances

$$v_{jk,i} = \frac{1}{n_j}p_{ji}(1 - p_{ji}) + \frac{1}{n_k}p_{ki}(1 - p_{ki})$$

have the same expression under H_0 and all alternatives. Let $\widehat{v}_{i,n}$ be the empirical estimator of v_i, the vector of the normalized observations

$$W_{j,n_j} = \{\widehat{v}_{i,n}^{-\frac{1}{2}}(n_j^{-1}N_{ji} - n_k^{-1}N_{ki})\}_{i=1,\ldots,l-1}$$

converges weakly to a l-dimensional vector of linearly independent normal variables $(Z_{ji})_{i=1,\ldots,l-1}$ since $\sum_{i=1,\ldots,l} Z_{ji} = 0$.

Proposition 5.16. *The test statistic defined as*

$$T_n = \sum_{j=1,\ldots,k-1} \sum_{i=1,\ldots,l-1} \widehat{v}_{ji,n}^{-1}(n_j^{-1}N_{ji} - n_k^{-1}N_{ki})^2$$

converges weakly to a $\chi^2_{(k-1)(l-1)}$ variable under H_0. Under fixed alternatives, T_n diverges.

Under fixed alternatives, among the k vectors defined by $j = 1, \ldots, k-1$, at least one component of a vector $(p_{ji} - p_{ki})_{i=1,\ldots,l-1}$ is nonzero and the vectors W_{j,n_j} diverge. The test based on the statistic T_n is therefore consistent and its asymptotic power is 1 for every fixed alternatives. Let K_n be a sequence of local alternatives with probabilities p_{ji,n_j} converging to p_{ji} and such that $r_{ji,n} = (n_j n_k)^{\frac{1}{2}}(n_j + n_k)^{-\frac{1}{2}}(p_{ji,n} - p_i)$ converges to a limit r_{ji} different from zero, for every $i = 1, \ldots, k$. Under K_n, their means are

$$E(n_j^{-1}N_{ji} - n_k^{-1}N_{ki}) = p_{ji,n} - p_{ki,n_j} = \left(\frac{n_j n_k}{n_j + n_k}\right)^{\frac{1}{2}}(r_{ji} - r_{ki})$$

and their variances are proportional to $(n_j + n_k)(n_j n_k)^{-1}$.

Proposition 5.17. *The statistic T_n converges weakly under K_n to a variable $T(r) = \sum_{j=1,\ldots,k-1} \sum_{i=1,\ldots,l-1} \{Z_{ji} + (r_{ji} - r_{ki})\{p_i(1 - p_i)\}^{-\frac{1}{2}}\}^2$.*

5.11.3 *Comparison of k regression curves*

Let $(X_j, Y_j)_{j=1,\ldots,k}$ be k independent sets of variables defined by the functions $m_j(x) = E(Y_j | X_j = x)$ for x in a subset $\mathcal{I}_X$ of $\mathbb{R}^d$. The hypothesis H_0 of identical regression curves $m_1 = \cdots = m_k$ in a sub-interval $\mathcal{I}$ of the support of the regression variables X_j is tested from the cumulated observations of the variables X_j in the sub-intervals of a common partition $(A_i)_{i=1,\ldots,l}$ of $\mathcal{I}$ for all curves. The cumulated observations of X_j are the sums of the indicators of X_j in the set A_i

$$N_{i,j} = \sum_{a=1}^{n} 1_{\{X_{ja} \in A_i\}}$$

and the variables Y_j are continuously observed in $m(\mathcal{I})$. The parameters of the model are $m_{i,j} = E(Y_j | X_j \in A_i)$, for $i = 1, \ldots, l$ and $j = 1, \ldots, k$. Their empirical estimators calculated from k independent samples of respective size n_j are defined by (3.26). With the notations of Section 4.8.3, a goodness of fit test statistic for the hypothesis $H_0 : m_1(A_i) = \cdots, m_k(A_i)$, for every class $i = 1, \ldots, l$, is defined as the normalized squared $l^2(\mathbb{R}^k)$-distance between the vectors with components $m_j(A_i)$ and respectively $m_k(A_i)$. Its empirical estimator is

$$S_{n,kl} = \sum_{j=1}^{k-1} \sum_{i=1}^{l} \frac{(\widehat{m}_{ji,n_i} - \widehat{m}_{ki,n_k})^2}{n_j^{-1}\widehat{v}_{ji,n_j} + n_k^{-1}\widehat{v}_{ki,n_k}}.$$

The mean of $\widehat{m}_{ji,n_j} - \widehat{m}_{ki,n_k}$ is zero for every $i = 1, \ldots, l$ under H_0 and there exists at least an interger $i_K \leq l-1$ such that they differ for i_K under an alternative K.

Proposition 5.18. *The statistic $S_{n,kl}$ converges weakly under H_0 to a $\chi^2_{(k-1)l}$ distribution as n_j tend to infinity with n. Under fixed alternatives, it diverges.*

By definition, the vector of components the variables $\{n_j^{-1}\widehat{v}_{ji,n_j} + n_k^{-1}\widehat{v}_{ki,n_k}\}^{-\frac{1}{2}}(\widehat{m}_{ji,n_j} - \widehat{m}_{ki,n_k})^2$, $i = 1, \ldots, l$, converges under H_0 to a vector $(Z_1, \ldots, Z_{k-1})$ of $k-1$ independent normal variables. Under alternatives, at least a component has a nonzero mean and it tends to infinity due to the normalization.

Let K_n be a sequence of local alternatives with conditional mean values m_{ji,n_j} for Y_j and m_{ki,n_k} for Y_k, such that m_{ji,n_j} and m_{ki,n_k} converge to the same limit m_i and $r_{ji,n_j} = (n_j n_k)^{\frac{1}{2}}(n_j + n_k)^{-\frac{1}{2}}(m_{ji,n_j} - m_i)$ converges to a limit r_{ji}, for all $i = 1, \ldots, l$ and $j = 1, \ldots, k-1$.

Proposition 5.19. *Under K_n, the statistic S_n converges weakly to a non-centered χ_k^2 variable $S(r) = \sum_{j=1}^{k-1} \sum_{i=1}^{l} \{X_{i,k} + v_i^{-\frac{1}{2}}(r_{ji} - r_{ki})\}^2$.*

5.12 Competing risks

In competing risks, the minimum of k possible variables occurs and censors the other $k-1$ variables. Let $T_1, \ldots, T_k$ be k dependent or independent positive random variables with respective marginal distribution functions F_j in $\mathbb{R}_+$ and with joint distribution function F in $\mathbb{R}_+^k$. The distribution function of the minimum $T_{m(k)}$ of k random variables T_k at a real value t is $F_{m(k)}(t) = 1 - \bar{F}(t, \ldots, t)$, where $\bar{F}$ is the joint survival function of the variables defined in the rectangles of $\mathbb{R}_+^k$. With independent competing risks, $F_{m(k)}(t) = 1 - \prod_{j=1,\ldots,k} \{1 - F_j(t)\}$. When the underlying variables $T_1, \ldots, T_k$ are unobserved, their marginal distribution functions are unidentifiable in a nonparametric setting.

In a semi-Markov jump process in a finite discrete state space $\mathcal{E}$, the direct transition functions between the states are described by the probabilities of direct transitions $p_{jj'}$ from a state j to another state j' and by the sojourn time distributions $F_{|jj'}$ as functions of the actual state j and the state j' reached from there at the end of the sojourn, for all j and j' in $\mathcal{E}$. The state j' is defined by competing risks as the state with the smallest sojourn time $W_{jj'}$ in j before going towards another state of $\mathcal{E}$. In a semi-Markov model as well as in a competing risks model, only the sub-distribution functions $F_{j'|j} = p_{jj'} F_{|jj'}$ are identifiable and $p_{jj'} = F_{j'|j}(\infty)$. Under independent competing risks

$$\overline{F}_{j'|j}(t) = \int_0^t \left\{ \prod_{j'' \neq j'} \overline{F}_{|jj''}(u) \right\} dF_{|jj'}(u) \tag{5.23}$$

$$= \int_0^t \frac{\overline{F}_{j'|j}^-}{\overline{F}_{j'|j}} \prod_{j''} \frac{\overline{F}_{j''|j}(u)}{p_{jj''}} d\Lambda_{j'|j}(u),$$

where $d\Lambda_{|jj'} = (\overline{F}_{|jj'}^-)^{-1}d\overline{F}_{|jj'}$, for all states j and j' of $\mathcal{E}$, with the convention $\frac{0}{0} = 0$ for the null probabilities.

The sub-distribution functions $F_{j'|j}(t)$ are estimated using Gill's empirical estimator (1980) from the observation of n independent sojourn times $(T_{i,j'|j})_{i=1,...,N_{j'|j}}$, where $N_{j'|j}$ is the random number of observed direct transitions from j to j'. Estimators $\widehat{p}_{jj',n}$ and $\widehat{p}_{jj',0n}$ of the probabilities of direct transitions from j to j' are deduced as $p_{jj'} = F_{j'|j}(\infty)$ under general conditions and under the independent competing risks assumption from (5.23). A test of independent competing risks is performed by a comparison of the estimators $\widehat{p}_n$ and $\widehat{p}_{0n}$ of the matrix of probabilities $(p_{jj'})_{j'\neq j\in\mathcal{E}}$. Both estimators are asymptotically centered Gaussian matrices and a bootstrap test can be performed with the statistic $n^{\frac{1}{2}}(\widehat{p}_{jj',n}-\widehat{p}_{jj',0n})$. The asymptotic power of the test against fixed alternatives is 1.

A direct nonparametric estimator of $\overline{F}_{|jj'}$ in a semi-Markov jump process is defined by Pons (2004) and from (5.23), a test statistic for H_0 is defined as

$$\sup_{t\in\mathbb{R}}\left|\widehat{F}_{j'|j,n}(t) - \widehat{p}_{jj',0n}\int_0^t\left\{\prod_{j''\neq j'}\widehat{\overline{F}}_{|jj'',n}(u)\right\}d\widehat{F}_{|jj',n}(u)\right|.$$

It converges under H_0 to the supremum of a centered Gaussian process and it yields a consistent test under local alternatives.

5.13 Tests for Markov renewal processes

The Markov renewal processes are generalized as processes with continuous marks which consist in predictable processes of covariates acting on the transition probabilities and functions. Consider a n-sample of independent sample-paths of a Markov renewal process observed up to a variable t_i from a known entry time $t_0 = 0$. For n independent sample-paths, the observation of the process on the interval $[0, t_i]$ consists in the sequence of states $J_i = (J_{i,0}, J_{i,1}, \ldots, J_{i,K_i})$ in a finite and irreducible m-dimensional state space $\mathcal{C}_m = \{1, \ldots, m\}$, where $J_{i,0}$ is the initial state and J_{i,K_i} the final state after a random number of transitions K_i. Let $T_i = (T_{i,1}, \ldots, T_{i,K_i})$ be the sequence of transition times, with $T_{i,k}$ the arrival time in state $J_{i,k}$ and $X_{i,k} = T_{i,k} - T_{i,k-1}$ the sojourn time in $J_{i,k-1}$, $i = 1, \ldots, n$. The covariate process Z_i is written as a sequence of its values on each sojourn intervals: $Z_i = (Z_{i,0}, \ldots, Z_{i,K_i})$, where $Z_{i,k}$ is a m-dimensional explanatory covariate for the distribution of the duration $X_{i,k}$, defined for x in $[0, X_{i,k+1}[$ by

$Z_{i,k}(x) = Z_i(T_{i,k} + x)$ from values of Z_i on the interval $[T_{i,k}, T_{i,k+1}[$. They are variables or processes with sample-paths $Z_{i,k}$ in a bounded subspace $\mathcal{Z}$ of $D([0,\tau])$.

Each sojourn time $X_{i,k}$ may be censored at a random time $C_{i,k}$, independent of $(T_{i,j})_{1 \leq j \leq k-1}$ and $(J_{i,k})_{0 \leq j \leq k-2}$ conditionally on $Z_{i,k}$, and depending on $J_{i,k-1}$ for $k = 1, \ldots, K_i + 1$ and $i = 1, \ldots, n$. Only the last sojourn time in a transient state J_{i,K_i} is censored, the last observed time t_i is then equal to $T_{i,K_i} + C_{i,K_i+1}$ and the last time C_{i,K_i+1} is denoted X_i^*.

Let $\delta_{i,k} = 1\{X_{i,k} \leq C_{i,k}\}$ for each sojourn time in a transient state and $\delta_{i,0} = 1$. If J_{i,K_i} is a transient state, the observation i is censored and the censoring indicator is $\delta_i = 0$, otherwise J_{i,K_i} is an absorbing state and the indicator is 1. The conditional probability of direct transition from state j to j', given the covariate, and the conditional distribution of the sojourn times, given the sojourn state j, the next state j' and the covariate, are denoted $p_{jj'}(z)$ and, respectively, $F_{|jj'z}$, and $F_{j'|jz}$ is deduced. The conditional hazard functions of the durations before a transition between two consecutive states j and j' are

$$\lambda_{|jj'z}(x) = \frac{d}{dx}P(x \leq X_{i,k} \leq x+dx \mid X_{i,k} \geq x, J_{i,k-1} = j, J_{i,k} = j', Z_{i,k-1}).$$

They are supposed to follow a multiplicative semi-parametric model

$$\lambda_{|jj'z}(x, \beta_{jj'}) = \lambda_{|jj'}(x)r_j(\beta_{jj'}, Z(x)) \tag{5.24}$$

where $\beta_{jj'}$ is a m-dimensional vector and $r_j(\beta, z)$ a known parametric function of the covariates at z, with parameter β. Maximum likelihood estimators of the parameter array and of the function $\lambda_{|jj'}(x)$ are defined by Pons (2008). They are asymptotically Gaussian under integrability and derivability conditions. Asymptotically Gaussian estimators of the conditional transition functions are deduced.

Goodness of fit tests and tests of equality of the transition intensities between the states are deduced from the functional estimators, they extend the tests of comparison of hazard functions and the tests for k-samples of variables. Tests about the parameters are similar to the tests in parametric models of densities.

5.14   Tests in $\mathbb{R}^{k_n}$ as k_n tends to infinity

Up to now, a finite number of sub-samples had increasing sizes of the same order as the sample size. Consider a n-sample of a random variable X in

$\mathbb{R}^{k_n}$, where the dimension k_n of the variable increases with n. A n-sample of X consists in an increasing number k_n of sub-samples and the sub-sample size n_j of the jth sub-sample increases at a slower rate than n. We assume that there exists a strictly positive constant a in $]0, \frac{1}{2}[$ such that

$$k_n = O(n^a), \qquad n_j = O(n^b),$$

with $b = 1 - a$, then $n^{-1} \sum_{j=1,\ldots,k_n} n_j = 1$. The marginal distribution functions have empirical estimators $\widehat{F}_{j,n}$ obtained from a sub-sample with the size $n_j = O(n^{1-a})$ and they converge to the marginal distribution functions F_j with the rate $n_j^{\frac{1}{2}} = O(n^{\frac{1-a}{2}})$, $j = 1, \ldots, k_n$. The joint empirical distribution function $\widehat{F}_n$ converges to the joint distribution function F with the rate $n^{\frac{1}{2}}$.

The asymptotic properties of the statistics 5.1 and 5.2 for the hypothesis of independence of components of X are modified. Under the hypothesis H_0 of independence of $k_n = O(n^a)$ components of a random variable, the expansion (5.3) diverges. The difference of the empirical estimators $\widehat{F}_n - \prod_{j=1,\ldots,k_n} \widehat{F}_{j,n}$ satisfies an expansion similar to (5.3) with a modified convergence rate. The process G_n of this expansion is asymptotically equivalent to ν_n and it is expanded as

$$G_n(x) = n^{\frac{1}{2}} \{ \widehat{F}_n(x) - \prod_{j=1}^{k_n} \widehat{F}_{j,n}(x_j) + \sum_{j=1}^{k_n} (\widehat{F}_{j,n} - F_j)(x_j) \prod_{l \neq k_n, l=1}^{k_n} F_l(x_l)$$

$$+ \prod_{j=1}^{k_n} F_j(x_j) - F(x) \} + o_p(1). \tag{5.25}$$

Since the marginal empirical distribution function in $\mathbb{R}^{k_n}$ have not the same convergence rates as ν_n, the tests statistics differs from those with a fixed number of components. They are based on the process $Z_n = n^{\frac{1-3a}{2}} \{ \widehat{F}_n(x) - \prod_{j=1}^{k_n} \widehat{F}_{j,n_j}(x_j) \}$.

Proposition 5.20. *Under H_0, the process Z_n is asymptotically equivalent to $Z = -F k_n^{-1} \sum_{j=1}^{k_n} F_j^{-1} W_j$ in $\mathbb{R}^{k_n}$. Under local alternatives defined by sequences of joint distribution functions $F_n = \prod_{j=1}^{k} F_{j,n} + n^{-\frac{1-3a}{2}} H_n$, where $(H_n)_n$ is a sequence functions H_n converging uniformly to a function $H \neq 0$ in $\mathbb{R}^{k_n}$, the process Z_n is asymptotically equivalent to $H + Z$.*

Proof. In the expansion (5.25), the processes $n^{\frac{1}{2}}(\widehat{F}_{j,n} - F_j)$ have the order $(nn_j^{-1})^{\frac{1}{2}} = n^{\frac{1-b}{2}} = n^{\frac{a}{2}}$, hence their sum has the order $k_n n^{\frac{1}{2}} n^{\frac{-b}{2}} = n^{\frac{3a}{2}}$ and it has to be normalized by $n^{-\frac{3a}{2}}$, then under H_0

$$n^{-\frac{3a}{2}} G_n(x) = n^{\frac{1-3a}{2}} \Big\{ \widehat{F}_n(x) - \prod_{j=1}^{k_n} \widehat{F}_j(x_j)$$

$$+ \sum_{j=1}^{k_n} F_j^{-1}(x_j)(\widehat{F}_{j,n} - F_j)(x_j) \prod_{l=1}^{k_n} F_l(x_l) \Big\} + o_p(1).$$

It follows that $n^{\frac{1-3a}{2}} \{ \widehat{F}_n(x) - \prod_{j=1}^{k_n} \widehat{F}_j(x_j) \}$ is asymptotically equivalent to $-n^{\frac{1-3a}{2}} \sum_{j=1}^{k_n} F_j^{-1}(x_j)(\widehat{F}_{j,n} - F_j)(x_j) \prod_{l=1}^{k_n} F_l(x_l)$, under H_0. Its limiting distribution is obtained as in Proposition 5.1. The local alternative to the hypothesis H_0 of independence of $k_n = O(n^a)$ components is defined by a sequence of joint distribution functions $F_n = \prod_{j=1}^{k_n} F_{j,n} + n^{-\frac{1-3a}{2}} H_n$, where $(H_n)_n$ is a sequence of functions H_n converging uniformly to a function $H \neq 0$ defined in $\mathbb{R}^{k_n}$, then $n^{\frac{1-3a}{2}} \{ \prod_{j=1}^{k_n} F_{j,n}(x_j) - F_n(x) \} = H_n$ converges to H. The expansion (5.3) is still valid under the alternative

$$n^{-\frac{3a}{2}} G_n(x) = n^{\frac{1-3a}{2}} \Big\{ \widehat{F}_n(x) - \prod_{j=1}^{k_n} \widehat{F}_{j,n}(x_j) - H$$

$$+ \sum_{j=1}^{k_n} (\widehat{F}_{j,n} - F_{j,n})(x_j) F_{j,n}^{-1}(x_j) \prod_{l=1}^{k_n} F_{l,n}(x_l) \Big\} + o_p(1),$$

its limit follows. □

If the order of the sub-sample sizes varies as $n_j = O(n^{b_j})$ with scalars b_j in $]0,1[$ for every $j = 1, \ldots, k_n$, and such that $a + \sum_{j=1,\ldots,k_n} b_j = 1$, the convergence rate of the sum $\sum_{j=1}^{k_n} F_j^{-1}(x_j)(\widehat{F}_{j,n} - F_j)(x_j)$ has the order $u_n = \sum_{j=1}^{k_n} n_j^{-\frac{1}{2}} = O(\sum_{j=1}^{k_n} n^{-\frac{b_j}{2}})$ instead of $n^{-\frac{3a}{2}}$. Under H_0, the process $Z_n = n^{\frac{1}{2}} u_n^{-\frac{1}{2}} \{ \widehat{F}_n(x) - \prod_{j=1}^{k_n} \widehat{F}_{j,n_j}(x_j) \}$ is asymptotically equivalent to $Z = -F k_n^{-1} \sum_{j=1}^{k_n} F_j^{-1} W_j$ in $\mathbb{R}^{k_n}$. Under local alternatives defined by sequences of joint distribution functions $F_n = \prod_{j=1}^{k} F_{j,n} + n^{-\frac{1}{2}} u_n^{\frac{1}{2}} H_n$, where $(H_n)_n$ is a sequence of functions converging uniformly to a function $H \neq 0$ in $\mathbb{R}^{k_n}$, the process Z_n is asymptotically equivalent to $H + Z$.

In tests of homogeneity of k_n sub-samples of a n-sample, the statistics

$$W_{j,n}(x) = \left(\frac{n_j n_{k_n}}{n} \right)^{\frac{1}{2}} (\widehat{F}_{j,n} - \widehat{F}_{k_n,n})(x) \tag{5.26}$$

have the respective orders $(n^{-1} n_j)^{-\frac{1}{2}} = O(n^{\frac{-a}{2}})$ which tends to zero as n tends to infinity. In Proposition 5.3, the statistic T_n has therefore the same order as $k_n n^{-a} = O(1)$.

Proposition 5.21. *Under H_0, the statistic*

$$T_n = \sum_{j=1,\ldots,k_n-1} \int_{\mathbb{R}} W_{j,n}^2(x)\, d\widehat{F}_n(x)$$

is asymptotically equivalent to $T = n^{-a} \sum_{j=1,\ldots,k_n-1} \int_{\mathbb{R}} W_{j,\lambda}^2(x)\, dF(x)$, *which is bounded in probability as* n *tends to infinity. Under fixed alternatives, it diverges with the rate* n^b.

Proof. The process $W_{j,n}$ defined by (5.26) is asymptotically equivalent to a Wiener process $W_{j,\lambda}$ and the limiting distribution of T_n under H_0 follows. Under fixed alternatives, the mean of the processes $W_{j,n}$ are $O(n^{\frac{1}{2}-a})$, $n^{-1} \sum_{j=1,\ldots,k_n-1} n_j n_{k_n} (F_j - F_{k_n})^2$ has the order $k_n n^{2b-1} = n^b$ and it tends to infinity. □

Proposition 5.22. *Under local alternatives* $K_n : F_{j,n} = F + n^{-\frac{b}{2}} H_{j,n}$ *and there exist nonzero functions* H_j *such that* $\lim_{n \to \infty} \sup_{\mathbb{R}} |H_{j,n} - H_j| = 0$, *for every* $j = 1,\ldots,k_n$, *the statistic* T_n *is asymptotically equivalent to*

$$T = n^{-a} \sum_{j=1,\ldots,k_n-1} \int_{\mathbb{R}} \{W_{j,\lambda}(x) + H_j(x) - H_{k_n}(x)\}^2\, dF(x).$$

This convergence is a consequence of the correction of the order n^b of the processes $W_{j,n}$ under alternatives by the order of the differences $F_{j,n} - F$ under K_n.

In the comparison of k_n regression curves, in the same setting as in Section 5.9, each term of the statistic (5.20) is a $O(1)$ and their sum is a $O(k_n)$. The Kolmogorov-Smirnov type statistic $k_n^{-1} S_n$ is asymptotically equivalent under H_0 to

$$k_n^{-1} \sum_{j=1,\ldots,k_n-1} \sup_{\mathcal{I}_{X,h}} \left| \left(\frac{\lambda_{k_n}}{\lambda_j + \lambda_{k_n}}\right)^{\frac{s}{2s+1}} B_j - \left(\frac{\lambda_j}{\lambda_j + \lambda_{k_n}}\right)^{\frac{s}{2s+1}} B_{k_n} \right|.$$

Under fixed alternatives, it tends to infinity and the under local alternatives of Proposition 5.13, it is asymptotically equivalent to the finite limit of $k_n^{-1} S_{k_n}$. The Cramer-von Mises type statistic T_n has a similar asymptotic behavior. With sub-sample sizes $n_j = O(n^{b_j})$, b_j in $]0,1[$, $j = 1,\ldots,k_n$, and such that $\sum_{j=1,\ldots,k_n} b_j = 1$, the convergence rate of the statistics $W_{j,n}$ is $O_p((n^{-1}n_{k_n})^{\frac{1}{2}}) + O_p((n^{-1}n_j)^{\frac{1}{2}}) = O(n^{\frac{b_{k_n} \wedge b_j - 1}{2}})$ under H_0 and it is a $O(n^{\frac{(b_{k_n}+b_j)-1}{2}})$ under alternatives. The normalization of the tests statistic is modified in consequence.

The χ^2 statistics designed for the tests with cumulated observations in k intervals are also modified as the fixed number of intervals is replaced by a sequence of k_n intervals increasing with the number n of observations, and as the dimension of the variables increases with n. The statistics are replaced by sums of k_n squared normal variables over the k_n sub-intervals and they are asymptotically equivalent to the related means over k_n terms under the local alternatives.

5.15   Exercises

5.15.1. Calculate the variance function of the empirical process of the dependence function, $n^{\frac{1}{2}}(\widehat{C}_n - C)$ in $\mathbb{R}^k$ and define its empirical estimator. Hints. The mean $E(\widehat{C}_n^2)$ is expanded in a sum of n terms according to the equal indices in the sums of the indicators defining the estimator.

5.15.2. Define tests of independence of $l < k$ components of a variable of $\mathbb{R}^k$ observed by intervals. Hints. This is a modification of Propositions 5.14 and 5.15.

5.15.3. Define tests of homogeneity of $l < k$ components of a variable of $\mathbb{R}^k$ observed by intervals. Hints. The asymptotic properties are proved by modification of Propositions 5.16 and 5.17.

5.15.4. Generalize the goodness of fit tests of Section 3.8.1 to a distribution function from observations by intervals of a variable in $\mathbb{R}^k$ and to a subset of its components. Hints. A statistic similar to T_n is defined with a maximum likelihood estimator of the parameter under the hypothesis, the variance v_i is modified according to the dimension k and the asymptotic behavior of this statistic is proved under local alternatives K_n as in Proposition 3.18.

5.15.5. Determine the limiting distributions of the test statistics of Section 3.8.1 with cumulated observations in k_n sub-intervals such that $k_n = O(n^a)$, with a constant a in $]0, \frac{1}{2}[$. Hints. Follow the same proofs as in Section 5.14.

Chapter 6

Nonparametric tests for processes

6.1 Introduction

The maximum likelihood ratio tests for processes are the most powerful tests. Generally they are not centered and normalized, the mean and variance of the likelihood ratio statistics for Brownian motions and point processes are calculated and estimated using arguments of the theory of the continuous martingales. Several transforms can be used such as bijections between Poisson processes and sequences of independent and identically distributed variables, and the tests of the previous sections apply to the independent variables. Point processes with a functional intensity can also be transformed in a sequence of independent variables, after the estimation of their intensity. The same transforms apply to all processes with independent increments.

Martingales properties of the difference between a point process N and its predictable compensator $\widetilde{N}$ provide useful tools for the calculus of estimators and their comparison, and for the calculus of the variance of the test statistics. The martingales properties and the central limit theorem for martingales are the main arguments for the convergence of the test statistics (Appendix A.2). The asymptotic properties of the statistics are studied in the general setting of ergodic and weakly dependent processes. The tests obtained for the transformed variables are compared to the maximum likelihood ratio tests and to other classes of tests built on weighted differences between nonparametric estimators of the intensities of processes under the null hypothesis. Optimal tests are defined from a comparison of the local asymptotic power of the tests and their efficiency, as defined in Chapter 1.8 for nonparametric hypotheses. All tests defined for the comparison of two processes are generalized to tests for k processes as in the previous chapter.

6.2   Goodness of fit tests for an ergodic process

Let $(\mathbb{X}, \mathcal{X}, \|\cdot\|)$ be a functional metric space and let X be a process defined from a probability space $(\Omega, \mathcal{A}, P)$ into $\mathbb{X}$. For a continuous process, $\mathbb{X}$ is the space of continuous functions in the support $\mathcal{I}_X$ of the sample paths of X. For a process X with a stationary distribution, for all $s < t$

$$P(X_s \leq x, X_t \leq y) = P(X_0 \leq x, X_{t-s} \leq y),$$
$$= \int_0^x P(X_{t-s} \leq y | X_0 = z)\, dP_{X_0}(z).$$

Then there exists an invariant distribution function F and a transition probability measure π in $\mathcal{I}_X^{\otimes 2}$ such that for every bounded and continuous function ψ on $\mathcal{I}_X^{\otimes 2}$

$$\lim_{T \to \infty} ET^{-1} \int_{[0,T]} \psi(X_t)\, dt = \int_{\mathcal{I}_X} \psi(x)\, dF(x), \qquad (6.1)$$

$$\lim_{T \to \infty} ET^{-1} \int_{[0,T]^{\otimes 2}} \psi(X_s, X_t)\, ds\, dt = \int_{\mathcal{I}_X^{\otimes 2}} \psi(x - y)\pi_x(dy)\, dF(x).$$

In particular

$$\lim_{T \to \infty} T^{-1} \int_0^T \int_0^T P(X_s \leq x, X_t \leq y)\, ds\, dt = \int_0^x \pi_{t-s}([0, u])\, dF(u).$$

A process with independent increment is ergodic and the limit l_2 of the second term is asymptotically equivalent to $T \int_{\mathcal{I}_X^{\otimes 2}} \psi(x, y - x)\, dF(x)\, dF(y)$. For a process with covariances tending to zero, $l_2 = \int_{\mathcal{I}_X^{\otimes 2}} \psi(x, x)\, dF(x)$.

An empirical estimator of the ergodic distribution function of the process is defined for every x in $\mathcal{I}_X$ by

$$\widehat{F}_T(x) = T^{-1} \int_0^T 1_{\{X_t \leq x\}}\, dt. \qquad (6.2)$$

Its mean is $T^{-1} \int_0^T P(X_t \leq x)\, dt$ and it converges to the invariant distribution function F defined by (6.1). For every compact subset C of $\mathcal{I}_X$, $\sup_{x \in C} |\widehat{F}_T(x) - F(x)|$ converges in L_1 to zero, as T tends to infinity. The asymptotic variance of the estimator $\widehat{F}_T(x)$ is deduced from the second equation of (6.1) as

$$v_F(x) = F(x) - F^2(x).$$

The asymptotic covariance of $\widehat{F}_T(x)$ and $\widehat{F}_T(y)$ such that $x < y$ is

$$Cov_F(x, y) = \int_0^x \pi_{y-x}([0, u])\, dF(u) - F(x)F(y).$$

If the process X has a second order stationary distribution function F, the empirical process

$$\nu_T(x) = T^{\frac{1}{2}}\{\widehat{F}_T(x) - F(x)\}$$

converges weakly, as T tends to infinity, to a centered Gaussian process ν_F with variance v_F. The asymptotic variance v_F is estimated by

$$\widehat{v}_{F,T}(x) = T^{-1} \int_{[0,T]^{\otimes 2}} 1_{\{X_t \leq x\}} 1_{\{X_s \leq x\}} \, ds \, dt - \widehat{F}_T^2(x).$$

A statistic for a test of a simple hypothesis $H_0 : F = F_0$ against the alternative $K : F \neq F_0$ is

$$W_T = T^{\frac{1}{2}} \sup_{x \in I_X} \widehat{v}_{F,T}^{-\frac{1}{2}}(x) |\widehat{F}_T - F_0|.$$

The variable W_T is the supremum of a Gaussian process with asymptotic variance 1 and covariance $v_F^{-\frac{1}{2}}(x) v_F^{-\frac{1}{2}}(y) cov_F(x, y)$. Under fixed alternatives that do not contain F_0, the asymptotic power of the test is one. Its local asymptotic power against H_0 is studied as in Section 3.2 for samples of independent variables.

Consider the hypothesis H_0 of a process X with a mean distribution function F_0 belonging to a regular parametric family of distribution functions $\mathcal{F}_\Theta$ indexed by a parameter of an open and bounded set Θ, $\mathcal{F}_\Theta = \{F_\theta \in C^2(\Theta) \cap C^1(I_X)\}$. A test of H_0 relies on a maximum likelihood estimator $\widehat{\theta}_T$ of the true parameter value θ_0 belonging to Θ. The mean log-likelihood process is defined as

$$l_T(\theta) = \int_0^T \log f_\theta(X_t) \, dt.$$

The maximum likelihood estimator of the parameter θ is defined under integrability and regularity conditions and its weak convergence also requires the following conditions.

Condition 6.1.

(1) The mean integral $\lim_{T \to \infty} T^{-1} E \int_0^T \log f_\theta(X_t) \, dt$ is finite and belongs to $C^2(\Theta)$.

(2) The following information matrix is finite in a neighborhood of θ_0

$$I_\theta = -\lim_{T \to \infty} T^{-1} \int_0^T \frac{f_\theta \ddot{f}_\theta - \dot{f}_\theta^2}{f_\theta^2}(X_t) \, dt = \int_{I_X} \frac{\dot{f}_\theta^2}{f_\theta}(x) \, dx.$$

Under Condition 6.1, the process $\vartheta_T = T^{\frac{1}{2}}(\widehat{\theta}_T - \theta_0)$ converges weakly to a centered Gaussian variable ϑ_0 with variance I_0^{-1}, at F_0 under H_0.

Proposition 6.1. *The statistic*

$$W_T = \sup_{I_X} |\nu_{T,F} - T^{\frac{1}{2}}(F_{\widehat{\theta}_T} - F_0)| \qquad (6.3)$$

converges weakly under H_0 to the supremum of the centered process $\nu_0 + \vartheta_0$, with the variance $v_{F_0} + [\int_{I_X} \dot{f}_{\theta_0}(u) \{1 - \pi_u([0,x])\} du]^t I_0^{-1} \dot{F}_{\theta_0}$.

Proof. By the weak convergence of ϑ_T, the process $\widehat{\nu}_T = T^{\frac{1}{2}}(F_{\widehat{\theta}_T} - F_0)$ develops as $\widehat{\nu}_T = \vartheta_T \dot{F}_{\theta_0} + o_p(1)$

$$\vartheta_T = T^{-\frac{1}{2}} l_T(\theta_0) I_0^{-1} + o_p(1),$$

$$l_T(\theta_0) = \int_0^T \frac{\dot{f}_\theta}{f_\theta}(X_t)\, dt.$$

The process $T^{-\frac{1}{2}} l_T(\theta_0)$ converges weakly under H_0 to a Gaussian variable with the mean zero and the variance I_0, it follows that $\widehat{\nu}_T$ converges weakly under H_0 to a centered Gaussian variable with the variance function $v_{\theta_0} = \dot{F}_{\theta_0}^t I_0^{-1} \dot{F}_{\theta_0}$. The asymptotic variance of $\nu_{T,F_0} - \widehat{\nu}_T$ depends on their variance and on the covariance of $\nu_{T,F}(x)$ and $l_T(\theta_0)$ which is the limit of $E_0 T^{-1}\{\int_0^T 1_{X_t \le x}\, dt \int_0^T f_{\theta_0}^{-1}(X_s) \dot{f}_{\theta_0}(X_s)\, ds\}$, as T tends to infinity, and it equals $c_0 = \int_{I_X} \dot{f}_{\theta_0}(u) \pi_u([0,x])\, du$. $\qquad \square$

Let Ξ be a set of measurable functions defined from $\mathbb{R}$ to $\mathbb{R}^*$ and let $\mathcal{G}_{T,\Xi}$ be a class of local alternatives such that for every F_T of $\mathcal{G}_{T,\Xi}$, there exist F in $\mathcal{F}_\Theta$ and ξ in Ξ such that $\xi = \lim_{T \to \infty} T^{\frac{1}{2}}(F_T - F)$. In $\mathcal{G}_{T,\Xi}$, the limiting distribution under $P_{F_T,\xi}$ of the process W_T is $v_0^{-\frac{1}{2}} \nu_0 + \vartheta_0 \dot{F}_\theta + \xi$, where $F = F_\theta$ belongs to $\mathcal{F}_\Theta$. At the level α, the test based on the statistic (6.3) has the asymptotic local power

$$\beta_{F_\theta,\xi} = \lim_{T \to \infty} \inf_{F_T \in \mathcal{G}_{T,\Xi}, \xi \in \Xi} P_{F_T,\xi}\left(\sup_{x \in I_X} |W_n(x)| > c_{\frac{\alpha}{2}} \right)$$

$$= 1 - \sup_{\xi \in \Xi,\, F_\theta \in \mathcal{F}} P\{-c_{-\frac{\alpha}{2}} - \xi \le \nu_\theta - \vartheta_\theta \dot{F}_\theta \le c_{\frac{\alpha}{2}} - \xi\}$$

and this is an unbiased test.

6.3   Poisson process

A Poisson process N in a measurable space $(\Omega, \mathcal{A})$ is a sequence of increasing and positive time variables $(T_i)_{i \ge 1}$ such that the differences $X_i = T_i - T_{i-1}$

are independent and identically distributed variables with an exponential distribution $\mathcal{E}(\lambda)$ under a probability distribution P_λ. The Poisson process has then a constant intensity $\lambda > 0$ and the likelihood ratio of the restriction of the process N to an interval $[0, t]$, under two probability distributions P_λ and P_{λ_0} is proportional to

$$L_t(\lambda) = e^{-(\lambda - \lambda_0)t} \left(\frac{\lambda}{\lambda_0}\right)^{N(t)},$$

the parameter λ is estimated at t by $\widehat{\lambda}_t = t^{-1} N_t$. This estimator is the maximum likelihood estimator and the variable $t^{\frac{1}{2}}(\widehat{\lambda}_t - \lambda)$ is a local martingale with respect to the filtration generated by the Poisson process. The moments of the process N_t is obtained from the generating function of the Poisson process, $g_t(x) = e^{-\lambda t(1-x)}$, hence $Var N_t = \lambda t$ and the variance of the estimator $\widehat{\lambda}_t$ is λ.

Goodness of fit tests for the intensity of a Poisson process on a time interval $[0, T]$ are deduced from the asymptotic behavior of the normalized process $\kappa(t) = N_t^{-\frac{1}{2}}(N_t - \lambda t)$, as t tends to infinity. The sequence of rescaled processes defined in $[0, 1]$ by $N_T(s) = T^{-1} N_{sT}$, are Poisson processes with the intensity λ and

$$\kappa_T(s) = T^{-1} \kappa(sT) = \frac{N_T(s) - \lambda s}{\sqrt{N_T}} = \frac{N_T(s) - \lambda s}{\sqrt{\lambda s}} + o_p(1) \qquad (6.4)$$

are such that κ_T converges weakly to a Gaussian process with mean zero, variance one and covariance $(st)^{-\frac{1}{2}}(s \wedge t)$ in $[0, 1]$.

Let N_1 and N_2 be two independent Poisson processes with constant intensities λ_1 and λ_2, respectively. A test of the hypothesis $H_0 : \lambda_1 = \lambda_2$ against an alternative $H_1 : \lambda_1 \neq \lambda_2$ relies on the difference of their estimators. The variable

$$S_T = \frac{N_{1T} - N_{2T}}{\sqrt{N_{1T} + N_{2T}}} \qquad (6.5)$$

converges weakly under H_0 to a centered Gaussian variable with variance 1. A test of level α for H_0 has the rejection domain $\{|S_T| > c_{\frac{\alpha}{2}}\}$ and its asymptotic power tends to 1 as T tends to infinity. Local alternatives H_{1T} are defined as $\lambda_{2T} = \lambda_{1T}\{1 + T^{-\frac{1}{2}}\gamma_{2T}\}$ and $\lambda_{1T} = \lambda_0\{1 + T^{-\frac{1}{2}}\gamma_{1T}\}$, where the scalars γ_{kT} converge to respective limits γ_k belonging to a subset Γ of $\mathbb{R}_+$, as T tends to infinity. The asymptotic power of the test of level α for H_0 is

$$\beta_\alpha = \sup_{\gamma \in \Gamma} P\left\{\left|S_T + \gamma\left(\frac{\lambda_1}{2}\right)^{\frac{1}{2}}\right| > c_{\frac{\alpha}{2}}\right\}.$$

Its Pitman efficiency for a test of $H_0 : \theta = \lambda_2 - \lambda_1 = 0$ is 1.

The Poisson processes is generalized to point processes N with a functional intensity $\lambda : \mathbb{R}_+ \mapsto \mathbb{R}_{*+}$. They have independent increments and mean processes are defined as

$$\Lambda_T(t) = T^{-1}\Lambda(Tt), \ N_T(t) = T^{-1}N(Tt), \ t \in [0,1].$$

For every continuous and bounded function φ, let $\Lambda(\varphi)(t) = T^{-1}\int_0^{Tt} \varphi \, d\Lambda$, the function Λ is supposed to satisfy the following ergodic conditions. There exist functions $\bar{\Lambda}$ and $\bar{\Lambda}(\varphi)$ such that

$$\lim_{T \to \infty} \sup_{t \in [0,1]} |\Lambda_T(t) - \bar{\Lambda}(t)| = 0, \tag{6.6}$$

$$\lim_{T \to \infty} \sup_{t \in [0,1]} \left| T^{-1}\int_0^T \varphi \, d\Lambda - \bar{\Lambda}(\varphi)(t) \right| = 0.$$

The likelihood of the Poisson process is expressed using the cumulative intensity $\Lambda(t) = \int_0^t \lambda(s) \, ds$

$$L_t(\lambda) = e^{-\Lambda(t)} \prod_{T_i \leq t} \lambda(T_i), \ t \in [0,T],$$

and the function Λ_T is estimated by

$$\widehat{\Lambda}_T(s) = T^{-1}N_T(s) = T^{-1}N(Ts), \ s \in [0,1].$$

The process $\widehat{\Lambda}_T(s) - \Lambda_T(s)$ defined as $T^{-1}\widehat{\Lambda}(Ts) - T^{-1}\Lambda(Ts)$ is a local martingale in $[0,1]$ and the variance of

$$M_{\Lambda T}(s) = T^{\frac{1}{2}}\{\widehat{\Lambda}_T(s) - \Lambda_T(s)\}$$

is $\Lambda_T(s)$, it converges uniformly to $\bar{\Lambda}$ in $[0,1]$, as T tends to infinity. In $[0,1]$, the empirical process

$$\mathcal{L}_T(t) = \frac{\widehat{\Lambda}_T(t) - \Lambda_T(t)}{\sqrt{\Lambda_T(t)}}$$

converges weakly to a centered Gaussian process $\mathcal{L}_0$ with variance 1 as T tends to infinity, with the usual convention $\frac{0}{0} = 0$.

The asymptotic covariance of the process $\mathcal{L}_0$ at s and t in $[0,1]$ is the limit of $C_T(s,t) = \bar{\Lambda}_T(s \wedge t)\{\bar{\Lambda}_T(s)\bar{\Lambda}_T(t)\}^{-\frac{1}{2}}$, that is

$$C_0(s,t) = \bar{\Lambda}(s \wedge t)\{\bar{\Lambda}(s)\bar{\Lambda}(t)\}^{-\frac{1}{2}}.$$

From the first ergodic property, it appears that $\widehat{\Lambda}_T(t)$ is an estimator of $\bar{\Lambda}(t)$ and this a consistent estimator of the mean function $\Lambda_T(t)$.

A test of a constant intensity is a goodness of fit test to an exponential distribution for the sample of variables $X_i = T_i - T_{i-1}$, $i \geq 1$. The unknown parameter of the exponential distribution is estimated by the mean $\widehat{\Lambda}_T$ and the alternative to H_0 is a composite alternative for functions $\Lambda(t)$.

Goodness of fit tests for the function Λ on an observation interval $[0, T]$ are based on the process $\mathcal{L}(t)$ and they are similar to the tests for the distribution function of a variable. Let $H_0 : \Lambda = \Lambda_0$ in $[0, T]$ be the hypothesis of a test and $H_1 : \Lambda$ and Λ_0 are not identical in $[0, T]$. The change of variable by the inverse of the scaled process Λ_{0T} provides an asymptotically free statistic

$$W_T = \sup_{t \in [0,1]} |\mathcal{L}_T \circ \Lambda_{0T}^{-1}(t)|. \tag{6.7}$$

Under H_0, W_T converges weakly, as T tends to infinity, to the supremum of a centered Gaussian process W_0 with variance 1 and covariance $(st)^{-\frac{1}{2}}(s \wedge t)$ at s and t. Under the alternative, W_T tends to infinity, as T tends to infinity.

Let $(H_{1T})_{T>0}$ be a sequence of local alternatives with deterministic cumulative intensities $\Lambda_{0T}(t)\{1 + T^{-\frac{1}{2}}\gamma_T(t)\}$, in $[0, 1]$, such that γ_T converges uniformly to a function γ belonging to a functional space Γ in $\mathbb{R}_+$, as T tends to infinity. Under H_{1T}, the process $\widehat{\Lambda}_T - \Lambda_T$ converges uniformly in $[0, 1]$ to zero in probability and the empirical process $\mathcal{L}_T$ is recentered as

$$\mathcal{L}_{KT}(t) = \frac{\widehat{\Lambda}_T(t) - \Lambda_T(t)}{\sqrt{\Lambda_{0T}(t)}} - T^{-\frac{1}{2}} \Lambda_{0T}^{\frac{1}{2}}(t)\gamma_T(t)$$

and it converges weakly to a Gaussian process $\mathcal{L}_K$ with mean function $\gamma \Lambda_0^{\frac{1}{2}}$ and with variance 1. The limit under the alternatives K_n of the process W_T defined by (6.7) is therefore $W_K = \lim_{T \to \infty} \sup_{t \in [0,1]} |\mathcal{L}_T \circ \Lambda_{0T}^{-1}(t) - t^{\frac{1}{2}}\gamma_T \circ \Lambda_{0T}^{-1}(t)|$. The local asymptotic power of the test of level α based on the statistic W_T is $\sup_{\gamma \in \Gamma} \beta_{\Lambda_0, \gamma}$, where

$$\beta_{\Lambda_0, \gamma} = \{ \sup_{t \in [0,1]} |\mathcal{L}_0 \circ \Lambda_0^{-1}(t) + t^{\frac{1}{2}}\gamma \circ \Lambda_0^{-1}(t)| > c_{\frac{\alpha}{2}} \}.$$

Tests of a parametric hypothesis $\Lambda \in \mathcal{H}_\Theta$ against an alternative $\Lambda \in \mathcal{H}_1$, such that $\mathcal{H}_1$ and $\mathcal{H}_\Theta$ do not overlap, rely on the asymptotic behavior of the difference $\widehat{\Lambda}_T - \Lambda_{\widehat{\theta}_T, T}$ of the parametric and nonparametric estimators of the cumulative intensity under H_0. The estimator $\widehat{\theta}_T$ of the unknown parameter value θ_0 under the hypothesis is the maximum likelihood estimator of the parametric Poisson process. The log-likelihood of the process N with intensity λ_θ is expressed in $[0, T]$ with the convention

$$\int_0^T \log \lambda_\theta(s) \, dN(s) = \int_{T_{1:N(T)}}^T \log \lambda_\theta(s) \, dN(s)$$

$$
\begin{aligned}
l_T(\theta) &= -\int_{T_{1:N(T)}}^T \lambda_\theta(s) \, ds + \int_{T_{1:N(T)}}^T \log \lambda_\theta(s) \, dN(s) \\
&= \int_{T_{1:N(T)}}^T \{\log \lambda_\theta(s) - 1\} \lambda_\theta(s) \, ds + \int_{T_{1:N(T)}}^T \log \lambda_\theta(s) \, dM(s),
\end{aligned}
$$

where $M(t) = N(t) - \int_0^t \lambda_0(s) \, ds$ is the local martingale of the compensated jumps of the Poisson process under H_0. For every t, $T^{-1}M(t)$ converges uniformly to zero in probability and $T^{-\frac{1}{2}}M(t)$ converges weakly to a centered Gaussian process $\bar{M}_0$ defined in $[0,1]$, with independent increments and with variance function $\lim_{T \to \infty} T^{-1} \int_0^t \lambda_0(s) \, ds$ asymptotically equivalent to $\bar{\Lambda}_0(T^{-1}t)$, at $T^{-1}t$ in $[0,1]$, from the definition (6.6). The derivatives at θ of the likelihood ratio are

$$\dot{l}_T(\theta) = -\int_{T_{1:N(T)}}^T \dot{\lambda}_\theta(s) \, ds + \int_{T_{1:N(T)}}^T \frac{\dot{\lambda}_\theta}{\lambda_\theta}(s) \, dN(s) = \int_{T_{1:N(T)}}^T \frac{\dot{\lambda}_\theta}{\lambda_\theta}(s) \, dM_\theta(s)$$

$$\ddot{l}_T(\theta) = -\int_{T_{1:N(T)}}^T \ddot{\lambda}_\theta(s) \, ds + \int_{T_{1:N(T)}}^T \left\{ \frac{\ddot{\lambda}_\theta}{\lambda_\theta}(s) - \left(\frac{\dot{\lambda}_\theta}{\lambda_\theta} \right)^{\otimes 2}(s) \right\} dN(s).$$

The maximum likelihood estimator $\widehat{\theta}_T$ satisfies the classical weak convergence property of $\vartheta_T = T^{\frac{1}{2}}(\widehat{\theta}_T - \theta_0)$ to a centered Gaussian variable ϑ_0 with variance I_0^{-1}, under the conditions that the information matrix $I_0 = \lim_{T \to \infty} T^{-1} \int_0^T \lambda_0^{-1} \dot{\lambda}_0^{\otimes 2} \, dt$ is finite and positive definite. The matrix I_0 is estimated by

$$I_T = T^{-1} \int_{T_{1:N(T)}}^T \lambda_{\widehat{\theta}_T}^{-2}(t) \dot{\lambda}_{\widehat{\theta}_T}^{\otimes 2}(t) \, dN(t).$$

Let $\sigma_0^2(s) = [\bar{\Lambda}_0^{\frac{1}{2}}(s) - \{\dot{\bar{\Lambda}}_0^t(s) I_0^{-1} \bar{\Lambda}_0(s)\}^{\frac{1}{2}}]^2$, for s in $[0,1]$, it is estimated by

$$\widehat{\sigma}_{0T}^2(s) = [\widehat{\bar{\Lambda}}_T^{\frac{1}{2}}(s) - T^{-2}\{\dot{\Lambda}_{\widehat{\theta}_T}^t(sT) I_T^{-1} \dot{\Lambda}_{\widehat{\theta}_T}(sT)\}^{\frac{1}{2}}]^2.$$

Let M_{Λ_0} denote the process in $[0,1]$ defined as the limit in distribution of the martingales $M_{\Lambda_0 T}$, under H_0.

Proposition 6.2. *The statistic*

$$W_T = \sup_{s \in [0,1]} \widehat{\sigma}_{0T}^{-1}(s) |\widehat{\Lambda}_T(s) - \Lambda_{\widehat{\theta}_T T}(s)| \tag{6.8}$$

converges weakly under H_0 to the supremum in $[0,1]$ of the centered Gaussian process $\sigma_0^{-1}(\bar{M}_0 + \vartheta_0^t \dot{\Lambda}_0)$ with variance 1.

Proof. For the Poisson processes, $\widehat{\theta}_T$ maximizes the log-likelihood ratio

$$l_T(\theta) - l_T(\theta_0) = \sum_{i=1}^{n} \log \frac{\lambda_\theta}{\lambda_0}(T_i) - \Lambda_\theta(T) + \Lambda_0(T) \tag{6.9}$$

$$= \int_{T_{1:N(T)}}^{T} \log \frac{\lambda_\theta}{\lambda_0} \, dM_0 + \int_{T_{1:N(T)}}^{T} \left\{ \log \frac{\lambda_\theta}{\lambda_0} - \frac{\lambda_\theta}{\lambda_0} + 1 \right\} d\Lambda_0$$

with the martingale difference $M_0 = N - \Lambda_{\theta_0}$, using the notation $\Lambda_0 = \Lambda_{\theta_0}$. The first derivative $\dot{l}_T(\theta)$ of log-likelihood process l_T is

$$\dot{l}_T(\theta) = \int_{T_{1:N(T)}}^{T} \lambda_\theta^{-1} \dot{\lambda}_\theta \, dM_0 + \int_{T_{1:N(T)}}^{T} (\lambda_\theta^{-1} - \lambda_0^{-1}) \dot{\lambda}_\theta \, d\Lambda_0,$$

therefore $T^{-\frac{1}{2}} \dot{l}_T(\theta_0)$ converges weakly to a centered variable X_0 with the finite variance $v_0 = \lim_{T \to \infty} T^{-1} \int_0^T \lambda_0^{-1} \dot{\lambda}_0^{\otimes 2} \, dt$. The second derivative $\ddot{l}_T(\theta)$ of l_T is such that

$$T^{-1} \ddot{l}_T(\theta) = -T^{-1} \int_0^T \left\{ \frac{\dot{\lambda}_\theta^{\otimes 2}}{\lambda_\theta^2} + \ddot{\lambda}_\theta \left(\frac{1}{\lambda_0} - \frac{1}{\lambda_\theta} \right) \right\} d\Lambda_0 + o_p(1)$$

converges by ergodicity to a finite limit denoted $-I_0$, as T tends to infinity. Then $\widehat{\vartheta}_T^t = I_0^{-1} T^{-\frac{1}{2}} \dot{l}_T(\theta_0) + o_p(1)$ and the asymptotic variance of $\widehat{\vartheta}_T^t$ is I_0^{-1}. The process $T^{\frac{1}{2}} \{ \widehat{\Lambda}_T(s) - \Lambda_{\widehat{\theta}_T, T}(s) \}$ develops as $\{ T^{-1} \Lambda_T(s) \}^{\frac{1}{2}} \mathcal{L}_T(s) + \widehat{\vartheta}_T^t \bar{\dot{\Lambda}}_0(s) + o_p(1)$, it converges weakly to a centered Gaussian process with variance

$$\sigma_0^2(s) = \bar{\Lambda}(s) + \bar{\dot{\Lambda}}_0^t(s) I_0^{-1} \bar{\dot{\Lambda}}_0(s)$$
$$- 2\bar{\Lambda}^{\frac{1}{2}}(s) T^{-1} E_0 \{ \dot{l}_T(\theta_0) M_{0T}(s) \}^t I_0^{-1} \bar{\dot{\Lambda}}_0(s).$$

Its limit as T tends to infinity depends on

$$\lim_{T \to \infty} T^{-1} E_0 \{ \dot{l}_T(\theta_0) M_{0T}(s) \} = \lim_{T \to \infty} T^{-1} E_0 \{ M_{0T}(s) \int_0^{sT} \lambda_\theta^{-1} \dot{\lambda}_\theta \, dM_0 \}$$

$$= \lim_{T \to \infty} T^{-1} \int_0^{sT} \lambda_0^{-1} \dot{\lambda}_0 \, d\Lambda_0 = \bar{\dot{\Lambda}}_0(s).$$

The asymptotic variance of the difference of the normalized estimators of the unknown function Λ_{θ_0} is σ_0. $\qquad\square$

Local alternatives H_{1T} to the parametric hypothesis H_0 are defined by sequences of functions $(\zeta_T)_{T \geq 1}$ in the tangent space $\mathcal{Z}$ of Θ, such that the parameter under H_{1T} is $\theta_T = \theta_0 + T^{-\frac{1}{2}} \zeta_T$ where ζ_T converges to a limit ζ. The cumulative intensity under H_{1T} has a parameter θ_T and there exists $\Lambda_0 = \Lambda_{\theta_0}$, with θ_0 in Θ, such that $T^{\frac{1}{2}} (\Lambda_{\theta_T T} - \Lambda_{0T}) = \zeta_T \dot{\Lambda}_{0T} + o_p(1)$. The process $T^{\frac{1}{2}}(\Lambda_{\widehat{\theta}_T T} - \Lambda_{0T})$ converges weakly under H_{1T} to $(\vartheta_0 + \zeta)^t \bar{\Lambda}_0$. The asymptotic power of the test of level α for H_0 is then

$$\beta_\alpha = \sup_{\Lambda_0} \inf_{\zeta \in \mathcal{Z}} P\Big\{ \sup_{s \in [0,1]} \big|S_T(s) - \sigma_0^{-1}\zeta\bar{\Lambda}_0\big| > c\Big\}.$$

We also consider local alternatives H_{1T} to H_0 defined by a larger parametric space that contains Θ. Under H_0, the parameter space Θ is a d_1-dimensional subset of a d_2-dimensional parameter space Γ. Under H_0, the parameter γ is a vector with the components of θ as its d_1 first components, and its other components are zero, like in Section 2.2. Under H_{1T}, the d_1 first components of the parameter γ_T are $\gamma_{kT} = \theta_{0k} + T^{-\frac{1}{2}}\zeta_{kT}$, $k = 1, \ldots, d_1$, and the last components of γ_T are $\gamma_{kT} = T^{-\frac{1}{2}}\zeta_{kT}$, $k = d_1 + 1, \ldots, d_2$, where ζ_T belongs to the tangent space $\mathcal{Z}$ and converges to a limit ζ. The cumulative intensity under H_{1T} has a parameter γ_T and there exists $\Lambda_0 = \Lambda_{\theta_0}$, with θ_0 in Θ, such that $T^{\frac{1}{2}}(\Lambda_{\gamma_T T} - \Lambda_{0T}) = \zeta_T^t \dot{\Lambda}_{0T} + o_p(1)$. The process $T^{\frac{1}{2}}(\Lambda_{\widehat{\gamma}_T T} - \Lambda_{0T})$ converges weakly under H_{1T} to $\vartheta_0^t \bar{\Lambda}_0 + \zeta^t \bar{\Lambda}_0$, using the d_1 and d_2-dimensional vector of derivatives of Λ in $C(\mathcal{Z})$. The asymptotic power of the test is written as in the previous case, with the supremum over θ_0 in Θ of Λ_{θ_0}.

Let N_1 and N_2 be two independent Poisson processes with functional intensities λ_1 and λ_2, respectively. A test of the hypothesis $H_0 : \lambda_1 \equiv \lambda_2$ on $[0, T]$ against an alternative H_1 of non identical intensities relies on the process defined in $[0, 1]$ as (6.5) by

$$S_T(t) = \frac{N_{1T} - N_{2T}}{\sqrt{N_{1T} + N_{2T}}}(t). \tag{6.10}$$

It converge weakly under H_0 to a centered Gaussian process with variance one and covariance

$$C(s, t) = \frac{\bar{\Lambda}_1(s \wedge t) + \bar{\Lambda}_2(s \wedge t)}{\{\bar{\Lambda}_1(s) + \bar{\Lambda}_2(s)\}^{\frac{1}{2}}\{\bar{\Lambda}_1(t) + \bar{\Lambda}_2(t)\}^{\frac{1}{2}}}.$$

A test of H_0 has the rejection domain $\{\sup_{s \in [0,1]}|S_T(s)| > c\}$ such that $P_0(\sup_{s \in [0,1]}|S_T(s)| > c\} = \alpha$ and its asymptotic power tends to 1 as T tends to infinity. Local alternatives H_{1T} are defined as $\lambda_{2T}(t) = \lambda_{1T}(t)\{1 + T^{-\frac{1}{2}}\gamma_T(t)\}$, where γ_T converges uniformly to a function γ belonging to a tangent space Γ of the space of the intensity functions, as T tends to infinity. The asymptotic power of the test of level α for H_0 is

$$\beta_\alpha = \sup_{\Lambda_1} \inf_{\gamma \in \Gamma} P\Big\{ \sup_{s \in [0,1]} \Big|S_T(s) + \frac{\bar{\Lambda}_1(\gamma)(s)}{\sqrt{2\bar{\Lambda}_1(s)}}\Big| > c\Big\}.$$

A test of the hypothesis $H_0 : \lambda(t)$ is a monotone function in $[0,T]$ relies on the difference between a kernel estimator $\widehat{\lambda}_{T,h}$ and a smooth monotone kernel estimator of the intensity λ defined by the isotonization Lemma 3.2 as for a density. Let K_h be a symmetric kernel density satisfying the usual Condition 1.1, with a bandwidth h_T converging at the optimal rate. For a function λ in $C^s(\mathbb{R}_+)$, the kernel estimator is

$$\widehat{\lambda}_{T,h}(t) = T^{-1} \sum_{T_i \leq sT} K_h(sT - T_i) = T^{-1} \int_0^T K_h(t - x) \, dN(x),$$

where $t = sT$ belongs to $[0,T]$. The isotonic estimator of a monotone intensity λ is obtained by isotonization lemma, like (1.8) from the integrated kernel estimator $A_{T,h}(t) = \int_0^t \widehat{\lambda}_{T,h}(x) \, dx$. The smooth increasing estimator of the intensity is

$$\lambda^*_{T,h}(x) = \inf_{v \geq x} \sup_{u \leq x} \frac{1}{v - u} \{A_{T,h}(v) - A_{T,h}(u)\}.$$

Lemma 3.2 applies and a test of an increasing intensity λ of $C^s(\mathcal{I}_X)$, $s > 1$, is defined as a weighted integrated squared difference of the estimators $\widehat{\lambda}_{\lambda,T,h}$ and $\widehat{\lambda}^*_{\lambda,T,h}$, with a bias correction

$$S_T = T^{\frac{2s}{2s+1}} \{ \int_0^T (\lambda^*_{T,h} - \widehat{\lambda}_{T,h} - \widehat{b}^*_{\lambda,T,h}(x) + \widehat{b}_{\lambda,T,h}(x))^2(x) w_T(x) \, dx\}, \quad (6.11)$$

where w_T is a positive weight function for which there exists a function w in $[0,T]$ such that $\lim_{T \to \infty} \sup_{t \in [0,T]} |w_T(t) - w(t)| = 0$. Both estimators converge to Gaussian processes and the asymptotic behavior of the statistic S_T under H_0 and under local alternatives converging with the rate $T^{\frac{s}{2s+1}}$ is studied like in Proposition 3.9.

A test of an unimodal intensity λ is performed with the same statistic replacing its increasing estimator by an unimodal estimator. The mode M_λ of an unimodal intensity is estimated by the mode $\widehat{M}_{T,\lambda}$ of its kernel estimator and $(Th^3)^{1/2}(\widehat{M}_{\lambda,T,h} - M_\lambda)$ converges weakly to a centered Gaussian variable. The isotonic estimator of a monotone decreasing density is defined as $\lambda^*_{T,h} = \lambda^*_{I,T}$ up to $\widehat{M}_\lambda$ where it is increasing, then $\lambda^*_{T,h} = \lambda^*_{D,T}$ where it is decreasing

$$\lambda^*_{D,T}(x) = \sup_{u \leq x} \inf_{v \geq x} \frac{1}{v - u} \{A_{T,h}(v) - A_{T,h}(u)\},$$

$$\lambda^*_{I,T}(x) = \inf_{v \geq x} \sup_{u \leq x} \frac{1}{v - u} \{A_{T,h}(v) - A_{T,h}(u)\},$$

and the test statistic S_T has the same asymptotic properties as in Proposition 3.9 for a unimodal density.

6.4   Poisson processes with scarce jumps

The intensities of the previous Poisson processes are $O(1)$ and their estimators are $O(T^{-\frac{1}{2}})$. In the setting of an increasing scarcity of its random points, a Poisson process observed in a time interval $[0, T]$ has an intensity λ_T of order T^{-1}

$$\lambda_T(t) = T^{-1}\mu(t) + o(T^{-1}), \tag{6.12}$$

uniformly in $[0, T]$. By the independence of the variables $X_i = T_i - T_{i-1}$ of the Poisson process N_T and the normalization of its intensity, the mean $EN_T(sT) = T^{-1}\int_0^{sT} \mu(x)\,dx$ has a finite limit denoted $\bar{\mu}(s)$, for every s in $[0, 1]$, as T tends to infinity. This is an ergodic property. The existence of this limit ensures the existence of a limit in probability for the process $N_T(sT)$, it satisfies also the ergodic property $\lim_{T\to\infty} N_T(sT) = \bar{\mu}(s)$. The log-likelihood process of N_T observed in $[0, T]$ is

$$l_T(t) = \int_0^t \log\lambda_T(s)\,dN_T(s) - \int_0^t \lambda_T(s)\,ds$$

$$= T^{-1}\int_0^t \{\log\mu(s) - 1\}\mu(s)\,ds - T^{-1}\log T \int_0^t \mu(s)\,ds$$

$$+ \int_0^t \log\lambda_T(s)\,dM_T(s) + o(T^{-1}\log T)$$

where

$$M_T(t) = N_T(t) - \int_0^t \lambda_T(x)\,dx,\ t \in [0, T].$$

The variance of $N_T(Ts)$ is $\int_0^{Ts} \lambda_T(x)\,dx$, it converges in probability to a finite limit $\bar{\mu}(s) = \lim_{T\to\infty} T^{-1}\int_0^{Ts} \mu(x)\,dx$, for every s in $[0, 1]$, as T tends to infinity. The process $M_T(sT)$ is the local martingale of the compensated jumps of the Poisson process $N_T(sT)$ in $[0, 1]$. Its asymptotic variance is finite and it converges weakly to a centered Gaussian process G_0 defined in $[0, 1]$, with independent increments and with variance function $v_\mu = \bar{\mu}$.

Proposition 6.3. *Under the assumption (6.12), the log-likelihood process* $(\log T)^{-1}l_T(sT) = -\{M_T(sT) + \bar{\mu}(s)\} + o_P(1)$ *converges weakly to a Gaussian process with independent increments and with mean and variance functions* $v_\mu = \bar{\mu}$ *in* $[0, 1]$.

Proof.   Since $M_T(sT) = O_p(1)$, the log-likelihood process of N_T has the approximation

$$l_T(sT) = \int_0^{sT} \log\lambda_T(x)\,dM_T(x) - T^{-1}\log T \int_0^{sT} \mu(x)\,dx + o_P(\log T)$$

$$= -\log T\{M_T(sT) + \bar{\mu}(s)\} + o_P(\log T) \qquad \square$$

and $l_T(sT) = O_p(\log T)$.

For every s in $[0,1]$, the variance function $v_\mu(s)$ is consistently estimated by

$$\widehat{v}_{\mu,T}(sT) = T^{-1} \int_0^{sT} \widehat{\mu}_T(s)\, ds = N_T(sT).$$

A goodness of fit likelihood ratio test of the hypothesis $H_0 : \mu = \mu_0$ under the assumption (6.12) relies on the process

$$S_{LR,T}(s) = \frac{\bar{\mu}_0(s) + (\log T)^{-1} l_T(sT)}{\mu_0^{\frac{1}{2}}(s)}.$$

Under H_0, S_T is asymptotically equivalent to the normalized martingale $v_\mu^{-1}(s) M_T(sT)$ and it converges weakly to the process $S_0 = v_0^{-1} G_0$ as T tends to infinity. This statistic is modified according to the parameter estimator in a test of a parametric model for the intensity function $\bar{\mu}$.

A comparison of the distributions of two independent sequences of Poisson processes N_{1T} and N_{2T} under the rarefaction assumption (6.12) is a comparison of their intensities λ_{1T} and, respectively, λ_{2T}. A test of the hypothesis $H_0 : \mu_1 = \mu_2$ against an alternative $H_1 : \mu_1 \neq \mu_2$ relies on the difference of the variables at T

$$S_T = \frac{N_{1T} - N_{2T}}{\sqrt{N_{1T} + N_{2T}}}(T) \qquad (6.13)$$

converge weakly under H_0 to a centered Gaussian variable S_0 with variance 1. Alternatives H_{1T} are defined by sequences of functions $\bar{\mu}_{1T}$ and $\bar{\mu}_{2T}$ converging uniformly to distinct functions $\bar{\mu}_1$ and $\bar{\mu}_2$ in $[0,1]$. The test statistic S_T converges under H_{1T} to $(S_0 + \bar{\mu}_1 - \bar{\mu}_2)(\bar{\mu}_1 + \bar{\mu}_2)^{-\frac{1}{2}}$, the test is therefore consistent against all fixed alternatives and it has the same limit as S_T under alternatives converging to H_0. Let $a = \bar{\mu}_1 - \bar{\mu}_2$ and $b = \bar{\mu}_1 + \bar{\mu}_2$. A test of level α for H_0 has the rejection domain $\{|S_0| > c_{\frac{\alpha}{2}}\}$ and its asymptotic power for H_0 against H_{1T} is

$$\beta_\alpha = 1 - P\{S_0 \in [-b^{-\frac{1}{2}} c_{\frac{\alpha}{2}} - a, b^{-\frac{1}{2}} c_{\frac{\alpha}{2}} - a]\}.$$

The likelihood ratio test for the comparison of the distributions of two Poisson processes with scarce jumps is based on the difference of the log-likelihood ratio statistics $l_{1T} - l_{2T}$. Under H_0, $\log T(l_{1T} - l_{2T})(sT)$ converges weakly to a Gaussian process with independent increments, with mean function $-\bar{\mu}_1(s) + \bar{\mu}_2(s)$ and variance function $\bar{\mu}_1(s) + \bar{\mu}_2(s)$. The normalized likelihood ratio test statistic is then written as

$$S_{LR,T} = \frac{(\log T)^{-1}(l_{1T} - l_{2T})}{\sqrt{N_{1T} + N_{2T}}}(T) = \frac{\bar{\mu}_{1T} - \bar{\mu}_{2T} + (M_{1T} - M_{2T})}{\sqrt{N_{1T} + N_{2T}}}(T) + o_p(1).$$

It is asymptotically equivalent to $-S_T$ under H_0 and under the consistent alternatives H_{1T}.

6.5   Point processes in $\mathbb{R}_+$

Let N be a time indexed point process in a measurable space $(\Omega, \mathcal{A})$, it is normalized in $[0, 1]$ as $N_T(t) = T^{-1}N(Tt)$. The predictable compensator of N_T is supposed to be continuous and multiplicative

$$\widetilde{N}_T(t) = \int_0^t Y_T(s)d\Lambda_T(s),\ t \in [0, 1]$$

under a probability distribution P_Λ, where the process Y_T is a positive left-continuous process with right-hand limits and the positive function Λ_T is the cumulative hazard function of the point process. The cumulative hazard function and the processes are normalized as $\Lambda_T(t) = T^{-1}\Lambda(Tt)$ so that the instantaneous hazard function is $\lambda_T(t) = \lambda(Tt)$, and

$$\begin{aligned} N_T(t) &= T^{-1}N(Tt), \\ Y_T(t) &= T^{-1}Y(Tt),\ t \in [0, 1], \end{aligned} \tag{6.14}$$

they are supposed to satisfy the following conditions of convergence.

Condition 6.2.

(1) Let $\tau = \sup\{t > 0; \Lambda(t) < \infty\}$ be the endpoint of the support of Λ, then $P(Y(\tau^-) > 0) = 1$.
(2) There exists a function y in $[0, 1]$ such that $\sup_{t \in [0,1]} |Y_T(t) - y(t)|$ converge a.s. to 0, as T tends to τ.
(3) The function Λ_T converges uniformly to a function $\bar{\Lambda}(t)$ in $[0, 1]$, as T tends to τ.

Under Condition 6.2, $EN_T(t) = T^{-1}\int_0^{Tt} Y(s)\,d\Lambda(s) = \int_0^t Y_T(s)\,d\Lambda_T(s)$ is asymptotically equivalent to $\int_0^t y(s)\,d\bar{\Lambda}_T(s)$. The function $\bar{\Lambda}$ is estimated from the observation of the processes N and Y in an interval $[0, T]$, with T large enough, by

$$\widehat{\Lambda}_T(t) = \int_0^t Y_T^{-1}(s)1_{\{Y_T(s) > 0\}}\,dN_T(s),\ t \in [0, 1], \tag{6.15}$$

with the convention $\frac{0}{0} = 0$. The difference $\widehat{\Lambda}_T - \Lambda_T$ is asymptotically equivalent to a local martingale with respect to the filtration $\mathbb{F}$ generated by the processes N_T and Y_T, by definition as the stochastic integral of a $\mathbb{F}$-predictable process with respect to the local martingale $M_T = N_T - \widetilde{N}_T$

$$\widehat{\Lambda}_T(t) - \Lambda_T(t) = \int_0^t Y_T^{-1}(s) 1_{\{Y_T(s) > 0\}} dM_T(s)$$
$$+ \int_0^t (1_{\{Y_T(s) > 0\}} - 1) d\Lambda_T(sT),$$

for every t such that Tt is strictly lower than the endpoint τ of the support of Λ. Under the ergodic Condition 6.2, the estimator $\widehat{\Lambda}_T$ is consistent, the function

$$v_T(t) = E \int_0^t Y_T^{-1}(s) 1_{\{Y_T(s) > 0\}} d\Lambda_T(s)$$

converges to a finite limit $v(t) = \int_0^t y^{-1} d\bar{\Lambda}$, for every t in $[0, 1[$, as T tends to τ, and the estimator $\widehat{v}_T(t) = \int_0^t Y_T^{-2}(s) 1_{\{Y_T(s) > 0\}} dN_T(t)$ is consistent for $v(t)$.

Proposition 6.4. *Under Condition 6.1, the process*

$$\mathcal{L}_T(t) = T^{\frac{1}{2}} \{\widehat{\Lambda}_T(t) - \Lambda_T(t)\}$$

converges weakly to a centered Gaussian process with the variance function $v(t)$ and the covariance $C(s,t) = v(s \wedge t)$, for all s and t in all compact sub-intervals of $[0, 1[$ where Λ_T is finite.

Proof. Under the conditions, the variance of $T^{\frac{1}{2}} M_T = T^{\frac{1}{2}} \{N_T(t) - T^{-1} \int_0^{Tt} Y d\Lambda\}$ is $\int_0^t Y_T d\Lambda_T$ which converges in probability to the function v uniformly in compact subsets of $[0, 1[$, as T tends to τ. The jump size of M_T is $T^{-1} = o(1)$ then by Rebolledo's convergence theorem for the local martingales related to the point processes, M_T converges weakly to a centered Gaussian process G with independent increments and variance v. The process $\mathcal{L}_T$ develops as a sum

$$\mathcal{L}_T(t) = T^{\frac{1}{2}} \int_0^t Y_T^{-1}(s) 1_{\{Y_T(s) > 0\}} dM_T(s) + T^{\frac{1}{2}} \int_0^t \{1_{\{Y_T(s) > 0\}} - 1\} d\Lambda_T(s)$$
$$= T^{\frac{1}{2}} \int_0^t Y_T^{-1}(s) 1_{\{Y_T(s) > 0\}} dM_T(s) + o(1),$$

and under the conditions, $\mathcal{L}_T$ has the limiting distribution $\int_0^t y^{-1}(s) dG(s)$ with variance $v_\Lambda(t) = \int_0^t y^{-1}(s) dv$. $\qquad\square$

A test for a simple hypothesis $H_0 : \Lambda = \Lambda_0$ in a subset $[0, T]$ of $[0, \tau[$, against local alternatives indexed by the length of the observation time of the process

$$H_{1T} : \lambda_T(t) = \lambda_0(t)\{1 + T^{-\frac{1}{2}}\theta_T(t)\},$$

where $(\theta_T)_{0 \leq T < \tau}$ is a positive sequence of functions converging uniformly to a functions $\theta \geq 0$ in $[0, T]$. A class of test statistics for H_0 against H_{1T} is defined from the integral of a process W with respect to the local martingale $\widehat{\Lambda}_T - \Lambda_0$, in the form

$$S_T(t) = T^{\frac{1}{2}} \int_0^t W_T(s)\, d\{\widehat{\Lambda}_T(s) - \Lambda_{0T}(s)\},\ t \in [0, 1].$$

Condition 6.3.
The process W_T is $\mathbb{F}$-predictable and positive, it converges uniformly to a function W, a.s. as T tends to τ.

Under H_0, the process S_T is a centered $\mathbb{F}$-martingale with the variance function

$$\sigma_T^2(t) = E \int_0^t W^2 Y_T^{-1} 1_{\{Y_T > 0\}}\, d\Lambda_T,$$

it converges to $\sigma_0^2(t) = \int_0^t W^2(s) y^{-1}(s) 1_{\{y(s) > 0\}}\, d\bar{\Lambda}_0(s)$. This limit is consistently estimated by $\widehat{\sigma}_T^2(t) = \int_0^t W_T^2 Y_T^{-2} 1_{\{Y_T > 0\}}\, dN_T$.

Let Z be a continuous and centered Gaussian process on $[0, 1]$, with independent increments and variance 1.

Proposition 6.5. *Under Conditions 6.2 and 6.3, the process $\widehat{\sigma}_T^{-1}(t)S_T(t)$ converges weakly under H_0 to Z and it converges under H_{1T} to $Z + m_1$, with mean $m_1(t) = \sigma_0^{-1}(t) \int_0^t W(s)\theta(s)\, d\bar{\Lambda}_0(s)$ and variance 1.*

Proof. The conditions imply that σ_T^2 converges in probability under H_0 to σ_0^2, as T tends to τ. Under the alternative, the mean of the process $S_T(t)$ is $\mu_{1T}(t) = T^{-1} \int_0^{Tt} W(s)\theta_T(s)\, d\Lambda_0(s)$ and its variance is $\sigma_{1T}^2 = \sigma_T^2 + o_p(1)$. The functions $\mu_{1T}(t)$ and $\sigma_{1T}^2(t)$ converge uniformly in probability to $\mu_1(t) = \int_0^t W(s)\theta(s)\, dd\bar{\Lambda}_0(s)$ and $\sigma_0^2(t)$, respectively. The result follows from Rebolledo's convergence theorem for the compensated martingales of point processes. $\square$

The hypothesis H_0 reduces therefore to the hypothesis that the normalized process $\sigma_T^{-1}S_T$ has the mean zero, against the alternative of a mean function $\sigma_T^{-1}\mu_{1T}$. The test statistic $\sup_{t \in [0,1]} \widehat{\sigma}_T^{-1}(t)|S_T(t)|$ converges to the supremum of the norm of a centered Gaussian process with variance 1 under H_0 and the local asymptotic power of the test with a critical value c for testing H_0 against H_{1T} is $\beta_\alpha = P(\sup_{t\in[0,1]} |Z(t) + m_1(t)| > c)$.

In a log-likelihood ratio test of goodness of fit for the hazard function of a point process N the hypothesis concerns the hazard function $H_0 : \lambda = \lambda_0$, their primitives are then equal $\Lambda = \Lambda_0$. Local alternatives are defined as $H_T : \lambda_T = \lambda_0\{1 + T^{-\frac{1}{2}}\theta_T\}$, in $[0,T]$, with a sequence θ_T of functions in the tangent space in the space of the hazard functions under H_T to λ_0 and such that θ_T converges uniformly to a function θ. The log-likelihood ratio test of the hypothesis H_0 against H_T is written $l_T = \int_0^T (\log \lambda_T - \log \lambda_0)\, dN - \int_0^T (\lambda_T - \lambda_0)(s)\, Y(s)ds$. Under H_0, it is written as a stochastic integral with respect to the local martingale $M_0 = N - \int_0^\cdot Y\, d\lambda_0$

$$l_T = \int_0^T \log \frac{\lambda_T}{\lambda_0}\, dM_0 + \int_0^T \left\{\log \frac{\lambda_T}{\lambda_0} - \frac{\lambda_T}{\lambda_0} + 1\right\} Y d\Lambda_0$$

and it develops as

$$l_T = \int_0^T \log\{1 + T^{-\frac{1}{2}}\theta_T\}\, dM_0 + \int_0^T [\log\{1 + T^{-\frac{1}{2}}\theta_T\} - T^{-\frac{1}{2}}\theta_T] Y\, d\Lambda_0$$

$$= T^{-\frac{1}{2}} \int_0^T \theta_T\, dM_0 - (2T)^{-1} \int_0^T \theta_T^2 Y\, d\Lambda_0 + o_p(T^{-1}).$$

By the ergodic property, the process l_T converges weakly to a Gaussian variable with mean $\mu_0 = -\frac{1}{2}\sigma_0^2$ and variance $\sigma_0^2 = \lim_{T\to\infty} T^{-1} \int_0^T \theta_T^2 Y\, d\Lambda_0$. The variance σ_0^2 is estimated by

$$\widehat{\sigma}_T^2 = T^{-1} \int_0^T \theta_T^2(s)\, dN(s).$$

The likelihood ratio test statistic $\widehat{\sigma}_T^{-1} l_T + \frac{1}{2}\widehat{\sigma}_T^2$ for the hypothesis H_0 against H_{1T} has therefore an asymptotically normal distribution.

Under the alternative, the variable l_T is centered using the martingale $M_T = N - \int_0^\cdot Y\, d\Lambda_T$ and its expansion becomes

$$l_T = \int_0^T \log \frac{\lambda_T}{\lambda_0}\, dM_T + \int_0^T \left\{\log \frac{\lambda_T}{\lambda_0} - 1 + \frac{\lambda_0}{\lambda_T}\right\} Y d\Lambda_T$$

$$= T^{-\frac{1}{2}} \int_0^T \theta_T\, dM_T + (2T)^{-1} \int_0^T \theta_T^2 Y\, d\Lambda_T + o_p(T^{-1}).$$

It converges weakly under H_T to a Gaussian variable with mean $\mu_1 = \frac{1}{2}\sigma_0^2$ and with variance σ_0^2. The local asymptotic power of the test of level α for

H_0 against H_T is therefore $1 - \phi(\sigma_0^{-1}c_\alpha - \sigma_0)$ where c_α is the $(1 - \alpha)$th quantile of the normal distribution and the test is consistent.

A test of comparison of two independent point processes N_1 and N_2 with predictable compensators $\tilde{N}_k(t) = \int_0^t Y_k(s)d\Lambda_k(s)$, satisfying Conditions 6.2 and 6.3, for $k = 1, 2$, is also defined from the integral of a process $W > 0$ with respect to the difference of the estimated cumulative hazard functions, in the form

$$S_{1T}(t) = T^{\frac{1}{2}} \int_0^t W_T(s) \{d\hat{\Lambda}_{2T}(s) - d\hat{\Lambda}_{1T}(s)\}, \ t \in [0, 1].$$

The variance of the process S_{1T} is the function

$$\sigma_{1T}^2(t) = T^{-1} \int_0^{Tt} W_T^2 \left(Y_1^{-1} 1_{\{Y_1 > 0\}} \, d\Lambda_1 + Y_2^{-1} 1_{\{Y_2 > 0\}} \, d\Lambda_2 \right).$$

Under H_0, $\Lambda_1 = \Lambda_2 = \Lambda_0$ and under the ergodic conditions, σ_T^2 converges in probability under H_0 and local alternatives to the function $\sigma_1^2 = \int_0^t W^2 \left(y_1^{-1} 1_{\{y_1 > 0\}} \, d\bar{\Lambda}_1 + y_2^{-1} 1_{\{y_2 > 0\}} \, d\bar{\Lambda}_0 \right)$, where Λ_0 is the common cumulative hazard function under H_0. The aymptotic variance σ_1^2 is consistently estimated by

$$\hat{\sigma}_{1T}^2(t) = T^{-1} \int_0^{Tt} W^2 \left(Y_1^{-2} 1_{\{Y_1 > 0\}} \, dN_1 + Y_2^{-2} 1_{\{Y_2 > 0\}} \, dN_2 \right).$$

Under the hypothesis H_0, the process S_{1T} is a local martingale with the mean zero and its variance also equals

$$\sigma_{0T}^2(t) = 2T^{-1} \int_0^{Tt} W(s)(Y_1^{-1} 1_{\{Y_1 > 0\}} + Y_2^{-1} 1_{\{Y_2 > 0\}}) \, d\Lambda_0,$$

the function σ_{0T}^2 is estimated by

$$\hat{\sigma}_{0T}^2(t) = (2T)^{-1} \int_0^{Tt} W^2 \, (Y_1 Y_2)^{-1} 1_{\{Y_1 > 0\}} 1_{\{Y_2 > 0\}} \, d(N_1 + N_2).$$

Under the alternative of non identical cumulative hazard functions, $S_{1T}(t)$ tends to infinity with T, for every t. Local alternatives H_{1T} of proportional hazard functions, $\lambda_{2T} = \lambda_1 \{1 + T^{-\frac{1}{2}} \zeta_T\}$ are defined by uniformly convergent functions $\zeta_T \geq 0$ of the tangent space to the space of the functions $\lambda_2 \lambda_1^{-1}$. Under H_{1T}, the mean of $S_{1T}(t)$ is $\int_0^t W_T(s) \zeta_T(s) \, d\Lambda_0(s)$ and it converges to a positive limit $\mu_1(t) = \int_0^t W(s) \zeta(s) \, d\Lambda_0(s)$, its variance converges to the same limit as under the hypothesis.

Proposition 6.6. *Under Condition 6.5, the statistic*

$$U_{1T} = \sup_{t \in [0,1]} \widehat{\sigma}_{1T}^{-1}(t)|S_{1T}(t)|$$

converges weakly under H_0 to the supremum U_1 of the norm of a centered Gaussian process with independent increments and variance 1. Under the local alternatives H_{1T}, the statistic U_{1T} converges weakly to $|U_1| + \sigma_0^{-1}(1)\mu_1(1)$.

Replacing the normalization of $S_{1T}(t)$ by the estimator of its variance under H_0, the statistic $U'_{1T} = \sup_{t \in [0,1]} \widehat{\sigma}_{0T}^{-1}(t)|S_{1T}(t)|$ has the same behavior under H_0 and its limit under the local alternatives H_{1T} is $|U_1| + \sigma_0^{-1}\mu_1$.

The statistics U_{1T} and U'_{1T} are therefore asymptotically equivalent and the local asymptotic power of the tests of level α based on U_{1T} and U'_{1T} are obtained with the $(1 - \frac{\alpha}{2})$-quantile $c_{\frac{\alpha}{2}}$ of the normal distribution as $\beta_{1,\alpha} = \inf_\zeta\{1 - \Phi(c_{\frac{\alpha}{2}} - \sigma_0^{-1}(1)\mu_1(1)) + \Phi(c_{-\frac{\alpha}{2}} - \sigma_0^{-1}(1)\mu_1(1))\}$.

Under the hypothesis H_0, the common cumulative hazard function of the point processes is defined in $[0,1]$ as

$$\widehat{\Lambda}_{0T}(t) = \int_0^{Tt} (Y_1 + Y_2)^{-1}(s)1_{\{Y_1+Y_2)(s)>0\}}\, d(N_1 + N_2)(s).$$

Another test statistic is deduced from the process

$$S_{2T}(t) = T^{\frac{1}{2}} \int_0^t W_T(s)\, d\{\widehat{\Lambda}_{2T}(s) - \widehat{\Lambda}_{0T}(s)\}$$

$$= T^{\frac{1}{2}} \int_0^t W_T(s)\frac{1_{\{Y_{2T}>0\}}}{(Y_{1T} + Y_{2T})Y_{2T}}(Y_{1T}\, dN_{2T} - Y_{2T}\, dN_{1T}),\ t \in [0,1].$$

It is centered under H_0 and its mean under alternatives is

$$\mu_{2T} = T^{\frac{1}{2}}E\int_0^t W_T(s)\frac{Y_{1T}1_{\{Y_{2T}>0\}}}{Y_{1T} + Y_{2T}}\, d(\Lambda_{2T} - d\Lambda_{1T}).$$

The process

$$T^{\frac{1}{2}}(\widehat{\Lambda}_{0T} - \Lambda_0)(t) = T^{\frac{1}{2}}\int_0^{Tt} (Y_1 + Y_2)^{-1}1_{\{Y_1+Y_2>0\}}\, d(M_1 + M_2)$$

is a centered local martingale and its variance is the function

$$\sigma_T^2(t) = T^{-1}E\int_0^t (Y_{1T} + Y_{2T})^{-1}1_{\{Y_{1T}+Y_{2T}>0\}}\, d(\Lambda_{1T} + \Lambda_{2T})$$

which converges to $\sigma^2(t) = (\bar{\Lambda}_1 + \bar{\Lambda}_2)((y_1 + y_2)^{-1})(t)$, uniformly in all compact subsets of $[0,1]$. Under H_0, the variance of $S_{2T}(t)$ is the mean of

$$\int_0^{Tt} W^2(s)\frac{Y_1^2 1_{\{Y_2>0\}}}{Y_1 + Y_2}\, d\Lambda_0,$$

it is consistently estimated by

$$\widehat{\sigma}^2_{S,T}(t) = \int_0^{Tt} W^2(s) \frac{Y_1^2}{(Y_1 + Y_2)^2} \, d(N_1 + N_2).$$

Under alternatives, the variance of $S_{2T}(t)$ is

$$\sigma^2_{2S,T}(t) = E \int_0^{Tt} W^2(s) \frac{Y_1^2 1_{\{Y_2 > 0\}}}{(Y_1 + Y_2)^2} \, (Y_2 \, d\Lambda_2 + Y_1 \, d\Lambda_1)$$

and $\widehat{\sigma}^2_{S,T}(t)$ is a consistent estimator of the variance S_{2T} under H_0 and alternatives which converges to a function $\sigma^2_{S_2}(t)$. Consider the test based on the the process

$$U_{2T}(t) = \widehat{\sigma}^{-1}_{S,T}(t) |S_{2T}(t)|$$

and local alternatives defined by sequences $\lambda_{2T} = \lambda_{1T}\{1 + T^{-\frac{1}{2}}\theta_T\}$ and $\lambda_{1T} = \lambda_1\{1 + T^{-\frac{1}{2}}\zeta_T\}$ in $[0, T]$, with a sequence of positive functions $(\zeta_T)_T$ in the tangent space in the space of the hazard functions under H_{1T}, at λ_0, converging uniformly to a non null function ζ and such that λ_{1T} converges uniformly in compacts subsets of $[0, 1]$ to λ_1.

Proposition 6.7. *Under the hypothesis H_0, the process U_{2T} converges weakly to the norm of a centered Gaussian process U_2 with independent increments and with variance 1. Under the local alternatives, the process U_{2T} converges weakly to $|U_2 + \sigma^{-1}_{S_2}\mu_S|$, with mean $\mu_S(t) = \int_0^t W y_1 1_{\{y_2 > 0\}} (y_1 + y_2)^{-1} \zeta d\Lambda_0$.*

The test based on $\sup_{t \in [0,1]} |U_{2T}(t)|$ is then consistent. Since the point processes are increasing, it is usual to consider the test statistic defined at the end of the observation interval, it converges to a normal variable. The local asymptotic power of the test with statistic $U_{2T}(1)$ is $\beta_\alpha = \inf_\zeta \{1 - \Phi(c_{\frac{\alpha}{2}} - \sigma^{-1}_{S_2}(1)\mu_2(1)) + \Phi(-c_{\frac{\alpha}{2}} - \sigma^{-1}_{S_2}(1)\mu_2(1))\}$. A comparison of the tests based on the normalized processes $U_{1T}(1)$ and $U_{2T}(1)$ depends the difference of the asymptotic means of the tests statistics $\mu_1(1)$ and $\mu_2(1)$ under the local alternatives, considering $\zeta_1 = 0$ in H_{2T}. Under Condition 6.5, the statistic $U_{2T}(1)$ has the same asymptotic behavior under H_0 as $U_{1T}(1)$. Under H_{1T}, the asymptotic relative efficiency of $U_{2T}(1)$ with respect to $U_{1T}(1)$ is $e = (\frac{\mu_2 \sigma_1}{\mu_1 \sigma_2})^2$. The means and variances differ, with $\mu_2 < \mu_1$ and $\sigma_2^2 < \sigma_1^2$. Generally, $U_{2T}(1)$ provides a most powerful test than $U_{1T}(1)$.

The optimal test is known to be the likelihood ratio test and a comparison of the tests is necessary. Let l_T be the log-likelihood ratio process for a test of the hypothesis H_0 of equality of the hazard functions of N_2 and N_1, $\lambda_2 \equiv \lambda_1$, denoted λ_0, against the local alternatives $H_T : \lambda_{kT} = \lambda_0\{1 + T^{-\frac{1}{2}}\zeta_{kT}\}$, in $[0,T]$, defined for $k = 1, 2$, by sequences ζ_{kT} of functions in the tangent spaces to the space of the hazard functions of N_2 and, respectively, N_1, and such that ζ_{kT} converges uniformly to a function ζ_k for $k = 1, 2$. It can be written as the log-likelihood ratio of the distributions of (N_1, N_2) under H_T and under H_0. Up to an additive constant, it is

$$l_T(t) = \int_0^{Tt} (\log \lambda_{2T} - \log \lambda_0)\, dN_2 + \int_0^{Tt} (\log \lambda_{1T} - \log \lambda_0)\, dN_1$$
$$- \int_0^{Tt} Y_2(s)\,(d\Lambda_{2T} - d\Lambda_0) - \int_0^{Tt} Y_1(s)\,(d\Lambda_{1T} - d\Lambda_0).$$

Under H_0, the martingales related to N_k is $M_{k0} = N_k - \int_0^\cdot Y_k\, d\Lambda_0$ and

$$l_T(t) = \int_0^{Tt} \log \frac{\lambda_{2T}}{\lambda_0}\, dM_{20} + \int_0^{Tt} \log \frac{\lambda_{1T}}{\lambda_0}\, dM_{10}$$
$$+ \int_0^{Tt} \left\{ Y_2 \log \frac{\lambda_{2T}}{\lambda_0} + Y_1 \log \frac{\lambda_{1T}}{\lambda_0} \right\} d\Lambda_0$$
$$- \int_0^{Tt} \left\{ (\lambda_{2T} - \lambda_0)Y_2 + (\lambda_{1T} - T^{-\frac{1}{2}}\lambda_0)Y_1 \right\} dt$$
$$= \sum_{k=1}^{2} \left[\int_0^{Tt} \zeta_{kT}\, dM_{k0} - (2T)^{-1} \int_0^{Tt} \zeta_{kT}^2 Y_k\, d\Lambda_0 \right] + o_p(T^{-\frac{1}{2}}).$$

By the ergodic property, the process l_T converges weakly to a Gaussian process with the mean $\mu_0 = -\frac{1}{2} \sum_{k=1}^{2} v_{0k}$, and the variance $v_0 = \sum_{k=1}^{2} v_{0k}$ defined by $v_{0k}(t) = \lim_{T \to \infty} T^{-1} \int_0^T \zeta_{kT}^2 Y_k\, d\Lambda_0 = \int_0^1 \zeta_k^2 y_k\, d\bar{\Lambda}_0$. The asymptotic variance v_0 is estimated by

$$\widehat{v}_T(t) = T^{-1} \sum_{k=1}^{2} \int_0^{Tt} \zeta_{kT}^2(s)\, dN_k(s).$$

The normalized likelihood ratio test statistic for the hypothesis H_0 against H_T is $\widehat{v}_T^{-1} l_T + \frac{1}{2}\widehat{v}_T$ and it has an asymptotically normal distribution. Under the alternative, the variable l_T is centered using the local martingales

$M_{kT} = N_k - \int_0^\cdot Y_k \, d\Lambda_{kT}$ and it has the following expansion

$$l_T(T) = \sum_{k=1}^{2} \left[\int_0^T \log \frac{\lambda_{kT}}{\lambda_0} \, dM_{kT} + \int_0^T \left\{ \log \frac{\lambda_{kT}}{\lambda_0} - 1 + \frac{\lambda_{kT}}{\lambda_0} \right\} Y_k d\Lambda_0 \right]$$

$$= \sum_{k=1}^{2} \left[T^{-\frac{1}{2}} \int_0^T \zeta_{kT} \, dM_{kT} + (2T)^{-1} \int_0^T \zeta_{kT}^2 Y_k \, d\Lambda_0 \right] + o_p(T^{-\frac{1}{2}}).$$

Under the alternative H_T, it converges weakly to a Gaussian variable with the mean $\mu_1 = \frac{1}{2}v_0$ and the variance v_0. The local asymptotic power of the test of level α for H_0 against H_T is therefore $1 - \phi(c_\alpha - v_0)$, where c_α is the $(1 - \alpha)$th quantile of the normal distribution. In this parametrization, the limiting distributions and the asymptotic power of the log-likelihood ratio test depend on the functions ζ_{1T} and ζ_{2T} defining the alternatives.

A comparison of the three tests is not straightforward because the functions ζ_{kT} defining the alternatives do not disappear by normalization of process $\sum_{k=1}^{2} \int_0^T \zeta_{kT} \, dM_{k0}$. It is proved in the next chapter that the log-likelihood ratio test is asymptotically equivalent to the previous tests based on S_{1T} and S_{2T} with a weighting process $W_T = Y_{1T} Y_{2T} (Y_{1T} + Y_{2T})^{-1}$ though the statistics are not equivalent.

6.6   Marked point processes

In marked point processes, a sequence of random variables $(Z_i)_{i \leq 1}$ is attached to the random time sequence $(T_i)_{i \leq 1}$ and the marked point process observed in the interval $[0, T]$ is

$$N(t) = \sum_{i \geq 1} Z_i 1_{\{T_i \leq t \wedge T\}}.$$

It is normalized as a point process in $[0, 1]$

$$N_T(t) = T^{-1} \sum_{i \geq 1} Z_i 1_{\{T_i \leq tT\}} \, t \in [0, 1].$$

The variables $(Z_i, T_i)_{i \leq 1}$ define a filtration $\mathbb{F} = \{\mathcal{F}_t, t \in [0, 1]\}$ such that $\mathcal{F}_t$ is the right-continuous σ-algebra generated by $\{N_T(s), s \leq t\}$. The predictable compensator of the process N_T is

$$\tilde{N}_T(t) = \sum_{i \geq 1} \int_0^t P\{s \leq T^{-1}(T_i - T_{i-1}) < s + ds | T_i - T_{i-1} \geq sT,$$

$$\sigma(Z_1, \ldots, Z_i), \sigma(T_1, \ldots, T_{i-1})\}.$$

The usual assumptions are the conditional dependence of $T_i - T_{i-1}$ to Z_i and its independence to the previous marks, the durations between the random times are independent in Poisson processes and in processes with independent increments.

Example. For insurance companies, the ith damage caused to an individual has a cost Z_i independent of the time T_i but depending of the calendar time where the higher risks are due to the bad weather occuring at the equinoxes and other periods, and to the car traffic during the holidays. Let w_j be the weeks in the year, for $j = 1, \ldots, k$. The mark Z_i is then supposed to depend on the week of the damage but not on the time T_i itself $E(Z_i|T_i) = \sum_{j=1}^{k} E(Z_i|w_j)1_{\{T_i \in w_j\}}$. Let $X_i = T_i - T_{i-1}$, assuming that $(Z_i)_{i \leq 1}$ and $(X_i)_{i \leq 1}$ are sequences of independent variables

$$EN_T(t) = T^{-1} \sum_{i \geq 1} \sum_{j=1}^{k} E(Z_i|w_j)P(\{T_i \leq tT\} \cap \{T_i \in w_j\})$$

$$= T^{-1} \sum_{i \geq 1} \sum_{j=1}^{k} E(Z_i|w_j)P\{X_i \in [0, tT - T_{i-1}] \cap (w_j - T_{i-1})\}.$$

Let $\Lambda_j(t, T_{i-1}) = \int_0^t P(T_i \in w_j, X_i \in (s, s+ds)|X_i \geq s, T_{i-1})$ be the cumulative intensities of X_i such that T_i belongs to w_j, conditionally on T_{i-1}. The predictable compensator of the marked point process N is

$$\widetilde{N}(t) = T^{-1} \sum_{i \geq 1} \sum_{j=1}^{k} E(Z_i|w_j) \int_0^\infty 1_{\{X_i \geq tT - T_{i-1}\}} \, d\Lambda_j(s - T_{i-1}).$$

For every t in $[0, 1]$ and for every $s < t$, $EN(t) = \widetilde{N}(t)$ and $M = N - \widetilde{N}$ is a L^2 local martingale with respect to the filtration $\mathbb{F}$.

Goodness of fit tests of semi-parametric models for the conditional intensity $\lambda(t; T_{i-1}, Z_i)$ are performed from the observation of the sequences N_T and $(Z_1, \ldots, Z_{N_T})$. For counting processes with independent increments, a proportional hazards model for the instantaneous intensity is written as

$$\lambda(t; Z_i, T_{i-1}) = \exp\{r_\theta(Z_i)\}\lambda_0(t - T_{i-1}).$$

The parameter θ is estimated by maximum likelihood and the estimator is $T^{\frac{1}{2}}$-consistent and asymptotically Gaussian and centered (Pons, 2008) and the mean baseline hazard function $\bar{\Lambda}_0(t) = \lim_{T \to \infty} T^{-1} \Lambda_{0T}(t)$ is estimated by

$$\widehat{\Lambda}_{0T}(t) = \lim_{T \to \infty} T^{-1} \int_0^{tT} \left[\sum_{i \geq 1} 1_{\{X_i \geq sT - T_{i-1}\}} \exp\{r_{\widehat{\theta}_T}(Z_i)\} \right]^{-1} dN_T(s).$$

A comparison of the cumulated hazard functions over several calendar periods or under different conditions relies on the difference of their asymptotically Gaussian estimators, since they are centered.

6.7 Spatial Poisson processes

In $\mathbb{R}^2$, the intensity of a Poisson process depends on a spatial measure. The process is sampled in rectangles, in balls or in patches of geometric forms which are not necessarily a partition of a region of $\mathbb{R}^2$. Let $N(t) = N(t_1, t_2)$ in $\mathbb{R}^2$ and let $N_j(t_j)$ denote its marginals, $j = 1, 2$. The increment of a process N in a rectangle $]s, t]$ of $\mathbb{R}^2$ is expressed as $N]s, t] = N(t) - N(s_1, t_2) - N(t_1, s_2) + N(s)$, in $\mathbb{R}^3$ this is

$$N]s, t] = N(t) - N(s_1, t_2, t_3) - N(t_1, s_2, t_3) - N(t_1, t_2, s_3) + N(t_1, t_2, s_3)$$
$$+ N(t_1, s_2, t_3) + N(s_1, t_2, t_3) - N(s)$$

and this formula extends to higher dimensions.

Under the hypothesis H_0 of two independent marginal Poisson processes, the intensity of a Poisson process $N(t) = \sum_{i \geq 1} 1_{\{T_{1i} \leq t_1, T_{2i} \leq t_2\}}$ is factorized like a density $\lambda = \lambda_1 \lambda_2$. With observations in a rectangle $[0, T] = [0, T_1] \times [0, T_2]$, the intensities are estimated by $\widehat{\lambda}_T = T^{-1} N(T)$ in the plane and by $\widehat{\lambda}_{kT} = T^{-1} N_k(T)$, $k = 1, 2$, for the marginals. The variable $U_T = T^{\frac{1}{2}} (\widehat{\lambda}_T - \widehat{\lambda}_{1T} \widehat{\lambda}_{2T})$ is centered under the hypothesis H_0 of independent processes. Under a local alternative such that $\lambda - \lambda_1 \lambda_2 = T^{-\frac{1}{2}} \zeta_T$, with a sequence $(\zeta_T)_{T>0}$ converging to $\zeta \neq 0$, a normalized statistic of independence is

$$U_T = \frac{\widehat{\lambda}_T - \widehat{\lambda}_{1T} \widehat{\lambda}_{2T}}{\sqrt{\zeta_T}}.$$

It converges to a centered Gaussian process with independent increments, and variance 1 under the alternatives as $\zeta_T \neq 0$ and under H_0, as ζ_T tends to zero. Under H_0, its mean $\sqrt{\zeta_T}$ is zero and it is different from zero under the alternative. Tests for the independence and for the homogeneity of Poisson processes are also based on the property of the independence of the difference variables $X_{k,i} = T_{k,i} - T_{k,i-1}$, with exponential distribution $\mathcal{E}(\lambda_k)$, for all $i \geq 1$ and $k = 1, 2$, denoting $T_0 = 0$. The same arguments apply to Poisson processes with a functional intensity $\lambda(t)$.

Let N be a counting process defined from $\Omega \times \mathbb{R}^d$ to $\mathbb{R}$ by the cardinal $N(A)$ of N in the subsets A of $\mathbb{R}^d$. A Poisson process N has independent rectangular increments and its cumulative intensity function Λ is a positive measure in $\mathbb{R}_+^d$ such that $N - \Lambda$ is a weak martingale with respect to the filtration generated by N on the rectangles of $\mathbb{R}_+^d$ (Cairoli and Walsh, 1975). Moreover, for every set A of $\mathbb{R}_+^d$, $N(A)$ is distributed as a Poisson variable with parameter $\Lambda(A)$. If $\sup_{A \in \mathbb{R}^d} |A|^{-1}\Lambda(A)$ is bounded, the normalized process

$$\Lambda(A)^{-\frac{1}{2}}\{N(A) - \Lambda(A)\}_{A \in \mathbb{R}^d}$$

converges weakly to a Gaussian process as the volume of A tends to infinity. The Poisson process N is homogeneous if there exists a strictly positive constant λ such that $\Lambda(A) = \lambda|A|$, for every set A with volume $|A|$. The parameter λ of its distribution is estimated on a set A where the process is observed by

$$\widehat{\lambda}_A = |A|^{-1}N(A).$$

The variable $N(A) - \lambda(A)|A|$ is then centered and its variance is $\lambda|A|$.

A test for homogeneity of a Poisson process is equivalent to a goodness of fit test for its intensity. For every set A of $\mathbb{R}^d$, $\widehat{\lambda}_A$ is an estimator of the constant parameter of the process under the hypothesis, independently of its volume $|A|$ and of its location in the plane. A comparison of independent estimators calculated in disjoint sets A_j, $j = 1, \ldots, n$ is performed by their comparison to the estimator the mean parameter $\bar{\lambda}_n = n^{-1}\sum_{j=1,\ldots,n} \lambda_{A_j}$ over the n sets, $\widehat{\bar{\lambda}}_n = n^{-1}\sum_{j=1,\ldots,n} \widehat{\lambda}_{A_j}$. The test statistic

$$S_n = n^{\frac{1}{2}}\bar{\lambda}_n^{-\frac{1}{2}} \sup_{j=1,\ldots,n} |\widehat{\lambda}_{A_j} - \bar{\lambda}_n|$$

is centered under the hypothesis of homogeneity and it tends to infinity under fixed alternatives of a functional intensity in $\mathbb{R}^d$. Its variance is 1 under H_0 and it is $(1 - 2n^{-1})\lambda_{A_j} + \bar{\lambda}_n$ under alternatives with different intensities λ_{A_j}. The test statistic S_n has an asymptotic normal distribution.

Goodness of fit test to multidimensional parametric models can be performed in $\mathbb{R}^d$ as in models for Poisson processes in $\mathbb{R}$. Two independent and non stationary Poisson processes, with functional intensities λ_1 and λ_2 in $\mathbb{R}^d$, are compared by Kolmogorov-Smirnov type tests and Cramer-von Mises type tests, they rely on the statistics

$$U_{1,n} = \sup_{j=1,\ldots,n} \frac{|\widehat{\lambda}_{A_j,1} - \widehat{\lambda}_{A_j,2}|}{(\widehat{\lambda}_{A_j,1} + \widehat{\lambda}_{A_j,2})^{\frac{1}{2}}},$$

$$U_{2,n} = \frac{1}{2}\int_{\mathbb{R}^d}\int_{\mathbb{R}^d \setminus A_{j,2}} \frac{(\widehat{\lambda}_{A_j,1} - \widehat{\lambda}_{A_j,2})^2}{\widehat{\lambda}_{A_j,1} + \widehat{\lambda}_{A_j,2}} \, d(\widehat{\lambda}_{A_j,1} + \widehat{\lambda}_{A_j,2}).$$

The second order characteristic $\mu_2(r)$ of stationary point processes is defined as the expected mean number of points of the process contained in the intersection of a ball centered at a point of the process and $B_r(0)$, with a correction to avoid multiple countings of the points

$$\mu_2(r) = (\lambda|B_r(0)|)^{-2} E N_2(r),$$

it is estimated by

$$N_2(r) = \sum_{X_i \in B_r(0)} \sum_{j \neq i} 1_{\{B_1(X_i) \cap B_r(0)\}\}}(X_j)$$

$$= \int_{B_r(0)} N(\{B_1(x) \cap B_r(0)\} \setminus \{x\}) \, dN(x).$$

For a Poisson process N with a functional intensity satisfying a second order stationarity, the variance of $N_2(r) - \mu_2(r)$ equals $\mu_2(r)$ and the normalized process

$$\kappa_{2r} = \frac{N_2(r) - \mu_2(r)}{\mu_2^{\frac{1}{2}}(r)}$$

converges weakly to a centered Gaussian process with independent increments and variance 1 as r tends to infinity. If the equality of the mean and the variance of the process fails, the asymptotic distribution of κ_{2r} is not free or it diverges as r tends to infinity. A bootstrap can then be performed to test the equality of the mean and the variance which is a property of a Poisson process. Similar statistics can be defined for the higher moments of the Poisson processes in order to test the validity of a Poisson distribution.

6.8   Tests of stationarity for point processes

In $\mathbb{R}$, the second order stationarity of a point process N in an observation interval $[0, T]$ is defined by the property

$$E\{N(t)N(s)\} - EN(t)\,EN(s) = R(t - s)$$

depending only on the difference $t - s$, for all s and t in $[0, T]$.

A non-stationary process has a covariance function $R(s, t)$ which is estimated using a discretization of the interval $[0, T]$ into n_T points $(t_i)_{i=1,\dots,n_T}$ such that $t_{i+1} - t_i = h_T$. Let $N_i = N(t_i)$, let s and t be in a sub-intervals $I_i =\,]t_i, t_{i+1}]$ and, respectively, $I_j =\,]t_j, t_{j+1}]$, and let k be an integer such that $i - k \geq 0$ and $j - k \geq 0$. The covariance of the process at s and t is estimated as the empirical covariance at $\pm k$ points of the partition around s and t by

$$\widehat{R}_k(s,t) = (2k+1)^{-1} \sum_{l=-k}^{k} (N_{i+l} - \bar{N}_{i,k})(N_{j+l} - \bar{N}_{j,k}),$$

where $\bar{N}_{i,k} = (2k+1)^{-1} \sum_{l=-k}^{k} N_{i+l}$ for every i. Under the hypothesis H_0 of stationarity, the covariance function is $R(t-s) = R(x, t-s+x)$ for every x in $]0, T-t+s]$, it is estimated by the mean covariance over all integers in $\{1, \ldots, n_T - j + i\}$

$$\widehat{R}_k(t-s) = (n_T - j + i)^{-1} \sum_{l=1}^{n_T - j + i} \widehat{R}_k(t_l, t - s + t_l),$$

with x in I_l. A bootstrap test based on the difference $\widehat{R}_k(s,t) - \widehat{R}_k(t-s)$ is consistent.

A second order stationary process N has the empirical mean function

$$\bar{N}(T) = \frac{1}{T} \int_{[0,T]} N(t)\, dt, \ T > 0$$

such that $\lim_{T\to\infty} T^{-1} \int_{[0,T]} N(t)\, dt = \bar{\Lambda}(t)$, with the expectation

$$\mu_T = T^{-1} E \int_{[0,T]} N(t)\, dt = T^{-1} \int_{[0,T]} \Lambda(t)\, dt.$$

The covariance at s and t is continuously estimated by

$$\widehat{C}_T(s,t) = \frac{1}{T} \int_{[0,T-t+s]} \{N(s+y) - \widehat{\Lambda}_T(s+y)\}\{N(t+y) - \widehat{\Lambda}_T(t+y)\}\, dy$$

with a mean estimator of Λ, $\widehat{\Lambda}_T(x) = (2a_T)^{-1} \int_{[x-a_T, x+a_T]} dN(t)$ in an interval centered at x. Under the hypothesis of a second order stationarity, the covariance at s and t such that $s < t$ is

$$\widehat{C}_T(t-s) = \frac{1}{T} \int_{[0,T-t+s]} \{N(y) - \widehat{\Lambda}_T(y)\}\{N(t-s+y) - \widehat{\Lambda}_T(t-s+y)\}\, dy.$$

The expectation of $U_T(s,t) = \widehat{C}_T(s,t) - \widehat{C}_T(t-s)$ is $\bar{U}_T(s,t)$ defined as

$$\bar{C}(s,t) - \bar{C}(t-s) = \lim_{T\to\infty} \frac{1}{T} \int_{[0,T-t+s]} \{R(s+y, t+y) - R(t-s)\}\, dy$$

it diverges under fixed alternatives and it is zero under the null hypothesis. The process U_T converges in probability to $\bar{U}_T$, uniformly in $[0, T \wedge T^*_{N(T)}]$. Bootstrap tests based on the statistics $\sup_{(s,t)\in[0,T^*_{N(T)}]^2, t-s>a} |U_T(s,t)|$ and

$$\sup_{(s,t)\in[0,T^*_{N(T)}]^2, t-s>a} |\widehat{R}_T(s,t) - \widehat{R}_T(t-s)|,$$

with an arbitrarily small constant $a > 0$, are consistent and their asymptotic power is 1.

The tests of stationarity are extended to ergodic point processes of $\mathbb{R}^d$, with $d \geq 1$. The second-order stationarity of point processes in $\mathbb{R}^d$ is characterized by a covariance function $R(A, B)$ between $N(A)$ and $N(B)$ satisfying

$$R(A, B) = VarN(A \cap B),$$

for all intersecting subsets A and B of $\mathbb{R}^d$. The increments of a Poisson process in disjoint spheres with the same radius constitute a sample of an exponential variable and a goodness of fit applies to such samples, for every radius and this property is used to test the hypothesis of a Poisson distribution.

The functions $R(A, B)$ is estimated by the empirical variance of the variables $N(A_x \cap B_x)$ as A_x and B_x are subsets of $\mathbb{R}^d$ having the same form as A and respectively B, with a translation of x belonging to $\mathbb{R}^d$ and such that all sets $A_x \cap B_x$ have the same length. The comparison of the estimators $N(A_x \cap B_x)$ with their mean $\bar{N}_n(A \cap B)$ over n points x_i yields a test of stationarity of the point process through the statistic

$$\max_{i=1\ldots,n} n^{\frac{1}{2}} |N(A_{x_i} \cap B_{x_i}) - \bar{N}_n(A \cap B)|.$$

The variance of the variables $N(A_{x_i} \cap B_{x_i}) - \bar{N}_n(A \cap B)$ is expressed by the means of the fourth moments of the process. For a stationary process N with independent increments and disjoints subsets $A_{x_i} \cap B_{x_i}$

$$E\{N(A_{x_i} \cap B_{x_i})N(A_{x_j} \cap B_{x_j})\} = E^2 N(A_{x_i} \cap B_{x_i}) = E^2 N(A_{x_j} \cap B_{x_j})$$

for every $j = 1, \ldots, n$, and the statistic is normalized by the empirical estimator of this fourth moment in order to test the stationary of the process. Under fixed alternatives, the normalized statistic diverges.

Other statistics defined from the properties of the moments of the spatial point processes have been considered, according to models.

6.9   Diffusion processes

Let X be a diffusion process on a probability space $(\Omega, \mathcal{A}, P_{\alpha, \sigma^2})$, indexed by continuous functions α and σ^2 defined in $\mathbb{R}_+$ as the drift and the variance of the process. A diffusion process with deterministic drift and variance is defined from a Wiener process $W_t = t^{\frac{1}{2}} W_1$ in $\mathbb{R}_+$ as

$$dX_t = \alpha(t)\, dt + \sigma(t)\, dW_t, \ t > 0.$$

Consider a goodness of fit test for a simple hypothesis $H_0 : \alpha = \alpha_0, \sigma = \sigma_0$ in $[0, T]$, against local alternatives indexed by the length of the observation time of the process

$$H_{1T} : \alpha_T(t) = \alpha_0(t)\{1 + T^{-\frac{1}{2}}\lambda_T(t)\}, \; \sigma_T(t) = \sigma_0(t)\{1 + T^{-\frac{1}{2}}\varsigma_T(t)\},$$

where $(\lambda_T)_{T\geq 0}$ and $(\varsigma_T)_{T\geq 0}$ are sequences of functions converging uniformly to functions λ and ς, respectively, in $[0, T]$. The next conditions ensure the ergodicity of the process X and its weak convergence. Let $\bar{\varphi}_T = T^{-1} \int_0^T \varphi(s) \, ds$ be the mean functions for a function φ, and let $\bar{W}_{\sigma T} = T^{-1} \int_0^T \sigma(s) \, dW_s$ be the mean diffusion process.

Condition 6.4.

(1) There exist constants $\bar{\alpha} = \lim_{T\to\infty} \bar{\alpha}_T$ and $\bar{\sigma}^2 = \lim_{T\to\infty} \bar{\sigma}_T^2$ and a process $\bar{W}_\sigma = W_{\bar{\sigma}T} > 0$, under H_0.

(2) Under H_{1n}, there exist constants limits of the following mean integrals

$$\bar{\lambda} = \lim_{T\to\infty} T^{-1} \int_0^T \alpha_0(s)\lambda_T(s) \, ds, \; \bar{\varsigma}^2 = \lim_{T\to\infty} T^{-1} \int_0^T \varsigma_T^2(s)\sigma_0^2(s) \, ds.$$

Under the hypothesis H_0 and the first Condition 6.4, the diffusion process satisfies

$$X_t = X_0 + \int_0^t \alpha_0(s) \, ds + \int_0^t \sigma_0(s) \, dW(s), \; t \in [0, T].$$

The variable $T^{-1}X_T$ is a Gaussian variable with mean $\bar{\alpha}_{0T}$ and variance $T^{-1}\bar{\sigma}_{0T}^2 = T^{-2} \int_0^T \sigma_0^2(s) \, ds$. It follows that, under H_0, $T^{-1}X_T$ converges in probability to the mean $\bar{\alpha}_0$ and

$$T^{-\frac{1}{2}}(X_T - T\bar{\alpha}_0) = T^{-\frac{1}{2}} \int_0^T \sigma_0(s) \, dW(s)$$

converges weakly to a centered Gaussian variable $\bar{W}_0$ with variance $\bar{\sigma}_0^2$. When α_0 is known and σ_0^2 is unknown, the mean variance $\bar{\sigma}_0^2$ is consistently estimated by

$$\hat{\sigma}_T^2 = T^{-1}(X_T - T\bar{\alpha}_{0T})^2.$$

Under the local alternatives H_{1T}, the diffusion process is

$$X_t = X_0 + \int_0^t \alpha_T(s) \, ds + \int_0^t \sigma_T(s) \, dW(s), \; t \in [0, T],$$

the process $T^{-1}X_T$ converges in probability to the variable $\bar{\alpha}_0$ and the mean variance $\bar{\sigma}_{1T}^2 = T^{-1}\int_0^T \sigma_T^2(s)\,ds$ converges to $\bar{\sigma}_0^2$. The process

$$T^{-\frac{1}{2}}(X_T - T\bar{\alpha}_0) = T^{-1}\int_0^T \alpha_0(s)\lambda_T(s)\,ds + T^{-\frac{1}{2}}\int_0^T \sigma_T\,dW$$

$$= \bar{\lambda}_T + T^{-\frac{1}{2}}\int_0^T \sigma_0\,dW + T^{-1}\int_0^T \sigma_0\varsigma_T\,dW + o(1)$$

converges weakly to a Gaussian process $\bar{W}_1$ with mean $\bar{\lambda}_T$ and variance $\bar{\sigma}_0^2$. This ergodicity is a consequence of the independence of the increments of the Wiener process W and of the above conditions.

A test statistic for a simple hypothesis $H_0 : \alpha = \alpha_0$ against the local alternative H_{1T} is the asymptotically normal variable

$$S_{0T} = \frac{X_T - X_0 - T\bar{\alpha}_{0T}}{T^{\frac{1}{2}}\bar{\sigma}_{0T}}. \tag{6.16}$$

Its asymptotic critical value at the level α satisfies $\alpha = P_0(|S_{0T}| > c_{\frac{\alpha}{2}})$ with the $(1 - \frac{\alpha}{2})$th normal quantile $c_{\frac{\alpha}{2}}$ and its local power is

$$\beta_T = P_1(|S_{0T}| > c_{\frac{\alpha}{2}}) = P_1\left\{\frac{X_T - T\bar{\alpha}_{1T}}{T^{\frac{1}{2}}\bar{\sigma}_{1T}} > \frac{\bar{\sigma}_{0T}}{\bar{\sigma}_{1T}}c_{\frac{\alpha}{2}} - T^{\frac{1}{2}}\left(\frac{\bar{\alpha}_{1T} - \bar{\alpha}_{0T}}{\bar{\sigma}_{1T}}\right)\right\}$$

$$+ P_1\left\{\frac{X_T - T\bar{\alpha}_{1T}}{T^{\frac{1}{2}}\bar{\sigma}_{1T}} < \frac{\bar{\sigma}_{0T}}{\bar{\sigma}_{1T}}c_{-\frac{\alpha}{2}} - T^{\frac{1}{2}}\left(\frac{\bar{\alpha}_{1T} - \bar{\alpha}_{0T}}{\bar{\sigma}_{1T}}\right)\right\},$$

where $T^{\frac{1}{2}}(\bar{\alpha}_{1T} - \bar{\alpha}_{0T}) = \bar{\lambda}_T$ and $\bar{\sigma}_{0T}\bar{\sigma}_{1T}^{-1}$ converges to 1 in probability under H_{1T}. The asymptotic local power of the test is therefore

$$\beta = 1 - \Phi\left(c_{\frac{\alpha}{2}} - \frac{\bar{\lambda}}{\bar{\sigma}_0}\right) + \Phi\left(c_{-\frac{\alpha}{2}} - \frac{\bar{\lambda}}{\bar{\sigma}_0}\right)$$

and the asymptotic efficiency of this test is $\bar{\lambda}_T\bar{\sigma}_0^{-1}$.

A test statistic for a composite hypothesis H_0, with unknown mean α_0 and variance σ_0, against the local alternative H_{1T} is the asymptotically normal variable

$$S_T = \frac{X_T - X_0 - T\bar{X}_{0T}}{T^{\frac{1}{2}}\hat{\sigma}_{0T}}, \tag{6.17}$$

where $\bar{X}_{0T}$ and $\hat{\sigma}_{0T}$ are the empirical estimators of $\bar{\alpha}_0$ and $\bar{\sigma}_0$ under the hypothesis

$$\bar{X}_{0T} = \frac{1}{T}\int_0^T X_s\,ds,$$

$$\hat{\sigma}_{0T}^2 = \frac{1}{T}\int_0^T X_s^2\,ds - \bar{X}_{0T}^2.$$

Its asymptotic level is $\alpha = P_0(S_T > c_\alpha)$ and its local power is

$$P_1\left(\frac{X_T - T\bar{X}_{1T}}{T^{\frac{1}{2}}\widehat{\sigma}_{1T}} > \frac{\widehat{\sigma}_{0T}}{\widehat{\sigma}_{1T}}c_{\frac{\alpha}{2}} - T^{\frac{1}{2}}\frac{\bar{X}_{1T} - \bar{X}_{0T}}{\widehat{\sigma}_{1T}}\right)$$

$$+ P_1\left(\frac{X_T - T\bar{X}_{1T}}{T^{\frac{1}{2}}\widehat{\sigma}_{1T}} < \frac{\widehat{\sigma}_{0T}}{\widehat{\sigma}_{1T}}c_{-\frac{\alpha}{2}} - T^{\frac{1}{2}}\frac{\bar{X}_{1T} - \bar{X}_{0T}}{\widehat{\sigma}_{1T}}\right),$$

where the empirical estimators of $\bar{\alpha}_T$ and $\bar{\sigma}_T$ under H_{1T} satisfy

$$T^{\frac{1}{2}}\bar{X}_{1T} = T^{\frac{1}{2}}\bar{X}_{0T} - \bar{\lambda}_T,$$

$$T^{\frac{1}{2}}\widehat{\sigma}_T^2 = T^{\frac{1}{2}}\widehat{\sigma}_{0T}^2 - 2\int_0^T \varsigma(s)\sigma_0^2(s)\,ds.$$

Then $\widehat{\sigma}_{0T}\widehat{\sigma}_{1T}^{-1}$ tends to 1 in probability under H_{1T} and the difference of the empirical means is such that $T^{\frac{1}{2}}(\bar{X}_{1T} - \bar{X}_{0T}) = \bar{\lambda}_T$. The asymptotic local power of the test is therefore

$$\beta_\alpha = \inf_\lambda\left\{1 - \Phi\left(c_{\frac{\alpha}{2}} - \frac{\bar{\lambda}_T}{\bar{\sigma}_0}\right) + \Phi\left(c_{-\frac{\alpha}{2}} - \frac{\bar{\lambda}_T}{\bar{\sigma}_0}\right)\right\}.$$

The log-likelihood ratio l_T of $X_T - X_0$ under H_{1T} and H_0 is the log-likelihood ratio of two normal variables $\mathcal{N}(T\bar{\alpha}_T, T\bar{\sigma}_T^2)$ and $\mathcal{N}(T\bar{\alpha}_{0T}, T\bar{\sigma}_{0T}^2)$. The log-likelihood ratio test statistic for hypotheses and alternatives such that X_T has the same variance under H_{1T} and H_0 is asymptotically equivalent to the previous statistic S_T given by (6.16). When σ_T^2 differs from σ_{0T}^2, up to an additive constant, l_T is asymptotically equivalent to

$$l_T = \frac{(X_T - T\bar{\alpha}_{0T})^2}{2T\bar{\sigma}_{0T}^2} - \frac{(X_T - T\bar{\alpha}_T)^2}{2T\bar{\sigma}_T^2} + o_p(1)$$

$$= T^{-\frac{3}{2}}X_T^2\frac{\bar{\varsigma}^2}{2\bar{\sigma}_{0T}^2\bar{\sigma}_T^2} - T^{-\frac{1}{2}}(X_T - T\bar{\alpha}_{0T}\bar{\sigma}_{0T})\frac{\bar{\alpha}_{0T}\bar{\varsigma}_T - \bar{\lambda}_T\bar{\sigma}_{0T}}{\bar{\sigma}_{0T}^2\bar{\sigma}_T^2} + o_p(1).$$

As T tends to infinity, $T^{-\frac{1}{2}}l_T$ converges in probability to

$$l_0 = \frac{\bar{\varsigma}^2}{2\bar{\sigma}_0^2\bar{\sigma}^2}\bar{\alpha}_0^2 - \bar{\alpha}_0(1 - \bar{\sigma}_0)\frac{\bar{\alpha}_0\bar{\varsigma} - \bar{\lambda}\bar{\sigma}_0}{\bar{\sigma}_0^2\bar{\sigma}^2}$$

and $l_T - T^{\frac{1}{2}}l_0$ is asymptotically equivalent to

$$\frac{\bar{\varsigma}^2}{2\bar{\sigma}_0^2\bar{\sigma}^2}T^{\frac{1}{2}}(T^{-2}X_T^2 - \bar{\alpha}_0^2) - T^{\frac{1}{2}}(T^{-1}X_T - \bar{\alpha}_{0T})\frac{\bar{\alpha}_{0T}\bar{\varsigma}_T - \bar{\lambda}_T\bar{\sigma}_{0T}}{\bar{\sigma}_{0T}^2\bar{\sigma}_T^2}$$

$$= T^{\frac{1}{2}}(\bar{X}_{0T} - \bar{\alpha}_{0T})\left\{\frac{\bar{\alpha}_{0T}\bar{\varsigma}_T^2}{\bar{\sigma}_{0T}^2\bar{\sigma}_T^2} - \frac{\bar{\alpha}_{0T}\bar{\varsigma}_T - \bar{\lambda}_T\bar{\sigma}_{0T}}{\bar{\sigma}_{0T}^2\bar{\sigma}_T^2}\right\} + o_p(1)$$

and it converges weakly to the process $\bar{W}_0\{\bar{\lambda}\bar{\sigma}_0^{-3} - \bar{\alpha}_0(\bar{\varsigma}^2 - \bar{\varsigma})\bar{\sigma}_0^{-4}\}$. The log-likelihood ratio test statistic for composite hypotheses against local alternatives is therefore asymptotically equivalent, up to location and scaling constants, to the statistic S_T given by (6.17).

A diffusion process with a linear drift and a deterministic variance is now defined in $[0, T]$ as

$$dX_t = \alpha_t X_t\, dt + \sigma_t\, dW_t. \tag{6.18}$$

The unique solution of this equation is

$$X_t - X_0 = e^{\int_0^t \alpha_s\, ds} \int_0^t e^{-\int_0^s \alpha_u\, du} \sigma_s\, dW_s = \int_0^t e^{-\int_s^t \alpha_u\, du} \sigma_s\, dW_s \tag{6.19}$$

for every t in $[0, T]$, so X_t is a transformed Brownian motion, with independent increments. Let $A_t = \int_0^t \alpha_s\, ds$, the rescaled process

$$Y_t = e^{-A_t}(X_t - X_0) = \int_0^t e^{-A_s} \sigma_s\, dW_s$$

is a transformed Wiener process satisfying the stochastic differential equation

$$dY_t = e^{-A_t} \sigma(t)\, dW_t := \beta(t)\, dW_t. \tag{6.20}$$

The processes $X_t - X_0$ and Y_t are centered martingales, the variance function of Y_t is $\int_0^t \beta^2(s)\, ds$ and the variance of X_t is $e^{-2A_t} \int_0^t \beta^2(s)\, ds$. The functions α and β are not identifiable from this expression. For every $x \neq 0$, the mean function

$$\bar{\alpha}_T = (xT)^{-1} \int_0^T \alpha_t E(X_t | X_t = x)\, dt = T^{-1} \int_0^T \alpha_t\, dt$$

has the kernel estimator

$$\widehat{\alpha}_{T,h}(x) = \frac{\int_0^T K_h(x - X_t)\, dX_t}{x \int_0^T K_h(x - X_t)\, dt}.$$

Its bias is obtained from the expansion

$$E \int_0^T K_h(x - X_t)\, dt = \int_0^T \left\{ f_{X_t}(x) + m_{2K} \frac{h^2}{2} f_{X_t}^{(2)}(x) \right\} dt + o(h^2),$$

$$:= T\bar{f}(x) + m_{2K} \frac{h^2}{2} \bar{f}^{(2)}(x)\} \, dt + o(h^2),$$

$$E \int_0^T K_h(x - X_t)\, dX_t = x^2 \int_0^T \alpha_t \left\{ f_{X_t}(x) + m_{2K} \frac{h^2}{2} x f_{X_t}^{(2)}(x) \right\} dt$$

$$+ m_{2K} h^2 \int_0^T f_{X_t}^{(1)}(x)\, dt + o(h^2).$$

Condition 1.1, as T tends to infinity and $h = h_T$ tends to zero, the convergences of the kernel estimators obtained from the process X_t are similar to those of regression functions. The bias of $\widehat{\alpha}_{T,h}(x)$ has the expansion $b_{\alpha,T,h}(x) = h^2 b_{\alpha,T}(x) + o(h^2)$, with

$$b_{\alpha,T}(x) = \frac{m_{2K}}{\bar{f}_T(x)} \left[\frac{1}{T} \int_0^T \{x\alpha_t\} f_{X_t}^{(2)}(x) + \alpha_t f_{X_t}^{(1)}(x)\} dt - \bar{\alpha}_T \bar{f}_T^{(2)}(x) \right]$$

and its variance is $v_{\alpha,T,h}(x) = (hT)^{-1} v_{\alpha,T}(x) + o((hT)^{-1})$, with

$$v_{\alpha,T}(x) = T^{-1} \int_0^T \int_0^T \alpha_t \alpha_s f_{X_s,X_t}(x,x) \, ds \, dt - \bar{\alpha}_T^2.$$

A goodness of fit test for the drift function can be performed with an Anderson-Darling type statistic

$$S_{\alpha,T} = \int_0^T \widehat{v}_{\alpha,T}^{-\frac{1}{2}}(x)\{Th(\widehat{\alpha}_{T,h} - \alpha_{0t}) - \widehat{b}_{\alpha,T,h}(x)\gamma\}^2 \, dt.$$

The variance is estimated as before by

$$\widehat{\sigma}_T^2 = T^{-1} \int_0^T \left\{ \bar{X}_T(t) - T^{-1} \int_0^T \widehat{\alpha}_{T,h}(t) \, dX_t \right\}^2 dt.$$

Adding a mean μ_t, the diffusion process is written as

$$dX_t = (m_t + \alpha X_t)dt + \sigma_t \, dW_t, \ t \in [0, T], \tag{6.21}$$

and the unique solution of this equation is

$$X_t - X_0 = e^{A_t} \int_0^t e^{-A_s}(m_s \, ds + \sigma_s \, dW_s).$$

The mean $z_t = EX_t$ and the variance v_t of $X_t - X_0$ are

$$z_t = \int_0^t e^{A_t - A_s} m_s \, ds, \quad v_t = \int_0^t e^{2(A_t - A_s)} \sigma_s \, ds.$$

The process $X_T(t) = T^{-1}X(Tt)$ estimates the integrated mean function $\bar{z}_T(t) = T^{-1} e^{A_{Tt}} \int_0^{Tt} e^{-A_s} m_s \, ds$, t in $[0, 1]$. The mean function

$$\theta_T(x) = T^{-1} \int_0^T E(m_t + \alpha_t X_t | X_t = x) \, dt = \bar{m}_T + x\bar{\alpha}_T$$

is estimated by the kernel estimator

$$\widehat{\theta}_{T,h}(x) = \frac{\int_0^T K_h(x - X_t) \, dX_t}{\int_0^T K_h(x - X_t) \, dt}.$$

Under Condition 1.1, the bias $b_{\theta,T,h}(x) = h^2 b_{\theta,T}(x) + o(h^2)$ of $\widehat{\theta}_{T,h}(x)$ and its variance $v_{\theta,T,h}(x) = (hT)^{-1} v_{\theta,T}(x) + o((hT)^{-1})$ have expansions

$$b_{\theta,T}(x) = \frac{m_{2K}}{\bar{f}_T(x)} \left[\frac{1}{T} \int_0^T \{(m_t + x\alpha_t) f_{X_t}^{(2)}(x) + \alpha_t f_{X_t}^{(1)}(x)\} \, dt - \theta_T(x) \bar{f}_T^{(2)}(x) \right]$$

$$v_{\theta,T}(x) = T^{-1} \int_0^T \int_0^T (m_t + x\alpha_t)(m_s + x\alpha_s) f_{X_s,X_t}(x,x) \, ds \, dt - \theta_T^2(x).$$

The primitive A_t of α_t is estimated by the primitive $\widehat{A}_t$ of the estimator of α_t and the variance of Y_t is estimated as the integrated squared $\widehat{Y}_t$. An estimator of the function α_t is deduced from the estimator of the linear function $\theta_t(x)$ and the function m_t is estimated from the estimators of θ_t and α_t. Goodness of fit tests about α_t and m_t are performed as above, using estimators of their bias and their variance.

More generally, a diffusion process satisfies a stochastic differential equation

$$dX_t = \alpha(X_t)dt + \beta(X_t) \, dW_t, \tag{6.22}$$

where α and β be two functions in a functional metric space $(\mathbb{X}, \|\cdot\|)$ where the sample paths of the process X_t are defined and the Brownian process W_t is such that $E(W_t - W_s | X_s) = 0$ for all $s < t$. From (6.22), the process X is the solution of the implicit equation

$$X_t - X_0 = \int_0^t \alpha(X_s)ds + \int_0^t \beta(X_s) \, dW_s.$$

The drift function $\alpha(x)$ is the derivative of the mean of X_t, conditionally on $\{X_t = x\}$, it is estimated by smoothing the sample-path of the process X in a neighborhood of $X_t = x$

$$\widehat{\alpha}_{T,h}(x) = \frac{\int_0^T K_h(x - X_s) \, dX_s}{\int_0^T K_h(x - X_s) \, ds}, \tag{6.23}$$

where K is a kernel satisfying the Condition 1.1 and its bandwidth $h = h_T$ tends to zero as T tends to infinity. By the martingale property of the Brownian motion, there exists an invariant distribution function F such that the ergodicity property (6.1) is satisfied. With the convergence rate $h_T = O(T^{-\frac{1}{5}})$, this property implies that the process $(Th_T^5)^{\frac{1}{2}}(\widehat{\alpha}_{T,h} - \alpha)$ converges weakly to a Gaussian process if α belongs to $L^2(I_X) \cap C^2(I_X)$ (Pons, 2011). Centering X_t with this estimator yields the process

$Z_t = X_t - X_0 - \int_0^t \widehat{\alpha}_{T,h}(X_s)$, it is expanded as

$$Z_t = \int_0^t (\alpha - \widehat{\alpha}_{T,h})(X_s)\, ds + \int_0^t \beta(X_s)\, dB_s = \int_0^t \beta(X_s)\, dB_s + o_p(1).$$

As t tends to infinity, the mean of the process $t^{-1}Z_t$ converges in probability to zero and the variance of $t^{-\frac{1}{2}}Z_t$ converges to $v = \int_{I_X} \beta^2(x)\, dx$, by the ergodic property (6.1) and since the difference $(Z_t)_{t \geq 0}$ is asymptotically equivalent to a centered martingale. If β belongs to $L^2(I_X)$, a consistent empirical estimator of the function $v(x) = \int_0^x \beta^2(y)\, dy$ is

$$\widehat{v}_T(x) = T^{-1}\left[\int_0^T 1_{\{X_s \leq x\}}\{dX_t - \widehat{\alpha}_{T,h}(X_t)\, dt\} \right]^2$$

$$= T^{-1}\left\{ \int_0^T \beta(X_s) 1_{\{X_s \leq x\}}\, dB_s \right\}^2 + o_p(1). \qquad (6.24)$$

The mean of $\widehat{v}_T(x)$ equals $T^{-1}E \int_0^T \beta^2(X_s) 1_{\{X_s \leq x\}}\, ds$ and by the ergodic property, it converges to $v(x)$ in probability. An estimator of the function $\beta^2(x)$, conditionally on $\{X_t = x\}$, is deduced by smoothing the estimator $\widehat{v}_T(x)$ with the kernel K

$$\widehat{\beta}_{T,h}^2(x) = 2\frac{\int_0^T Z_s K_h(X_s - x)\, dZ_s}{\int_0^T K_h(X_s - X_t)\, ds}. \qquad (6.25)$$

Tests of hypotheses for functions α and β of $L^2(I_X) \cap C^2(I_X)$ rely on the weak convergence of their kernel estimators which was proved by Pons (2011). The asymptotic behavior of the tests is the same as the similar tests about the regression functions in Propositions (3.11)-(3.13). Goodness of fit tests for the variance are performed as tests about the cumulative function v_t, with the empirical estimator $\widehat{v}_t$.

Tests for parametric models of the drift or the variance in the differential equation (6.22) rely on estimators of the parameters of the models. A minimum distance estimator of the parameter θ in a model of the variance $\{v_\theta; \theta \in \Theta\}$ can be defined, under H_0, as the parameter value $\widehat{\theta}_T$ that minimizes the empirical distance of the function β_θ to the function β_0, at the actual parameter value θ_0 in Θ

$$d_T(\theta) = T^{-1}\int_0^T \{dZ_t - \beta_\theta(X_t)\, dB_t\}^2 \qquad (6.26)$$

$$= T^{-1}\int_0^T \{\beta_0(X_t) - \beta_\theta(X_t)\}^2\, dt + o_p(1).$$

As T tends to infinity, $d_T(\theta)$ tends to zero under H_0 if and only if $\theta = \theta_0$ and its first two derivatives are

$$\dot{d}_T(\theta) = -2T^{-1} \int_0^T \{dZ_t - \beta_\theta(X_t)\,dB_t\}\dot{\beta}_\theta(X_t)\,dB_t$$

$$= -2T^{-1} \int_0^T \{\beta_0(X_t) - \beta_\theta(X_t)\}\dot{\beta}_\theta(X_t)\,dt + o_p(1),$$

$$\ddot{d}_T(\theta) = 2T^{-1} \int_0^T [\{dZ_t - \beta_\theta(X_t)\,dB_t\}\ddot{\beta}_\theta(X_t)\,dB_t - \dot{\beta}_\theta^{\otimes 2}(X_t)\,dt].$$

The variables $\dot{d}_T(\theta_0)$ converge in probability to zero and $-\ddot{d}_T(\theta_0)$ converges in probability to the limit $s_0 = 2\int_{I_X} \dot{\beta}_\theta^{\otimes 2}(x)\,dx$. Moreover, the weak convergence of $A_{T,h} = (Th_T)^{\frac{1}{2}}(\widehat{\alpha}_{T,h} - \alpha_0)$ to a Gaussian process W_α, with mean $-b_\alpha$ and variance σ_α^2, implies that the process

$$W_{1T} = (Th_T)^{\frac{1}{2}}\dot{d}_T(\theta_0) = (Th_T)^{\frac{1}{2}} \int_0^T (\alpha_0 - \widehat{\alpha}_{T,h})(X_t)\dot{\beta}_{\theta_0}(X_t)\,dt$$

converges weakly to a Gaussian variable $W_1 = -2\int_{I_X} W_\alpha(x)\dot{\beta}_{\theta_0}(x)\,dx$ with a mean $b_0 = \int_{I_X} b_\alpha\dot{\beta}_{\theta_0}(x)\,dx$ with a finite variance v_0. It follows that $\widehat{\vartheta}_{T,h} = (Th_T)^{\frac{1}{2}}(\widehat{\theta}_{T,h} - \theta_0)$ converges weakly to a Gaussian process $\widehat{\theta}_0$, with a mean $b_0 = s_0^{-1}\int_{I_X} b_\alpha\dot{\beta}_\theta(x)\,dx$ and a variance $\sigma_0^2 = s_0^{-1}v_0s_0^{-1}$. The limit of $d_T(\theta)$ defined by (6.26) is

$$\lim_{T\to\infty} d_T(\theta) = \int_{I_X} \{\beta_{\widehat{\theta}_T}(x) - \beta_{\theta_0}(x)\}^2\,dx + \int_{I_X} \{\widehat{\alpha}_{T,h}(x) - \alpha_0(x)\}^2\,dx$$

$$= (\widehat{\theta}_T - \theta_0)^t\Big\{\int_{I_X} \dot{\beta}_{\theta_0}^{\otimes 2}(x)\,dx\Big\}(\widehat{\theta}_T - \theta_0)$$

$$+ \int_{I_X} (\widehat{\alpha}_{T,h} - \alpha_0)^2(x)\,dx.$$

Therefore $Th_T d_T(\theta)$ converges weakly to the sum $\mathcal{D}_0 = \int_{I_X} \{W_\alpha^2(x) + W_\beta^2(x)\}\,dx$ of two variables corresponding to the sum of the integrated squared Brownian errors of estimation for the drift and the variance of the diffusion.

A test for the hypothesis of a parametric model for α and β against separate alternatives is performed in the same way replacing the nonparametric estimator of the drift by a parametric distance. They have rejection domains $\{Th_T d_T(\theta) > c\}$, for a level $P(\mathcal{D}_0 > c)$. Local alternatives for α and β in neighborhoods of their parametric models are

$$\alpha_T(x) = \alpha_{\theta_0}(x)\{1 + (Th_T)^{-\frac{1}{2}}a_T(x)\},$$

$$\beta_T(x) = \beta_{\theta_0}(x)\{1 + (Th_T)^{-\frac{1}{2}}b_T(x)\},$$

where the functions a_T and b_T belong to the tangent spaces H_α and H_β of the models and converge uniformly to functions a and b, respectively, and θ_0 is an unknown parameter value in Θ. The mean of the limit $\mathcal{D}_0$ of $d_T(\theta)$ is modified as $\mathcal{D}_{a,b} = \int_{I_X} \{(W_\alpha + a)^2(x) + (W_\beta + b)^2(x)\} \, dx$ and the limiting variance is unchanged. The asymptotic local power of the test is therefore $\sup_{\alpha,\beta} \inf_{a \in H_\alpha} \inf_{b \in H_\beta} P(\mathcal{D}_{a,b} > c)$.

When the diffusion processes are only observed in a grid $(t_{iT})_{i \leq n_T}$ of an interval $[0, T]$ such that $h_T = t_{i,T} - t_{i-1,T}$ is constant and h_n tends to zero as T and $n = n_T$ tend to infinity, the nonparametric estimators of their drift and variance functions are modified. Let $Y_i = X_{i+1} - X_{t_i}$ and let $Z_i = Y_i - h_n \widehat{\alpha}_{n,h}(X_{t_i}) = h_n\{\alpha(X_{t_i}) - \widehat{\alpha}_{n,h}(X_{t_i})\} + \beta(X_{t_i})\varepsilon_i$, where ε_i is a Gaussian error $\varepsilon_i = B_{t_{i+1}} - B_{t_i} \sim \mathcal{N}(0, h_n)$. The estimators are modified

$$\widehat{\alpha}_{n,h}(x) = \frac{\sum_{i=1}^{n} Y_i K_h(x - X(t_i))}{\sum_{i=1}^{n} K_h(x - X(t_i))},$$

$$\widehat{\beta}_{n,h}^2(x) = \frac{\sum_{i=1}^{n} Z_i^2 K_h(x - X(t_i))}{h_n \sum_{i=1}^{n} K_h(x - X(t_i))}.$$

Test statistics are defined for each parameter like with for a continuous diffusion.

6.10 Comparison of diffusion processes

A comparison of two independent diffusion processes defined by model (6.22) with unspecified drift and diffusion functions is based on the difference of the estimated sample paths, with nonparametric estimators of their drift and variance, through the distance L^2-distance d_T. This is equivalent to consider the L^2-distances of the drifts and of the variances

$$d_T(X_{1t}, X_{2t}) = T^{-1} \int_0^T \{\widehat{\alpha}_{1t}(X_t) - \widehat{\alpha}_{2t}(X_2)\}^2 \, dt$$

$$+ T^{-1} \int_0^T \{\widehat{\beta}_{1t}^2(X_t) - \widehat{\beta}_{2t}^2(X_2)\}^2 \, dt,$$

with the kernel estimators (6.23) and (6.25) of the functions α and β^2. As the processes X_{kt} are determined by the functions α_{kt} and β_{kt}, the integrals over the observed time interval is equivalent to integrals over the subset $\mathcal{I}_{X,T}$ of the sample paths of the processes $(X_{kt})_{t \in [0,T]}$

$$d_T(X_{1t}, X_{2t}) = \int_{\mathcal{I}_{X,T}} \{\widehat{\alpha}_{1t}(x) - \widehat{\alpha}_{2t}(x)\}^2 \, dx$$

$$+ \int_{\mathcal{I}_{X,T}} \{\widehat{\beta}_{1t}^2(x) - \widehat{\beta}_{2t}^2(x)\}^2 \, dx.$$

The null hypothesis of equal functions for the diffusion processes of model (6.22) is $H_0 : \alpha_1 = \alpha_2$ and $\beta_1 = \beta_2$. For $k = 1, 2$, we consider the sum of the L^2-distances of the drift and diffusion functions to their estimators

$$d_T(X_{kt}) = \int_{\mathcal{I}_{X,T}} \{\widehat{\alpha}_{kt}(x) - \alpha_{kt}(x)\}^2 \, dx + \int_{\mathcal{I}_{X,T}} \{\widehat{\beta}_{kt}^2(x) - \beta_{kt}^2(x)\}^2 \, dx.$$

Under H_0 and by the independence of the processes X_{1t} and X_{2t}, $Ed_T(X_{1t}, X_{2t}) = E\{d_T(X_{1t}) + d_T(X_{2t})\}$ and the variance of $d_T(X_{1t}, X_{2t})$ is also the sum of the variances of $d_T(X_{1t})$ and $d_T(X_{2t})$. Assuming that the functions α_{kt} and β_{kt} belong to $C^s(\mathcal{I}_{X,T}) \cap L^2(\mathcal{I}_{X,T})$, as $h = h_T$ tends to zero and T tends to infinity, $\{\widehat{\alpha}_{kt}(x) - \alpha_{kt}(x)\}^2$ and $\{\widehat{\beta}_{kt}^2(x) - \beta_{kt}^2(x)\}^2$ tend to zero with the rate $T^{\frac{2s}{2s+1}}$. The optimal bandwidth minimizing the L^2-norm of the difference between the estimators and the function they estimate has the order $T^{-\frac{s}{2s+1}}$. With a bandwidth of optimal order and under the hypothesis of the same model for both processes X_{1t} and X_{2t}, it follows that $T^{\frac{2s}{2s+1}} d_T(X_{1t}, X_{2t})$ converges to the sum of integrated squared Gaussian processes $\int_{\mathcal{I}_{X,T}} (W_{\alpha,1}^2 + W_{\alpha,2}^2 + W_{\beta,1}^2 + W_{\beta,2}^2)(x) \, dx$.

The first order moments of $d_T(X_{1t}, X_{2t})$ are

$$Ed_T(X_{1t}, X_{2t}) = O(h^{2s}) + O((hT)^{-1}),$$

$$Vard_T(X_{1t}, X_{2t}) = O(h^{4s}) + O((hT)^{-1}),$$

and they have the same order with the optimal bandwidth, then $d_T(X_{1t}, X_{2t})$ has to be centered and normalized by an estimator of its variance. A bootstrap test is advised. Under alternatives in a neighborhood of the differences $\alpha_1 - \alpha_2$ or $\beta_1^2 - \beta_2^2$, with a radius $T^{-\frac{1}{2s+1}}$, the statistic $T^{\frac{2s}{2s+1}} d_T(X_{1t}, X_{2t})$ converges to a limit of the same form with noncentered Gaussian processes, so the test is consistent and its asymptotic power is 1.

A comparison of models is performed with the same statistic, using the differences of the estimators of the drifts and of the variances in both models. A test for the comparison of the variances of two diffusions is more simply defined from the empirical estimator (6.24) of the function $v(x) = \int_0^x \beta^2(y) \, dy$. A test of the equality of the variances β_1 and β_2 of two diffusions is defined from the maximal difference of the empirical estimators $\widehat{v}_{1T}(x)$ and $\widehat{v}_{2T}(x)$.

6.11   Exercises

6.11.1. Generalize the goodness of fit statistic of a composite hypothesis H_0 for the functional intensity of a Poisson process with parameter θ in Θ, against a parametric alternative K : the parameter γ belongs to Γ such that Θ defines a sub-model of the alternative.

Hints. A sequence of Poisson processes N_T is observed in a time interval $[0,1]$, with intensity $\lambda_{T,\theta}$ indexed by a parameter θ in an open convex set Θ under H_0. Under the alternative, the parameter of the model is $\gamma = (\theta^t, \xi^t)^t$ and the hypothesis is $H_0 : \xi = 0$. The maximum likelihood estimators of the parameters are denoted $\widehat{\theta}_T$ under H_0 and $\widehat{\xi}_T$ under K, they maximize the log-likelihood ratio $l_T(\gamma)$ given by (6.9) in Γ under K and, respectively, Θ under H_0. The variables $\vartheta_T = T^{\frac{1}{2}}(\widehat{\theta}_T - \theta_0)$ and $z_T = T^{\frac{1}{2}}(\widehat{\gamma}_T - \gamma_0)$ converge to centered Gaussian variables ϑ_0 and, respectively, z. The variance of ϑ_0 is $I_0 = \lim_{T\to\infty} T^{-1} \int_0^T \lambda_0^{-1} \dot{\lambda}_0^2 \, dt$ and the variance of z is $I = \lim_{T\to\infty} T^{-1} \int_0^T \lambda_\xi^{-1} \dot{\lambda}_\xi^2 \, dt$. The process $\widehat{\Lambda}_{T,\widehat{\gamma}_T}(s) - \Lambda_{T,\widehat{\theta}_{0T}}(s)$ is asymptotically equivalent to $\widehat{\xi}_T^t \dot{\Lambda}_{T,\theta_0}(s)$, it follows that the log-likelihood ratio statistic converges to a χ_d^2 where d is the difference between the dimensions of Γ and Θ.

6.11.2. Generalize the reparametrization (6.7) to a test statistic of the hypothesis of two functional Poisson processes with identical distributions against local alternatives, determine its asymptotic distributions under the hypothesis and under the alternatives.

Hints. Let Λ_0 be the unknown intensity under H_0, the statistic

$$W_T = \sup_{t\in[0,1]} t^{-\frac{1}{2}} |\{\widehat{\Lambda}_{1T}(t) - \widehat{\Lambda}_{2T}(t)\} \circ \widehat{\Lambda}_{0T}^{-1}(t)|$$

defined with the nonparametric estimator $\widehat{\Lambda}_{0T}$ under H_0 has an asymptotic free Gaussian distribution. Under a local alternative such that $T^{-\frac{1}{2}}(\Lambda_T - \Lambda_{0T})$ converges to a bounded limit, its mean is bounded and its variance has the same limit.

6.11.3. Define a test of the hypothesis $H_0 : N = N_1 + N_2$, for two independent Poisson processes N_1 and N_2 and for a Poisson process N.

Hints. Under H_0, the difference $\Delta_T = N_T - (N_{1T} + N_{2T})$ has the variance $v_{0T} = \Lambda_T - (\Lambda_{1T} + \Lambda_{2T})$ and a test is defined by the statistic $v_{0T}^{-\frac{1}{2}} \Delta_T$, it converges to a centered Gaussian variable under H_0 and it has a bounded mean under local alternatives.

6.11.4. Define a goodness of fit statistic for a parametric model of functional intensities of Poisson processes under the rarefaction assumption (6.12) and determine its asymptotic behavior.
Hints. The log-likelihood process $(\log T)^{-1}l_T(sT)$ Proposition 6.3 and the statistic (6.13) define two statistics which are modified to goodness of fit tests. Their limits are studied as above.

6.11.5. Generalize the likelihood ratio tests to point processes in $\mathbb{R}_+$ under the rarefaction assumption (6.12).
Hints. Use Proposition 6.3.

6.11.6. Generalize the likelihood ratio test of goodness of fit to a Poisson process in $\mathbb{R}_+^2$.
Hints. The log-likelihood is defined in rectangles determined by the random jump points of the process or in balls centered at these points. It is expanded as T tends to inifinity.

6.11.7. Determine a goodness of fit test to a diffusion with linear drift and variance (6.18).
Hints. The nonparametric estimators must be constants under the hypothesis, apply the results concerning their asymptotic behavior to the statistics $\sup_x |\widehat{\alpha}_T(x) - \sup_y \widehat{\alpha}_T(y)|$ and $\sup_x |\widehat{\beta}_T(x) - \sup_y \widehat{\beta}_T(y)|$.

6.11.8. Determine tests of equality of the drift functions of k diffusion processes.
Hints. The test statistic is defined as in Chapter 4.9 for the comparison of regression curves and its asymptotic behavior is studied along the same lines.

6.11.9. Determine nonparametric tests of equality of the variance functions of k diffusion processes.
Hints. As in Exercise 8.

Chapter 7

Nonparametric tests under censoring or truncation

7.1 Introduction

On a probability space $(\Omega, \mathcal{A}, P)$, let X and C be two independent positive random variables such that $P(X < C)$ is strictly positive, and let

$$T = X \wedge C, \quad \delta = 1_{\{X \leq C\}}$$

denote the observed variables when X is right-censored by C. Let F be the distribution function of X, its survival function is $\bar{F}(x) = 1 - F(x)$ such that $\bar{F}^-(x) = P(X \geq x)$.

Let G be the distribution function of the censoring variable C, its survival function is denoted $\bar{G}$ and $\bar{H} = \bar{G}\bar{F}$ is the survival function of the variable T. The end points of their support $\tau_F = \sup\{t : \bar{F}(t) > 0\}$, $\tau_G = \sup\{t : \bar{G}(t) > 0\}$ and $\tau = \sup\{t : \bar{H}(t) > 0\}$ satisfy $\tau = \tau_F \wedge \tau_G$. The cumulated hazard function related to F is defined for every $t < \tau$ by

$$\Lambda(t) = \int_0^t \frac{dF}{\bar{F}^-}, \tag{7.1}$$

conversely

$$\bar{F}(t) = \exp\{-\Lambda^c(t)\} \prod_{s \leq t} \Delta\Lambda(s),$$

where $\Lambda^c = \int_0^t (\bar{F}^-)^{-1} dF^c$ is the continuous part of Λ and the sum of its jumps is $\Lambda - \Lambda^c = \sum_{s \leq t} \Delta\Lambda(s)$. For every $t < \tau$, the cumulated hazard function of the censored variable T is written

$$\Lambda(t) = \int_0^t \frac{\bar{G}\, dF}{\bar{H}^-}. \tag{7.2}$$

Let X be a time variable with a density f. From (7.1), the hazard function $\lambda(t) = \bar{F}^{-1}(t)f(t)$ of X defines its density via the bijection between the hazard function λ and

$$f(t) = \lambda(t) \exp\{-\Lambda(t)\}. \tag{7.3}$$

Consider a sample of independent and identically distributed right-censored variables and their censoring indicators, $(T_i, \delta_i)_{i \leq n}$. Let $\mathbb{F}_n = (\mathcal{F}_{nt})_{t \in \mathbb{R}_+}$ denote the history generated by the observations before t, i.e. $\mathcal{F}_{nt}$ is the σ-algebra generated by the events $\{\delta_i 1_{\{T_i \leq s\}}, 1_{\{T_i \leq s\}}$, for $0 < s \leq t$ and $i = 1, \ldots, n\}$. The counting process of the censored times until t is

$$N_n(t) = \sum_{1 \leq i \leq n} \delta_i 1_{\{T_i \leq t\}}$$

and the number of individuals at risk at t is

$$Y_n(t) = \sum_{1 \leq i \leq n} 1_{\{T_i \geq t\}}.$$

The process N_n is the empirical estimator of $\int_0^{\cdot} \bar{G} \, dF$ and Y_n is the empirical estimator of $\bar{H}$, it is $\mathbb{F}_n$-predictable and the processes

$$M_n(t) = N_n(t) - \int_0^t Y_n(s) \, d\Lambda(s), \ t \in]0, \tau[$$

are local $\mathbb{F}_n$-martingales. An empirical estimator of Λ is defined as

$$\widehat{\Lambda}_n(t) = \int_0^t \frac{1_{\{Y_n(s) > 0\}}}{Y_n(s)} \, dN_n(s) \tag{7.4}$$

$$= \sum_{1 \leq i \leq n} \delta_i \frac{1_{\{Y_n(X_i) > 0\}}}{Y_n(X_i)} 1_{\{X_i \leq t\}}, \ t < \tau,$$

with the convention $0/0$. This is Nelson's estimator (Nelson 1969, 1984, Aalen 1978) and it satisfies

$$\widehat{\Lambda}_n(t) - \Lambda(t) = \int_0^t \frac{1_{\{Y_n(s) > 0\}}}{Y_n(s)} \, dM_n(s) + \int_0^t (1_{\{Y_n(s) > 0\}} - 1) \, d\Lambda(s), \tag{7.5}$$

for every t in $]0, X_{n:n}[$. The largest value $X_{n:n}$ converges in probability to τ and $Y_n(X_{n:n}) = 1$, it follows that $P(Y_n(X_{n:n}) > 0) = 1$ and $\widehat{\Lambda}_n(t) - \Lambda(t)$ is approximated by a martingale in $]0, \tau[$.

Condition 7.1.

(1) $\sup_{s \leq \max_{i=1,\ldots,n} \{X_i; \delta_i = 1\}} |n^{-1} Y_n(s) - \bar{H}(s)|$ converges a.s. to zero, as n tends to infinity.
(2) The integral $\int_0^t \bar{H}^{-1} \, d\Lambda = \int_0^t \bar{H}^{-2} \, dF$ is finite for every $t < \tau$.

Under Condition 7.1, the process $n^{\frac{1}{2}}(\widehat{\Lambda}_n - \Lambda)$ converges weakly to a continuous Gaussian process with mean zero and covariance function

$$C(s, t) = \int_0^{s \wedge t} (\bar{H}^-)^{-1} (1 - \Delta\Lambda) \, d\Lambda \tag{7.6}$$

in every compact sub-interval of $[0, \tau]$.

Let $0 < T_{n:1} \leq \ldots \leq T_{n:n}$ be the ordered sequence of the observation times T_i and $\delta_{n:k}$ be the censoring indicator for the time $T_{n:k}$. The classical estimator of the survival function $\bar{F} = 1 - F$ is the product-limit Kaplan-Meier estimator, it is right-continuous with left-hand limits step function, constant between two observations times such that $\delta_i = 1$

$$\widehat{\bar{F}}_n(t) = \prod_{j:T_j \leq t} \left\{ \frac{Y_n(T_j) - 1}{Y_n(T_j)} \right\}^{\delta_j}$$

$$= \prod_{T_{n:k} \leq t} \left\{ 1 - \Delta\widehat{\Lambda}_n(T_{n:k}) \right\} = \widehat{\bar{F}}_n(t^-)\left\{ 1 - \Delta\widehat{\Lambda}_n(t) \right\}.$$

The estimator $\widehat{\bar{F}}_n$ only depends on the censoring variables through the size of its jumps. For every $t \geq T_{n:n}$, $\widehat{\bar{F}}_n(t) = 0$ if $\delta_{n:n} = 1$ and it is strictly positive if $\delta_{n:n} = 0$. When the last observed time is a censoring time, the total mass of $\widehat{F}_n = 1 - \widehat{\bar{F}}_n$ is therefore less than 1. The Kaplan-Meier estimator is a.s. uniformly consistent in $[0, T_{n:n}[$. Gill (1983) proved that

$$\frac{\widehat{\bar{F}}_n - \bar{F}}{\bar{F}}(t) = \int_0^{t \wedge T_{n:n}} \frac{1 - \widehat{F}_n(s^-)}{1 - F(s)} \left\{ d\widehat{\Lambda}_n(s) - d\Lambda(s) \right\}$$

and $n^{1/2}(\widehat{\bar{F}}_n - \bar{F})\bar{F}^{-1}$ converges weakly to a centred Gaussian process B_F on $[0, \tau[$, with a covariance function defined from (7.6) as

$$K(s,t) = \int_0^{s \wedge t} (\bar{F}^{-1}\bar{F}^-)^2 \, dC.$$

Due to the equivalence between the expression of the likelihood of the observations from the distribution function F and its cumulative hazard function Λ_F, tests about the distribution functions of time variables are expressed as tests about their cumulative hazard functions and tests about the density f of time variables are tests about their hazard function. In particular, likelihood ratio tests are written with their hazard functions. Similar tests are established from the estimators of the distribution functions and the cumulative hazard functions of time variables under left-censoring and under right or left-truncations, their estimators were defined and studied by Pons (2008).

These methods are generalized to sequences of point processes $(N_n)_{n \geq 1}$ such that N_n has a predictable compensator with respect to the natural filtration $\{\mathcal{F}_n(t), t > 0\}$ of N_n, in the form $\widetilde{N}_n(t) = \int_0^t Y_n \lambda \, ds$, where $(Y_n)_{n \geq 1}$ is a sequence of predictable processes and λ is a positive hazard function. In $\mathbb{R}_+^k$, we consider a vector $N_n = (N_{1n}, \ldots, N_{kn})$ of independent or dependent point processes with marginal predictable processes $\widetilde{N}_{kn}(t) = \int_0^t Y_{kn}\lambda_k\, ds$. Tests of equality of the hazard functions of the processes are defined, their asymptotic behavior under the hypothesis and local alternatives are studied and they are compared to local log-likelihood ratio test. This approach is extended to other tests for censored observations. All results apply to uncensored variables with the notations $n^{-1}N_n = \widehat{F}_n$ and $n^{-1}Y_n = 1 - \widehat{F}_n^-$.

7.2   Comparison of right-censored distributions

On a probability space $(\Omega, \mathcal{A}, P)$, let T_1 and T_2 be two independent time variables with distribution functions F_1 and F_2 and hazard functions λ_1 and λ_2, respectively. The comparison of the distribution functions F_1 and F_2 from two samples of censored variables is based on the observations $(X_{ki}, \delta_{ki})_{i=1,\dots,n_k}$, for $k = 1, 2$. Let τ be the stopping time of the observations for both samples, such that $\bar{H}_1(\tau) > 0$ and $\bar{H}_2(\tau) > 0$, the hypothesis is also written as $H_0 : \Lambda_1(t) = \Lambda_2(t), t \in [0, \tau]$. The alternative is first unspecified H_1 : there exist sub-intervals of $[0, \tau]$ where the functions Λ_1 and Λ_2 differ. Under H_0, the hazard function is denoted Λ_0. The total sample size is $n = n_1 + n_2$ such that $\lim_{n\to\infty} n^{-1}n_k = \rho_k$ is strictly positive for $k = 1, 2$.

The notations of the introduction are indexed by k for the kth sample, $N_{kn} = N_{kn_k}$ and $Y_{kn} = Y_{kn_k}$, for $k = 1, 2$. Under the hypothesis H_0, the martingale sequences $M_{kn}(t) = N_{kn}(t) - \int_0^t Y_{kn}(s)1_{\{Y_{kn}(s)>0\}}\, d\Lambda_0(s)$ are independent and the cumulative hazard functions Λ_k have the estimators $\widehat{\Lambda}_{kn}(t) = \int_0^t Y_{kn}^{-1}(s)1_{\{Y_{kn}(s)>0\}} N_{kn}(s)$, for $k = 1, 2$. Let $N_n = N_{1n} + N_{2n}$ and $Y_n = Y_{1n} + Y_{2n}$ be the cumulated counting processes of the total sample. Under H_0, the common cumulative hazard function is then estimated by the process $\widehat{\Lambda}_n(t)$ given by (7.4). A family of test statistics is defined from a weighting predictable process $W_n = W(Y_{1n}, Y_{2n})$, in the form

$$U_n = n^{\frac{1}{2}} \int_0^\tau W_n 1_{\{Y_{1n}(s)>0\}} 1_{\{Y_{2n}(s)>0\}} (d\widehat{\Lambda}_{2n} - d\widehat{\Lambda}_{1n}).$$

Proposition 7.1. *Let us assume that Condition 7.1 holds for both samples and there exists a function $w \neq 0$ such that $\sup_{t\in[0,\tau]} |W_n(t) - w(t)|$ converges to zero in probability as n tends to infinity. Then Under H_0, the statistic $n^{\frac{1}{2}}U_n$ converges weakly to a centered normal variable with variance*

$$\sigma_\tau^2 = \int_0^\tau w^2(t)\left\{\frac{1}{\rho_1 \bar{H}_1(t)} + \frac{1}{\rho_2 \bar{H}_2(t)}\right\} d\Lambda_0(t)$$

and σ_τ^2 is estimated by

$$V_n = n \int_0^\tau W_n^2(t) \left\{ 1_{\{Y_{1n}(t)>0\}} \frac{1}{Y_{1n}(t)} + 1_{\{Y_{2n}(t)>0\}} \frac{1}{Y_{2n}(t)} \right\} d\widehat{\Lambda}_n(t).$$

An alternative estimator of the variance σ_τ^2 still valid under alternatives is

$$V_n = n \int_0^\tau W_n^2(t) 1_{\{Y_{1n}(t)>0\}} \frac{d\widehat{\Lambda}_{1n}(t)}{Y_{1n}(t)} + n \int_0^\tau 1_{\{Y_{2n}(t)>0\}} \frac{d\widehat{\Lambda}_{2n}(t)}{Y_{2n}(t)}.$$

The normalized statistic $V_n^{-\frac{1}{2}} U_n$ is therefore asymptotically free.

Local alternatives H_{1n} are defined by sequences of real functions $(\theta_n)_{n\geq 1}$ and $(\gamma_n)_{n\geq 1}$ in $[0,\tau]$ converging uniformly in $[0,\tau]$ to functions θ and γ, respectively, and such that the hazard functions are

$$\lambda_{n,1}(t) = \lambda_{n,2}(t)\{1 + n^{-\frac{1}{2}}\theta_n(t)\},$$
$$\lambda_{n,2}(t) = \lambda_0(t)\{1 + n^{-\frac{1}{2}}\gamma_n(t)\}. \tag{7.7}$$

Under H_{1n}

$$n^{\frac{1}{2}}(\widehat{\Lambda}_{1n} - \Lambda_{n,1})(t) = n^{\frac{1}{2}}(\widehat{\Lambda}_{1n} - \Lambda_0)(t) - \sqrt{\rho_1} \int_0^t (\theta_n + \gamma_n)\,d\Lambda_0 + o(1),$$

$$n^{\frac{1}{2}}(\widehat{\Lambda}_{2n} - \Lambda_{n,2})(t) = n^{\frac{1}{2}}(\widehat{\Lambda}_{2n} - \Lambda_0)(t) - \sqrt{\rho_2} \int_0^t \gamma_n\,d\Lambda_0 + o(1),$$

then the mean of the statistic U_n under the local alternative H_{1n} is asymptotically equivalent to $\mu_{1n} = -\int_0^\tau W_n(t)\theta_n(t)\,d\Lambda_0(t)$, its limit is $\mu = -\int_0^\tau w(t)\theta(t)\,d\Lambda_0(t)$. The asymptotic variance of the statistic U_n under the local alternative H_{1n} is still σ_τ^2, it is deduced from the limit of

$$Var\{n^{\frac{1}{2}}(\widehat{\Lambda}_{kn} - \Lambda_{n,k})(t)\} = \int_0^t 1_{\{Y_{kn}(s)>0\}} \frac{n_k}{Y_{kn}(s)}\,d\Lambda_{kn}(s)$$

$$= \int_0^t \frac{1}{\bar{H}_k(s)}\,d\Lambda_0(s) + o(1).$$

It follows that under H_{1n}, $n^{\frac{1}{2}}U_n$ converges weakly to a Gaussian variable with mean μ and variance σ_τ^2. The asymptotic local power at the level α of the test defined by the statistic $V_n^{-\frac{1}{2}}|U_n|$ is $\inf_\theta\{1 - \Phi(c_{\frac{\alpha}{2}} + \sigma_\tau^{-1}\mu) + \Phi(c_{-\frac{\alpha}{2}} + \sigma_\tau^{-1}\mu)\}$.

Consider two dependent variables T_1 and T_2 having a joint distribution function F, right-censored at a random time $C = (C_1, C_2)$ with joint distribution function G and survival function $\bar{G}$. The covariance of the local martingales M_{1n} and M_{2n} conditionally on $\mathcal{F}_{s \wedge t}^-$ is

$$E\{M_{1n}(s)M_{2n}(t)\} = \int_0^s \int_0^t \bar{G}(u,v)\{F(du,dv) + S(u,dv)\,d\Lambda_1(u)$$

$$+ S(du,v)\,d\Lambda_2(v) + S(u,v)\,d\Lambda_1(u)\,d\Lambda_2(v)\}.$$

The asymptotic covariance of $n_1^{\frac{1}{2}}(\widehat{\Lambda}_{n1} - \Lambda_1)(t)$ and $n_2^{\frac{1}{2}}(\widehat{\Lambda}_{n2} - \Lambda_2)(t)$ is then

$$C_{(s,t)} = (n_1 n_2)^{\frac{1}{2}} \int_0^s \int_0^t S^{-1}(u,v)\{F(du,dv) + S(u,dv)\,d\Lambda_1(u)$$

$$+ S(du,v)\,d\Lambda_2(v) + S(u,v)\,d\Lambda_1(u)\,d\Lambda_2(v)\}, \tag{7.8}$$

and the asymptotic variance of the statistic U_n is $v_\tau = \sigma_\tau^2 - 2(\rho_1 \rho_2)^{-\frac{1}{2}} C_\tau$. The test statistic normalized by an empirical estimator of $v_\tau^{-\frac{1}{2}}$ is asymptotically normal under the hypothesis of identical marginal distributions and the local behavior of this statistic is as above.

A comparison of J independent groups relies on statistics of the same form with the notation N_{jn} and Y_{jn} for the counting processes of the j-th group and $N_n(t) = \sum_{j=1}^J N_{jn}$, $Y_n(t) = \sum_{j=1}^J Y_{jn}$ for the sums of the counting processes over all groups. The cumulative intensity function under the hypothesis of homogeneous groups is denoted Λ_0 and it is estimated by $\widehat{\Lambda}_n(t) = \int_0^t 1_{\{Y_n(s)>0\}} Y_n^{-1}(s)\,dN_n(s)$. In group j, a weighting function has the form $W_{jn} = L_n Y_{jn}$, where L_n is a predictable process converging uniformly to a function L and a marginal statistic compares the cumulative intensity functions of the group j to the global function Λ_0 through weighted differences of these estimators

$$K_{jn} = n^{-\frac{1}{2}} \int_0^\tau W_{jn}\,(d\widehat{\Lambda}_{jn} - d\widehat{\Lambda}_n) = n^{-\frac{1}{2}} \int_0^\tau L_n\Big(dM_{jn} - \frac{Y_{jn}}{Y_n}\,dM_n\Big).$$

The covariance matrix $\Sigma_n = (\Sigma_{jj'n})_{j,j'=1,\ldots,J}$ of the vector $n^{-\frac{1}{2}}(M_{jn} - \int_0^\tau Y_{jn} Y_n^{-1}\,dM_n)_{j=1,\ldots,J}$ under H_0 has the components

$$\Sigma_{jjn} = n^{-1} Var\Big(\int_0^\tau \frac{Y_n - Y_{jn}}{Y_n}\,dM_{jn} - \sum_{j' \neq j} \int_0^\tau \frac{Y_{jn}}{Y_n}\,dM_{j'n}\Big)$$

$$= n^{-1} E_0\Big\{\int_0^\tau \Big(\frac{Y_n - Y_{jn}}{Y_n}\Big)^2 Y_{jn}\,d\Lambda_0 + \sum_{j' \neq j} \int_0^\tau \Big(\frac{Y_{jn}}{Y_n}\Big)^2 Y_{j'n}\,d\Lambda_0\Big\}$$

$$= n^{-1} E_0 \int_0^\tau \frac{Y_n - Y_{jn}}{Y_n} Y_{jn}\,d\Lambda_0,$$

$$\Sigma_{jkn} = n^{-1}E_0\left(\int_0^\tau \frac{Y_n - Y_{jn}}{Y_n}\,dM_{jn} - \sum_{j'\neq j}\int_0^\tau \frac{Y_{jn}}{Y_n}\,dM_{j'n}\right)$$

$$\left(\int_0^\tau \frac{Y_n - Y_{kn}}{Y_n}\,dM_{kn} - \sum_{j'\neq k}\int_0^\tau \frac{Y_{kn}}{Y_n}\,dM_{j'n}\right)$$

$$= -n^{-1}E_0\left(\int_0^\tau \frac{Y_n - Y_{jn}}{Y_n^2}Y_{jn}Y_{kn}\,d\Lambda_0 - \int_0^\tau \frac{Y_n - Y_{kn}}{Y_n^2}Y_{jn}Y_{kn}\,d\Lambda_0\right.$$

$$\left. + \sum_{j'\neq j,k}\int_0^\tau \frac{Y_{jn}}{Y_n}\frac{Y_{kn}}{Y_n}Y_{j'n}\,d\Lambda_0\right)$$

$$= -n^{-1}E_0\int_0^\tau \frac{Y_{jn}Y_{kn}}{Y_n}\,d\Lambda_0, j\neq k\in\{1,\ldots,J\}.$$

Under the hypothesis of homogeneity, the variable $\sqrt{n}(K_{1n},\ldots,K_{Jn})$ converges weakly to a J-dimensional Gaussian variable with mean zero and variance matrix $(\Sigma_{jj'})_{j,j'=1,\ldots,J}$ such that

$$\Sigma_{jj} = \rho_j\int_0^\tau L^2(1 - \rho_j y_j y^{-1})y_j\,d\Lambda_0,$$

$$\Sigma_{jk} = -\rho_j\rho_k\int_0^\tau L^2 y_j y_k y^{-1}\,d\Lambda_0, j\neq k\in\{1,\ldots,J\}.$$

The components of Σ are estimated under H_0 by

$$V_{jj,n} = n^{-1}\int_0^\tau L_n^2(1 - Y_{jn}Y_n^{-1})Y_{jn}\,d\widehat{\Lambda}_n,$$

$$V_{jk,n} = -n^{-1}\int_0^\tau L_n^2 Y_{jn}Y_{kn}Y_n^{-1}\,d\widehat{\Lambda}_n, j\neq k\in\{1,\ldots,J\}$$

and a test statistic for the homogeneity of the population is defined as

$$(K_{1n},\ldots,K_{Jn})^t V_n^{-1}(K_{1n},\ldots,K_{Jn}),$$

it has an asymptotic χ_{k-1}^2 distribution and it tends to infinity under fixed alternatives. Under local alternatives such as (7.7), the asymptotic variance matrix and its estimator are still valid and the mean of $(K_{1n},\ldots,K_{Jn})$ is a vector $\mu_n = -\int_0^\tau W_n(t)\theta_n(t)\,d\Lambda_0(t)$, where $\theta_n = (\theta_{1n},\ldots,_{1n},\ldots,\theta_{Jn})^t$ converges uniformly to a vector θ of J functions.

If the components of the time variable are dependent, the covariance of the local martingales M_{kn} is still given by (7.8) and the matrix Γ is modified by taking into account the covariances C_{jkn} of M_{jn} and M_{kn}, for all $j \neq k$ in $\{1, \ldots, J\}$

$$\Sigma_{jj} = \int_0^\tau L^2 (1 - y_j y^{-1}) y_j \, d\Lambda_0$$

$$- 2 \sum_{j' \neq j} \int_{[0,\tau]^2} L(s) L(t) \left(1 - \frac{y_j}{y} \right) (s) y_j(t) \, dC_{jj'}(s,t),$$

$$\Sigma_{jk} = \int_{[0,\tau]^2} L(s) L(t) \left(1 - \frac{y_j}{y} \right) (s) \left(1 - \frac{y_k}{y} \right) (t) \, dC_{jk}(s,t)$$

$$- \sum_{j' \neq j,k} \int_{[0,\tau]^2} L(s) L(t) \left(1 - \frac{y_j}{y} \right) (s) \frac{y_k}{y}(t) \, dC_{jj'}(s,t)$$

$$- \sum_{j' \neq j,k} \int_{[0,\tau]^2} L(s) L(t) \left(1 - \frac{y_k}{y} \right) (s) \frac{y_j}{y}(t) \, dC_{kj'}(s,t)$$

$$+ \sum_{j' \neq j, k' \neq k} \int_{[0,\tau]^2} L(s) L(t) \frac{y_k}{y}(s) \frac{y_j}{y}(t) \, dC_{k'j'}(s,t)$$

$$- \int_0^\tau L^2 y_j y_k y^{-1} \, d\Lambda_0.$$

7.3   Likelihood ratio test of homogeneity

Lecam's theory (1956) of tests provides the limiting distribution of the local log-likelihood ratio test statistics under the hypothesis and the alternative and their asymptotic equivalence with asymptotically normal test statistics, according to the alternatives. The same idea was used to prove the asymptotic equivalence of the log-likelihood ratio tests and tests based on weighted differences of the empirical estimators of the cumulative hazard functions of point processes with a multiplicative intensity (Pons 1980, 1981) and the main results in this domain are detailed below.

The proportionality of hazard functions $\lambda_1 = c\lambda_2$ in the interval $[0, T]$ is equivalent to the relationship $\bar{F}_1 = \bar{F}_2^c$ in $[0, T]$ between the survival functions. The observations are n-sample drawn from a global population of individuals with hazard function λ_1 or λ_2. Under the hypothesis H_0 of equality of the distribution functions, the population is homogeneous and under the local alternatives K_n, two sub-populations with proportional hazard functions are sampled, each with a positive probability in the observation interval $[0, T]$. The local alternatives are defined by

$$K_n : \begin{cases} \lambda_{1n}(t) = \lambda_{2n}(t)(1 + n^{-\frac{1}{2}} \rho_n), \\ \lambda_{2n}(t) = \lambda_2(t)\{1 + n^{-\frac{1}{2}} \gamma_n(t)\}, \ t \in [0, T], \end{cases}$$

where $(\rho_n)_{n\geq 1}$ is a real sequence converging to a strictly positive limit ρ and $(\gamma_n)_{n\geq 1}$ is a sequence of functions converging uniformly in $[0,T]$ to a strictly positive function γ, as n tends to infinity. Under the hypothesis, the hazards functions are also denoted λ_0. The log-likelihood ratio statistic for H_0 against K_n is

$$l_n = \sum_{j=1}^{2} \left\{ \int_0^T \log \frac{\lambda_{jn}}{\lambda_j} \, dN_{jn} + \int_0^T (\lambda_j - \lambda_{jn}) Y_{jn} \, ds \right\}.$$

Let $M_{jn,0}(t) = N_{jn}(t) - \int_0^t Y_{jn}\lambda_0 \, ds$ and $M_{jn}(t) = N_{jn}(t) - \int_0^t Y_{jn}\lambda_{jn} \, ds$, for t in $[0,T]$, be the local martingales related to the counting processes N_{jn}, $j = 1,2$, under H_0 and K_n, respectively. The Cox statistic (1972) for H_0 against an alternative of proportional hazard functions in $[0,T]$ is

$$S_n(T) = V_n^{-\frac{1}{2}}(T) W_n(T), \tag{7.9}$$

where the variables $W_n(T)$ and $V_n(T)$ are defined by

$$W_n(T) = \int_0^T \frac{Y_{1n}Y_{2n}}{Y_{1n} + Y_{2n}} \left(1_{\{Y_{1n}>0\}} \frac{dN_{1n}}{Y_{1n}} - 1_{\{Y_{2n}>0\}} \frac{dN_{2n}}{Y_{2n}} \right), \tag{7.10}$$

$$= \int_0^T \frac{Y_{2n}}{Y_{1n} + Y_{2n}} 1_{\{Y_{1n}>0\}} \, dN_{1n} - \int_0^T \frac{Y_{1n}}{Y_{1n} + Y_{2n}} 1_{\{Y_{2n}>0\}} \, dN_{2n},$$

$$V_n(T) = \int_0^T \frac{Y_{1n}Y_{2n}}{(Y_{1n} + Y_{2n})^2} 1_{\{Y_{1n}+Y_{2n}>0\}} (dN_{1n} + dN_{2n}), \tag{7.11}$$

with the convention $\frac{0}{0} = 0$. According to the martingale property of the processes $M_{1n,0}$ and $M_{2n,0}$, the variable $V_n(T)$ is an estimator of the variance of $W_n(T)$ in the general model and under H_0

$$\sigma_{nT}^2 = E \int_0^T \left(\frac{Y_{1n}Y_{2n}}{Y_{1n} + Y_{2n}} \right)^2 (1_{\{Y_{1n}>0\}} Y_{1n}^{-1}\lambda_1 + 1_{\{Y_{2n}>0\}} Y_{2n}^{-1}\lambda_2) \, ds$$

$$= E \int_0^T \frac{Y_{1n}Y_{2n}}{Y_{1n} + Y_{2n}} \lambda_0 \, ds.$$

The asymptotic behavior of the log-likelihood ratio and Cox's test statistic relies on convergence of the processes, with integrability and boundedness conditions.

Condition 7.2.

(1) The endpoint of the observation interval T is strictly smaller than the end point of the support of the distribution function of the censored time.

(2) There exist functions $y_j : \mathbb{R}_+ \mapsto \mathbb{R}_+$, such that $n^{-1}Y_{jn}$ converges in probability to y_j uniformly in $[0, T]$ and $\int_0^T y_1 y_2 \{y_1 + y_2\}^{-1}\lambda_0 \, ds$ is finite.

(3) The integrals $\int_0^T \gamma^2 y_j \lambda_0 \, ds$, $j = 1, 2$, and $\int_0^T y_1 \lambda_0 \, ds$ are finite.

Under Condition 7.2, the estimator $V_n(T)$ converges in probability to

$$\sigma_T^2 = \int_0^T \frac{y_1 y_2}{y_1 + y_2} \lambda_0 \, ds \qquad (7.12)$$

and Rebolledo's weak convergence theorem (1977) for the local martingales related to the counting processes N_{1n} and N_{2n} in $\mathbb{R}_+$ applies to the test statistic $W_n(T)$.

Theorem 7.1. *Under Condition 7.2, the statistic $S_n(T)$ converges weakly under the hypothesis H_0 to a normal variable S. Under the local alternatives K_n, it converges weakly to $S + \mu$ où $\mu = \sigma_T^{-1} \rho \int_0^T y_1 y_2 (y_1 + y_2)^{-1} \, d\Lambda_0$.*

Proof. The variable $n^{-\frac{1}{2}} W_n(T)$ is centered under H_0 and its variances under H_0 and K_n are both asymptotically equivalent to σ_T^2. Under K_n, $W_n(T)$ is the sum of a variable with mean zero and the variable $n^{-1} \rho_n \int_0^T Y_{1n} Y_{2n} (Y_{1n} + Y_{2n})^{-1} \lambda_0 \, ds$, it converges in probability to $\mu \sigma_T$. □

Let X be a normal variable, the test of H_0 against K_n has the asymptotic level $\alpha = P(X > c) = 1 - \Phi(c)$ and the asymptotic power $\beta(\alpha) = P(X > c - \mu) = 1 - \Phi(c - \mu)$.

Let $\sigma_{\rho,\gamma}^2 = \int_0^T (\rho + \gamma)^2 y_1 \lambda_0 + \int_0^T \gamma^2 y_2 \lambda_0$. Expanding $\log \lambda_{jn} - \log \lambda_j$ in the expression of l_n and applying Rebolledo's weak convergence theorem provides the analoguous results for the log-likelihood test statistic, with the scaling $\sigma_{\rho,\gamma}^2$ and the location $\pm \frac{1}{2}\sigma_{\rho,\gamma}^2$.

The following asymptotic results are a consequence of Lecam's theory of tests (1956) for adapted to the independent and non identically distributed processes (Pons 1980).

Theorem 7.2. *Under Condition 7.2, the log-likelihood ratio statistic has an expansion $l_n(T) = Y_n - \frac{1}{2}Z_n + o_p(1)$ where Y_n is a centered process under H_0. Under H_0 and K_n, the variance of Y_n is asymptotically equivalent to Z_n, as n tends to infinity. The statistic $n^{-\frac{1}{2}} l_n(T)$ converges weakly under H_0 to a Gaussian variable Y with mean $-\frac{1}{2}\sigma_{\rho,\gamma}^2$ and variance $\sigma_{\rho,\gamma}^2$. Under the local alternatives K_n, it converges weakly to $Y + \sigma_{\rho,\gamma}^2$.*

Proof. The log-likelihood ratio statistic for H_0 against K_n is written as

$$l_n = \int_0^T \log\{(1 + n^{-\frac{1}{2}}\rho_n)(1 + n^{-\frac{1}{2}}\gamma_n)\} \, dN_{1n} - \int_0^T (\lambda_{1n} - \lambda_1) Y_{1n} \, ds$$

$$+ \int_0^T \log(1 + n^{-\frac{1}{2}}\gamma_n) \, dN_{2n} - n^{-\frac{1}{2}} \int_0^T \gamma_n \lambda_2 Y_{2n} \, ds.$$

A second order Taylor expansion of the logarithm in this expression leads to $\log\{(1 + n^{-\frac{1}{2}}\rho_n)(1 + n^{-\frac{1}{2}}\gamma_n)\} = \log\{1 + n^{-\frac{1}{2}}(\rho_n + \gamma_n) + n^{-1}\rho_n\gamma_n\}$, which equals $n^{-\frac{1}{2}}(\rho_n + \gamma_n) - \frac{1}{2}n^{-1}(\rho_n^2 + \gamma_n^2) + o(n^{-1})$, hence

$$l_n = n^{-\frac{1}{2}}\left\{ \int_0^T (\rho_n + \gamma_n) \, dN_{1n} + \int_0^T \gamma_n \, dN_{2n} \right\}$$

$$- (2n)^{-1}\left\{ \int_0^T (\rho_n^2 + \gamma_n^2) \, dN_{1n} + \int_0^T \gamma_n^2 \, dN_{2n} \right\}$$

$$- \int_0^T (\lambda_{1n} - \lambda_1) Y_{1n} \, ds - n^{-\frac{1}{2}} \int_0^T \gamma_n \lambda_2 Y_{2n} \, ds + R_n$$

$$= n^{-\frac{1}{2}}\left\{ \int_0^T (\rho_n + \gamma_n) \, dM_{1n} + \int_0^T \gamma_n \, dM_{2n} \right\}$$

$$- (2n)^{-1}\left\{ \int_0^T (\rho_n + \gamma_n)^2 Y_{1n} \lambda_0 \, ds + \int_0^T \gamma_n^2 Y_{2n} \lambda_0 \, ds \right\} + R_n,$$

where R_n converges in probability to zero under H_0. It follows that l_n converges weakly under H_0 to a Gaussian variable with mean $\mu = -\frac{1}{2}\sigma^2(\rho, \gamma)$ and variance $\sigma^2(\rho, \gamma) = \int_0^T (\rho + \gamma)^2 y_1 \lambda_0 \, ds + \int_0^T \gamma^2 y_2 \lambda_0 \, ds$. The probability distributions of l_n under H_0 and K_n are therefore contiguous and this implies that R_n converges in probability to zero under the alternative (Ex. 2.9.1 and 2.9.2). Under K_n, the first order term of the approximation of l_n is $n^{-\frac{1}{2}}\{\int_0^T (\rho_n + \gamma_n) \, dM_{1n} + \int_0^T \gamma_n \, dM_{2n}\} + n^{-\frac{1}{2}}\{\int_0^T (\rho_n + \gamma_n)(\lambda_{1n} - \lambda_0) Y_{1n} \, ds + \int_0^T \gamma_n (\lambda_{2n} - \lambda_0) Y_{2n} \, ds\}$ and its asymptotic mean is $\frac{1}{2}\sigma_{\rho,\gamma}^2$. $\quad\square$

Theorem 7.3. *The log-likelihood ratio test is asymptotically equivalent to the test defined by the statistic $S_n(T)$.*

Proof. From the previous expansion and the weak convergence of its first order term, the log-likelihood ratio test H_0 against K_n is asymptotically equivalent to the test based on

$$T_n = \sigma^{-1}(\rho, \gamma) n^{-\frac{1}{2}} l_n. \tag{7.13}$$

The functions γ_n are uniformly approximated in $[0, T]$ by a sequence of step functions $\gamma_{n,m}$ in a partition $\pi_m = ([t_{m,k-1}, t_{m,k}[)_{k=1,\dots,m}$ such that $t_{m,k-1} = 0$ and $t_{m,m} = T$, and the value of γ_n in the interval $I_k = [t_{m,k-1}, t_{m,k}[$ is denoted $\gamma_{n,m,k}$. As m tends to infinity, l_n is approximated under H_0 by

$$l_{n,m} = \theta_{n,m}^t \Delta_{n,m} - \frac{1}{2}\sigma_{n,m}^2,$$

where

$$\theta_{n,m} = (\rho_n, \gamma_{n,m,1}, \ldots, \gamma_{n,m,m})^t,$$

$$\Delta_{n,m} = n^{-\frac{1}{2}}\left(M_{1n,0}(T), \sum_{j=1,2} M_{jn,0}(t_{m,1}), \sum_{j=1,2}\{M_{jn,0}(t_{m,2}) - M_{jn,0}(t_{m,1})\}, \right.$$

$$\left. \ldots, \sum_{j=1,2}\{M_{1n,0}(t_{m,m}) - M_{1n,0}(t_{m,m-1})\}\right),$$

and $\sigma_{n,m}^2 = \theta_{n,m}^t \Gamma_m \theta_{n,m}$, with a symmetric matrix $\Gamma_{n,m}$ of $\mathbb{R}^{m+1} \times \mathbb{R}^{m+1}$ with terms defined from the constants $C_{j,m,k} = \int_{t_{j,m,k-1}}^{t_{m,k}} y_j \lambda_0 \, ds$ as

$$\Gamma_{m,11} = \sum_{k=1}^{m} C_{1,m,k},$$

$$\Gamma_{m,1k} = C_{1,m,k},$$

$$\Gamma_{m,kk} = C_{1,m,k} + C_{2,m,k},$$

and the other terms of Γ_m are zero. The hypothesis H_0 is $\rho_n = 0$ and γ_n are nuisance functions for the test of H_0. Let

$$e_m = \left(1, \frac{-C_{1,m,1}}{C_{1,m,1} + C_{2,m,1}}, \ldots, \frac{-C_{1,m,m}}{C_{1,m,m} + C_{2,m,m}}\right)^t,$$

then $\Gamma_m e_m$ is proportional to the unit vector $e_1 = (1, 0, \ldots, 0)$ of $\mathbb{R}^{m+1}$, with first component

$$(\Gamma_m e_m)_1 = \Gamma_{m,11} - \sum_{k=1}^{m} \frac{C_{1,m,k}^2}{C_{1,m,k} + C_{2,m,k}} = \sum_{k=1}^{m} \frac{C_{1,m,k} C_{2,m,k}}{C_{1,m,k} + C_{2,m,k}},$$

therefore $e_m^t \Gamma_m e_m = e_1^t \Gamma_m e_m$.

Under H_0, the projection of the parameter vector $\theta_{n,m}$ onto e_1 must be zero. The other components of $\theta_{n,m}$ being nuisance parameters, one can omit them in the test statistic, hence H_0 reduces to $\theta_{n,m}^t e_1 = 0$, equivalently $\theta_{n,m}^t \Gamma_m e_m$ must be zero. It follows that $\theta_{n,m}$ is Γ_m-orthogonal to e_m under H_0 and it is possible to reduce the question to test whether the statistic is zero as $\theta_{n,m} = e_m$. The log-likelihood ratio test statistic is asymptotically equivalent to T_n given by (7.13) and by projection along e_m, it becomes asymptotically equivalent

$$U_{n,m} = (e_m^t \Gamma_m e_m)^{-\frac{1}{2}} e_m^t \Delta_{n,m},$$

as the bandwidth of the partition π_m tends to zero. Under H_0, $U_{n,m}$ is written as

$$U_{n,m} = n^{-\frac{1}{2}} \Big(\sum_{k=1}^{m} \frac{C_{1,m,k} C_{2,m,k}}{C_{1,m,k} + C_{2,m,k}} \Big)^{-1} \Big\{ M_{1n,0}(T)$$
$$- \sum_{k=1}^{m} \frac{C_{1,m,k}}{C_{1,m,k} + C_{2,m,k}} \int_{t_{j,m,k-1}}^{t_{m,k}} (dM_{1n,0} + dM_{2n,0}) \Big\}.$$

The limit as m and n tend to infinity of $U_{n,m}$ is

$$U_n = \sigma_T^{-1} \Big\{ \int_0^T \frac{y_2}{y_1 + y_2} Y_{1n} \lambda_0 \, ds - \int_0^T \frac{y_1}{y_1 + y_2} Y_{2n} \lambda_0 \, ds \Big\}.$$

By (7.12) and the uniform convergence of the processes Y_{jn} to y_{jn} for $j = 1, 2$, the statistic (7.9) is such that $U_n - S_n(T)$ converges to zero in probability under H_0 and under K_n. □

The asymptotic variance σ_T of the likelihood ratio statistic $W_n(T)$ is minimal among the variances of the test statistic for H_0. For example, the statistic

$$W_{2n}(T) = \int_0^T \frac{Y_{1n}}{Y_{1n} + Y_{2n}} dN_{1n} - \int_0^T \frac{Y_{2n}}{Y_{1n} + Y_{2n}} dN_{2n}$$

has, under H_0, the variance

$$\sigma_{2n}^2(T) = E_0 \int_0^T \frac{Y_{1n}^3 + Y_{2n}^3}{(Y_{1n} + Y_{2n})^2} \lambda_0 \, ds = E_0 \int_0^T \frac{Y_{1n}^2 + Y_{2n}^2 - Y_{1n} Y_{2n}}{Y_{1n} + Y_{2n}} \lambda_0 \, ds$$

which is always larger than the variance of $W_n(T)$.

The tests are modified when the sizes n_1 and n_2 of the two sub-samples are different. The terms of the difference defining the variable $n^{-\frac{1}{2}} W_n(T)$ is weighted by the ratios $n^{-1} n_1$ and, respectively, $n^{-1} n_2$ which converge respectively to strictly positive constants c and $1 - c$ as n tends to infinity. Under H_0

$$n^{-\frac{1}{2}} W_n(T) = \Big(\frac{n_1}{n} \Big)^{\frac{1}{2}} \int_0^T \frac{Y_{2n}}{Y_{1n_1} + Y_{2n_2}} 1_{\{Y_{1n_1} > 0\}} n_1^{-\frac{1}{2}} (dN_{1n_1} - d\Lambda_0)$$
$$- \Big(\frac{n_2}{n} \Big)^{\frac{1}{2}} \int_0^T \frac{Y_{1n_1}}{Y_{1n_1} + Y_{2n_2}} 1_{\{Y_{2n_2} > 0\}} n_2^{-\frac{1}{2}} (dN_{2n_2} - d\Lambda_0).$$

It is asymptotically equivalent to

$$n^{-\frac{1}{2}} W_n(T) = c^{\frac{1}{2}} \int_0^T \frac{Y_{2n}}{Y_{1n_1} + Y_{2n_2}} 1_{\{Y_{1n_1} > 0\}} n_1^{-\frac{1}{2}} dM_{1n_1}$$
$$- (1-c)^{\frac{1}{2}} \int_0^T \frac{Y_{1n_1}}{Y_{1n_1} + Y_{2n_2}} 1_{\{Y_{2n_2} > 0\}} n_2^{-\frac{1}{2}} dM_{2n_2}.$$

Under the condition of convergence in probability to zero of $\sup_{t \in [0,T]} |n_k^{-1} Y_{kn_k}|$, as n tends to infinity, for $k = 1, 2$, the local martingales $M_{kn_k} = N_{kn_k} - \int_0^T Y_{kn_k} \, d\Lambda_0$ are such that $n_k^{-\frac{1}{2}} M_{kn_k}$ converge weakly under H_0 to centered Gaussian processes G_k with independent increments and with variance $\int_0^T y_k \, d\Lambda_0$. It follows that $n^{-\frac{1}{2}} W_n(T)$ converges weakly under H_0 to a centered Gaussian variable with variance

$$\sigma_0^2(T) = c(1 - c) \int_0^T \frac{y_1 y_2}{y_1 + y_2} \, d\Lambda_0$$

and this is estimated by

$$V_n(T) = \int_0^T \frac{Y_{1n_1} Y_{2n_2}}{(Y_{1n_1} + Y_{2n_2})^2} 1_{\{Y_{1n_1} + Y_{2n_2} > 0\}} \, d(N_{1n_1} + N_{2n_2}).$$

The test statistic $S_n(T)$ is now defined with these weighting variables in $W_n(T)$.

With dependent point processes N_{1n} and N_{2n}, the variance of $W_n(T)$ under H_0 modifies the test statistic (7.9). Their covariance is supposed to satisfy the next conditions.

Condition 7.3. There exists a continuous and bounded function ϕ from $\mathbb{R}_+$ to $\mathbb{R}_+$ such that

$$E_0(N_{1n} N_{2n}) = E_0 \int_0^T \phi(Y_{1n}, Y_{2n}) 1_{\{Y_{1n} Y_{2n} > 0\}} \lambda_0 \, ds.$$

Under Condition 7.2, the process $n^{-1}\phi(Y_{1n}, Y_{2n})$ converges in probability to a function $\phi(y_1, y_2)$, uniformly in $[0, T]$, as n tends to infinity. The covariance under H_0 of the local martingales $n^{-\frac{1}{2}} M_{1n,0}$ and $n^{-\frac{1}{2}} M_{2n,0}$ is

$$n^{-1} E_0(N_{1n}(s) N_{2n}(t)) = n^{-1} \int_0^{s \wedge t} \phi(Y_{1n}, Y_{2n}) \lambda_0 \, ds$$

and it converges in probability to the same limit $\int_0^{s \wedge t} \phi(y_1, y_2) \lambda_0 \, ds$. The variance of the variable $W_n(T)$ is then estimated by

$$V_{n,\phi}(T) = V_n(T) - 2 \int_0^T \frac{Y_{1n} Y_{2n}}{(Y_{1n} + Y_{2n})^2} \frac{\phi(Y_{1n}, Y_{2n})}{Y_{1n} + Y_{2n}} (dN_{1n} + dN_{2n})$$

and the Cox statistic is normalized by this estimator. Theorem 7.3 is modified in the same way and the log-likelihood ratio test is asymptotically equivalent to the test defined by the statistic $V_{n,\phi}^{-\frac{1}{2}}(T) W_n(T)$.

7.4 Tests of homogeneity against general local alternatives

The alternative of proportional hazard functions is restrictive and it is extended to a test of H_0 against general local alternatives

$$K_n : \begin{cases} \lambda_{1n}(t) = \lambda_{2n}(t)\{1 + n^{-\frac{1}{2}}\eta_n(t)\}, \\ \lambda_{2n}(t) = \lambda_2(t)\{1 + n^{-\frac{1}{2}}\gamma_n(t)\}, \ t \in [0, T], \end{cases} \tag{7.14}$$

where the sequence of positive functions $(\eta_n)_{n \geq 1}$ converge uniformly to a limit η in $[0, T]$. It can be performed using the log-likelihood ratio statistic or an asymptotically equivalent Student type statistic. The log-likelihood ratio statistic satisfies asymptotic expansions similar to those of Theorems 7.2 and 7.3, with weighting processes depending on the functions η_n defining the alternative. Replacing the constant ρ by the function η, the Student statistic and Theorem 7.1 are not modified.

Theorem 7.4. *Under Condition 7.2, the log-likelihood ratio statistic for H_0 against the local alternatives K_n (7.14) is asymptotically equivalent to the statistic $S_n(T)$. It converges weakly to a normal variable under H_0 and to the sum of a normal variable and σ_T^2 under K_n.*

Proof. The functions η_n and γ_n are uniformly approximated in $[0, T]$ by sequences of step functions $\eta_{n,m}$ and, respectively, $\gamma_{n,m}$ in a partition $\pi_m = (I_{n,k})_{k=1,\dots,m}$ such that $I_{n,k} = [t_{m,k-1}, t_{m,k}[$ with $t_{m,k-1} = 0$ and $t_{m,m} = T$, and the values of the functions in the interval $I_{n,k}$ are constants denoted $\eta_{n,m,k}$ and, respectively, $\gamma_{n,m,k}$. For $j = 1, 2$, the variations of the processes $A_{j,n} = n^{-\frac{1}{2}}M_{j,n,0}$ in $I_{n,k}$ are denoted $\Delta A_{j,n}(I_{n,k})$. Under H_0, the log-likelihood ratio statistic has an expansion, as m tends to infinity, in the form

$$l_{n,m} = \theta_{n,m}^t \Delta_{n,m} - \frac{1}{2}\theta_{n,m}^t \Gamma_m \theta_{n,m},$$

where

$$\theta_{n,m} = (\eta_{n,m,1}, \dots, \eta_{n,m,m}, \gamma_{n,m,1}, \dots, \gamma_{n,m,m})^t,$$

$$\Delta_{n,m} = (\Delta A_{1,n}(I_{n,1}), \Delta A_{1,n}(I_{n,2}), \dots, \Delta A_{1,n}(I_{n,m}), \sum_{j=1,2} \Delta A_{j,n}(I_{n,1}),$$

$$\sum_{j=1,2} \Delta A_{j,n}(I_{n,2}), \dots, \sum_{j=1,2} \Delta A_{m,n}(I_{n,m})),$$

and $\Gamma_m = (\Gamma_{m,k,l})_{k,l=1,\dots,m}$ is a symmetric matrix of $\mathbb{R}^{2m} \times \mathbb{R}^{2m}$ defined by the constants $C_{j,m,k} = \int_{t_{j,m,k-1}}^{t_{m,k}} y_j \lambda_0 \, ds, \ k = 1, \dots, m$, as

$$\Gamma_{m,k,k} = C_{1,m,k},$$

$$\Gamma_{m,k,m+k} = C_{1,m,k}, \ l = 1, \dots, m,$$

$$\Gamma_{m,m+k,m+k} = C_{1,m,k} + C_{2,m,k},$$

and the other terms of Γ_m are zero. The hypothesis H_0 is expressed as $\eta_n(x) = 0$ in $[0, T]$ and γ_n is a nuisance function for the test of H_0. Under the hypothesis H_0, the first k components of the parameter $\theta_{n,m}$ are zero and the other components are nuisance parameters. Let $e_1 = (1, 0, \ldots, 0)$, $e_2 = (0, 1, \ldots, 0), \ldots, e_m = (0, \ldots, 0, 1, 0, \ldots, 0)$ be unit vectors of $\mathbb{R}^{2m}$ with one component 1 among the first m components and the other components 0. The Γ_m-orthogonal projection of e_k in the space Θ_m is explicitly written as

$$u_{k,m} = \Big(0, \ldots, 0, 1, 0, \ldots, 0, \frac{-C_{1,m,k}}{C_{1,m,k} + C_{2,m,k}}, 0, \ldots, 0\Big)^t,$$

with nonzero kth and $(m+k)$th components, and

$$u_{k,m}^t \Gamma_m u_{k,m} = \frac{C_{1,m,k} C_{2,m,k}}{C_{1,m,k} + C_{2,m,k}}, \quad k = 1, \ldots, m.$$

The log-likelihood ratio test statistic is asymptotically equivalent to T_n given by (3.4) and to its approximation

$$U_{n,m} = \Big(\sum_{k=1}^m u_{k,m}^t \Gamma_m u_{k,m}\Big)^{-1} \sum_{k=1}^m u_k^t \Delta_{n,m}$$

using the partition π_m and the Γ_m-orthonormal basis of $\mathbb{R}^{2m} \setminus \Theta_m$. It must be zero under H_0, it is written as

$$U_{n,m} = n^{-\frac{1}{2}} \Big(\sum_{k=1}^m \frac{C_{1,m,k} C_{2,m,k}}{C_{1,m,k} + C_{2,m,k}}\Big)^{-\frac{1}{2}} W_{n,m}(T)$$

where the variable $W_{n,m}(T)$ is the sum

$$
\begin{aligned}
W_{n,m}(T) = &\sum_{k=1}^m \Big(\int_{t_{j,m,k-1}}^{t_{m,k}} Y_{1n} \lambda_0 \, ds \\
&- \sum_{k=1}^m \frac{C_{1,m,k}}{C_{1,m,k} + C_{2,m,k}} \int_{t_{j,m,k-1}}^{t_{m,k}} (Y_{1n} + Y_{2n}) \lambda_0 \, ds \Big) \\
= &\sum_{k=1}^m \Big(\frac{C_{2,m,k}}{C_{1,m,k} + C_{2,m,k}} \int_{t_{j,m,k-1}}^{t_{m,k}} Y_{1n} \lambda_0 \, ds \\
&- \sum_{k=1}^m \frac{C_{1,m,k}}{C_{1,m,k} + C_{2,m,k}} \int_{t_{j,m,k-1}}^{t_{m,k}} Y_{2n} \lambda_0 \, ds \Big).
\end{aligned}
$$

The limit as m and n tend to infinity of $U_{n,m}$ is the same as the limit of the variable

$$U_n = \sigma_T^{-1} n^{-\frac{1}{2}} \Big(\int_0^T \frac{y_2}{y_1 + y_2} \, dN_{1n} - \int_0^T \frac{y_1}{y_1 + y_2} \, dN_{2n} \Big),$$

where σ_T^2 is the limit as n tends to infinity of $V_n(T)$ defined in (7.9). By the uniform convergence of the processes Y_{jn} to y_{jn} for $j = 1, 2$, the Cox statistic is such that $U_n - S_n(T)$ converges to zero in probability under H_0 and under K_n. □

The test (7.9) for homogeneity of the population against local alternatives of proportional functions has been generalized to the alternative (7.14) using a weighted statistic and an estimator of its variance. They are defined from a positive predictable process H_n in $[0, T]$, as

$$W_n(T) = \int_0^T H_n \left(1_{\{Y_{1n} > 0\}} \frac{dN_{1n}}{Y_{1n}} - 1_{\{Y_{2n} > 0\}} \frac{dN_{2n}}{Y_{2n}} \right),$$

$$V_n(T) = \int_0^T H_n^2 1_{\{Y_{1n} + Y_{2n} > 0\}} \frac{dN_{1n} + dN_{2n}}{Y_{1n} + Y_{2n}}.$$

The process H_n is supposed to converge in probability to a function h, uniformly in $[0, T]$, and the integrability conditions are modified by the multiplicative function h or h^2 under the integral. The efficiency of these tests is measured by the ratio

$$\Delta = \sigma_{h,T}^{-1} \int_0^T h \eta \lambda_0 \, ds$$

where $\sigma_{h,T}^2 = \int_0^T h^2 \varphi \lambda_0 \, ds$ is finite and

$$\varphi(t) = \frac{y_1 y_2}{y_1 + y_2}(t).$$

The function h is supposed of the form $h = \beta \eta \varphi + v$, with a constant β and with a function v satisfying the orthogonality property $\int_0^T \eta v \lambda_0 \, ds = 0$. The limit of $n^{-1} W_n(T)$ is then $\int_0^T h \eta \lambda_0 \, ds = \beta \int_0^T \eta^2 \varphi \lambda_0 \, ds$. The coefficient Δ is such that

$$\Delta^2 = \frac{\beta^2 \{\int_0^T \eta^2 \varphi \lambda_0 \, ds\}^2}{\int_0^T h^2 \varphi \lambda_0 \, ds} = \frac{\{\int_0^T \beta \eta^2 \varphi \lambda_0 \, ds\}^2}{\int_0^T (\beta^2 \eta^2 \varphi + v^2) \lambda_0 \, ds}$$

and it is maximum as the function v is zero. An optimal weighting process H_n is therefore proportional to $\eta_n Y_{1n} Y_{2n} (Y_{1n} + Y_{2n})^{-1}$.

7.5 Goodness of fit for the hazard functions ratio

Let ξ be a positive real function in $\mathbb{R}_+$, the statistics of the previous section have been modified in order to test the hypothesis $H_0 : \lambda_1 = \xi \lambda_2$ in the observation interval $[0, T]$ against alternatives of a ratio different from the function ξ in $[0, T]$ (Pons 1981). A sequence of local alternatives K_n to the hypothesis H_0 is

$$K_n : \begin{cases} \lambda_{1n}(t) = \xi(t)\lambda_{2n}(t)(1 + n^{-\frac{1}{2}}\rho_n), \\ \lambda_{2n}(t) = \lambda_2(t)\{1 + n^{-\frac{1}{2}}\gamma_n(t)\}, \ t \in [0, T], \end{cases}$$

and the common hazard function under H_0 is denoted λ_0. Its integral over $[0, t]$ is estimated under H_0 and the local alternatives by

$$\widehat{\Lambda}_{n\xi}(t) = \int_0^t 1_{\{Y_{1n}+Y_{2n}>0\}} \frac{dN_{1n} + dN_{2n}}{\xi Y_{1n} + Y_{2n}}, \tag{7.15}$$

using the martingale property of the processes $M_{1n,0}$ and $M_{2n,0}$, $\widehat{\Lambda}_{n\xi}$ is unbiased under H_0 and consistent under K_n. A test statistic for H_0 against K_n

$$S_{n\xi}(T) = V_{n\xi}^{-\frac{1}{2}}(T)W_{n\xi}(T) \tag{7.16}$$

is defined by a variable $W_{n\xi}(T)$ and an estimator of its variance $V_{n\xi}(T)$ depending on the function ξ of the hypothesis (Pons, 1981)

$$W_{n\xi}(T) = \int_0^T \frac{Y_{2n}1_{\{Y_{1n}>0\}}}{\xi(\xi Y_{1n} + Y_{2n})} dN_{1n} - \int_0^T \frac{Y_{1n}1_{\{Y_{2n}>0\}}}{\xi Y_{1n} + Y_{2n}} dN_{2n},$$

$$= \int_0^T \frac{Y_{1n}Y_{2n}}{\xi Y_{1n} + Y_{2n}}\left(1_{\{Y_{1n}>0\}}\frac{dN_{1n}}{\xi Y_{1n}} - 1_{\{Y_{2n}>0\}}\frac{dN_{2n}}{Y_{2n}}\right),$$

$$V_{n\xi}(T) = \int_0^T \frac{Y_{1n}Y_{2n}}{\xi(\xi Y_{1n} + Y_{2n})^2} 1_{\{Y_{1n}+Y_{2n}>0\}}(dN_{1n} + dN_{2n}).$$

The variance of $W_{n\xi}(T)$ under H_0 is

$$\sigma_{n\xi,T}^2 = E \int_0^T \left(\frac{Y_{1n}Y_{2n}}{\xi Y_{1n} + Y_{2n}}\right)^2 \left(\frac{1_{\{Y_{1n}>0\}}}{\xi Y_{1n}} + \frac{1_{\{Y_{2n}>0\}}}{Y_{2n}}\right)\lambda_0 \, ds$$

$$= E \int_0^T \frac{Y_{1n}Y_{2n}}{\xi(\xi Y_{1n} + Y_{2n})}\lambda_0 \, ds + o(1)$$

and $V_{n\xi}(T)$ is an unbiased estimator of $\sigma_{n\xi,T}^2$ based on (7.15). The log-likelihood ratio test statistic is similar to statistic for the test of propotional hazards and it depends on ξ, it is denoted $l_{n\xi}$. The functions y_1 and y_2 are supposed to satisfy the following conditions which modify Condition 7.2.

Condition 7.4.

(1) The variance $\sigma_{\xi,T}^2 = \int_0^T y_1 y_2 \{\xi(\xi y_1 + y_2)\}^{-1}\lambda_0 \, ds$ is finite.

(2) The integral $\int_0^T \xi \gamma^2 y_1 \lambda_0 \, ds$ is finite.

Under Conditions 7.2 and 7.4, the variable $V_{n\xi}(T)$ converges in probability to $\sigma^2_{\xi,T}$ under H_0 and K_n.

Proposition 7.2. *Under the condition of finite integrals $\int_0^T \gamma\lambda_0\,ds$ and $\int_0^T(\xi y_1 + y_2)^{-1}\lambda_0\,ds$, the process $n^{\frac{1}{2}}(\widehat{\Lambda}_{n\xi} - \Lambda_0)$ converges weakly in $D_{[0,T]}$ to a centered continuous Gaussian process with independent increments and with variance function $\sigma^2_{\Lambda,\xi}(t) = \int_0^t(\xi y_1 + y_2)^{-1}\lambda_0\,ds$, under H_0. Under K_n, it converges weakly to a Gaussian process with independent increments and with mean function*

$$m_{\Lambda,\xi}(t) = \int_0^t \gamma\lambda_0\,ds + \rho\int_0^t \frac{\xi y_1}{\xi y_1 + y_2}\lambda_0\,ds,$$

and variance $\sigma^2_{\Lambda,\xi}$.

This is a direct consequence of the weak convergence under H_0 of the local martingales $n^{-\frac{1}{2}}(N_{jn} - \int_0^\cdot Y_{jn}\lambda_{jn}\,ds)$, $j = 1, 2$, to centered Gaussian processes with independent increments and with respective variances $\int_0^\cdot \xi y_1\lambda_0\,ds$ and $\int_0^\cdot y_2\lambda_0\,ds$, when they are finite. Under K_n and under the same conditions, the processes converge weakly to Gaussian processes with respective means $m_1 = \int_0^\cdot(\rho+\gamma)\xi y_1\lambda_0\,ds$ and $m_2 = \int_0^\cdot \gamma y_2\lambda_0\,ds$, for $j = 1, 2$. The asymptotic variances are the same as under H_0.

Theorem 7.5. *Under Conditions 7.2 and 7.4, the statistic $S_{n\xi}(T)$ converges weakly under the hypothesis H_0 to a normal variable S. Under the local alternatives K_n, it converges weakly to $S+\mu$ où $\mu = \sigma^{-1}_{\xi,T}\rho\int_0^T y_1 y_2(\xi y_1 + y_2)^{-1}\,d\Lambda_0$.*

This result is proved like Theorem 7.1 for the limiting distributions of the statistic (7.9) and the asymptotic power of the one-sided test of H_0 against K_n based on the statistic $S_{n\xi}(T)$ is deduced as $\beta(\alpha) = 1 - \Phi(c_\alpha - \mu)$, with the normal quantile. The local asymptotic power of the two-sided test at the level α is $\beta(\alpha) = 1 - \Phi(c_{\frac{\alpha}{2}} - \mu) + \Phi(c_{1-\frac{\alpha}{2}} - \mu)$.

Let $\sigma^2_{\xi,\rho,\gamma} = \int_0^T(\rho + \gamma)^2 y_1\xi\lambda_0\,ds + \int_0^T \gamma^2 y_2\lambda_0\,ds$. Expanding $\log\lambda_{jn} - \log\lambda_j$ in the expression of $l_{n,\xi}$ and applying Rebolledo's weak convergence theorem provides the analoguous results for the log-likelihood test statistic, with the variances and means depending only of $\sigma^2_{\xi,\rho,\gamma}$.

Theorem 7.6. *Under Conditions 7.2 and 7.4, the statistic $n^{-\frac{1}{2}}l_{n,\xi}(T)$ converges weakly under the hypothesis H_0 to a Gaussian variable Y with mean $-\frac{1}{2}\sigma^2_{\xi,\rho,\gamma}$ and variance $\sigma^2_{\xi,\rho,\gamma}$. Under the local alternative K_n, it converges weakly to $Y + \frac{1}{2}\sigma^2_{\xi,\rho,\gamma}$.*

The asymptotic equivalence of the likelihood ratio and the statistic weighted difference of estimated cumulative hazard functions extends Theorem 7.3 to the goodness of fit tests (Pons 1981).

Theorem 7.7. *Under Conditions 7.2 and 7.4, the log-likelihood ratio statistic has an expansion $l_{n,\xi} = Y_{n,\xi} - \frac{1}{2}Z_{n,\xi}$ where $E_0 Y_{n,\xi} = 0$. Under H_0 and K_n, the limit of $\mathrm{Var}\,Y_{n,\xi}$ is the limit in probability of $Z_{n,\xi}$, as n tends to infinity and the tests performed with $Z_{n,\xi}^{-\frac{1}{2}} Y_{n,\xi}$ and $S_{n,\xi}(T)$ are asymptotically equivalent.*

The proof of Theorem 7.3 is modified by the multiplicative hazard function $\xi\lambda_0$ for λ_1 in the expression of the integrals $C_{j,m,k}$ defining the orthogonal basis e_m.

The theorems of this section extend to goodness of fit tests of the hazard functions ratio against general local alternatives

$$K_n : \begin{cases} \lambda_{1n}(t) = \xi(t)\lambda_{2n}(t)\{1 + n^{-\frac{1}{2}}\eta_n(t)\}, \\ \lambda_{2n}(t) = \lambda_2(t)\{1 + n^{-\frac{1}{2}}\gamma_n(t)\}, \ t \in [0,T], \end{cases} \tag{7.17}$$

where the function ξ is known and the sequence of positive functions $(\eta_n)_{n\geq 1}$ converge uniformly to a limit η in $[0,T]$. The proofs rely on the same arguments as that for Theorem 7.3.

More generally, the function ξ belongs to a parametric class of regular functions in a compact set Θ, $\Xi = \{\xi_\theta : [0,T] \mapsto \mathbb{R}, \theta \in \Theta\}$. The true parameter value under the hypothesis H_0 is denoted θ_0 and the unknown ratio ξ_0 of the hazard functions is estimated by maximizing the likelihood process

$$l_{1n}(\theta) = \int_0^T \log(\xi_\theta)\,dN_{1n} - \int_0^T \xi_\theta Y_{1n}\lambda_0\,ds.$$

The process

$$C_n(\theta) = n^{-1}\{l_{1n}(\theta) - l_{1n}(\theta_0)\} = n^{-1}\Big\{\int_0^T \log\frac{\xi_\theta}{\xi_0}\,(dN_{1n} - Y_{1n}\xi_0\lambda_0\,ds)$$

$$- \int_0^T \Big(\frac{\xi_\theta}{\xi_0}\log\frac{\xi_\theta}{\xi_0} - 1\Big)Y_{1n}\xi_0\lambda_0\,ds\Big\}$$

converges in probability to the function

$$C(\theta) = \int_0^T \Big(\log\frac{\xi_\theta}{\xi_0} - \frac{\xi_\theta}{\xi_0} + 1\Big)\xi_0\lambda_0\,ds,$$

uniformly in Θ. Its first two derivatives are

$$\dot{C}(\theta) = \int_0^T \left(\frac{\dot{\xi}_\theta}{\xi_\theta} - \frac{\dot{\xi}_\theta}{\xi_0} \right) \xi_0 \lambda_0 \, ds,$$

$$\ddot{C}(\theta) = \int_0^T \left(\frac{\ddot{\xi}_\theta}{\xi_\theta} - \frac{\ddot{\xi}_\theta}{\xi_0} - \frac{\dot{\xi}_\theta^2}{\xi_\theta^2} \right) \xi_0 \lambda_0 \, ds,$$

the function C is therefore concave with a maximum at ξ_0 and it follows that the maximum likelihood estimator $\widehat{\theta}_n$ of ξ_0 is consistent. By a Taylor expansion of C_n in a neighborhood of ξ_0, it follows that

$$n^{\frac{1}{2}}(\widehat{\theta}_n - \xi_0) = -\{\ddot{C}(\theta_0)\}^{-1} n^{\frac{1}{2}} \dot{C}_n(\theta_0) + o_p(1).$$

By Rebolledo's convergence theorem, the process $n^{\frac{1}{2}} \dot{C}_n(\theta_0)$ converges weakly to a centered Gaussian process with variance I_0^{-1}, where I_0 is the matrix defined as $I_0 = -\ddot{C}(\theta_0) = \int_0^T (\dot{\xi}_\theta \xi_\theta^{-1})^{\otimes 2} \xi_0 \lambda_0 \, ds$.

Let $\widehat{\xi}_n = \xi_{\widehat{\theta}_n}$, the process $n^{\frac{1}{2}}(\widehat{\xi}_n - \xi_0)(s)$, s in $[0, T]$, converges weakly to a centered Gaussian process with variance $J_0 = \dot{\xi}_0^t I_0^{-1} \dot{\xi}_0$, where $\dot{\xi}_0$ is the derivative of the function ξ at θ_0. The variable $W_{n\widehat{\xi}_n}(T)$ is defined by plugging the estimator of the unknown parameter in the statistic (7.16). Let $\dot{W}_{n\xi}$ be the derivative of the process $W_{n\xi}$ defined by (7.16), with respect to the parameter ξ. A first order expansion of $W_{n\widehat{\xi}_n}(T)$ at ξ_0 has the form

$$n^{\frac{1}{2}} \{ W_{n\widehat{\xi}_n}(T) - W_{n\xi_0}(T) \} = n^{\frac{1}{2}} \int_0^T (\widehat{\xi}_n - \xi_0) d\dot{W}_{n\xi_0} + R_n$$

where

$$dW_{n\xi_0}(s) = -\frac{Y_{1n}^2 Y_{2n}}{(\xi_0 Y_{1n} + Y_{2n})^2} \left(1_{\{Y_{1n}>0\}} \frac{dN_{1n}}{\xi_0 Y_{1n}} - 1_{\{Y_{2n}>0\}} \frac{dN_{2n}}{Y_{2n}} \right)$$

$$- \frac{Y_{2n}}{\xi_0 Y_{1n} + Y_{2n}} 1_{\{Y_{1n}>0\}} \frac{dN_{1n}}{\xi_0^2}. \tag{7.18}$$

Under H_0, $d\dot{W}_{n\xi_0}$ reduces to $Z_n = \int_0^T Y_{2n} (\xi_0 Y_{1n} + Y_{2n})^{-1} 1_{\{Y_{1n}>0\}} \xi_0^{-2} dN_{1n}$, it converges in probability to $z_0 = \int_0^T y_1 y_2 \xi_0^{-1} (\xi_0 y_1 + y_2)^{-1} \lambda_0 \, ds$ under Condition 7.4. It follows that the variable $n^{-1} \dot{W}_{n\xi_0}(T) = n^{-1} Z_n + o_p(1)$ converges in probability to z_0 and the statistic $S_{n\widehat{\xi}_n}(T)$ has to be centered in order to converge. Finally, let us consider the variables

$$A_n = n^{-\frac{1}{2}} \{ \dot{W}_{n\widehat{\xi}_n}(T) - Z_n \}, \tag{7.19}$$

$$V_{nA} = \int_0^T \frac{Y_{1n}^3 Y_{2n}}{\xi_0 (\xi_0 Y_{1n} + Y_{2n})^3} \lambda_0 \, ds,$$

under the next conditions.

Condition 7.5.
The variance $\sigma_A^2 = \int_0^T y_1^3 y_2 \{ \xi_0 (\xi_0 y_1 + y_2)^3 \}^{-1} \lambda_0 \, ds$ is finite.

Theorem 7.8. *Under Conditions 7.2, 7.4 and 7.5, the test statistic*

$$S_{nA}(T) = V_{nA}^{-\frac{1}{2}} A_n$$

converges weakly under H_0 to a normal variable and it converges weakly under K_n to the sum of a normal variable and $\sigma_A^{-1} \int_0^T y_1^2 y_2 (\xi_0 Y_1 + Y_2)^{-2} \eta \lambda_2 \, ds$.

Proof. From (7.18) and by Rebolledo's convergence theorem, A_n converges weakly under H_0 to a centered Gaussian variable with variance $\sigma_A^2 = EV_{nA}$. Under K_n, the martingales $\int_0^\cdot \{1_{\{Y_{1n}>0\}} (\xi_0 Y_{1n})^{-1} dN_{1n} - 1_{\{Y_{2n}>0\}} Y_{2n}^{-1} dN_{2n}\}$ have the mean $n^{-\frac{1}{2}} \int_0^\cdot \eta_n \lambda_{2n} \, ds + o_p(1)$, it follows that

$$A_n = n^{-1} \int_0^T \frac{Y_{1n}^2 Y_{2n}}{(\xi_0 Y_{1n} + Y_{2n})^2} \eta_n \lambda_{2n} \, ds + o_p(1)$$

and it converges to $y = \int_0^T y_1^2 y_2 (\xi_0 y_1 + y_2)^{-2} \eta \lambda_2 \, ds$. The limit of the variance V_{nA} is the same under H_0 and K_n. □

The asymptotic equivalence between the log-likelihood ratio test statistic and the Student statistic are no longer satisfied and Theorems 7.6 and 7.7 concerning the expansion and the limiting distributions of the log-likelihood ratio test under H_0 and K_n cannot be extended to a functional class of alternatives. The log-likelihood ratio statistic for H_0 against K_n is written as $l_n = l_n(\widehat{\xi}_n)$ and the limiting distributions of $l_n(\xi_0)$ under H_0 and K_n are similar to those of Theorem 7.1. By an expansion of $l_n(\widehat{\xi}_n)$ near ξ_0

$$l_n(\widehat{\xi}_n) - l_n(\xi_0) = n^{\frac{1}{2}} \int_0^T (\widehat{\xi}_n - \xi_0) \, n^{-\frac{1}{2}} d\dot{l}_n(\xi_0) + o_p(1) \qquad (7.20)$$

and the right-hand side of this equation does not tend to zero. The previous approximations of log-likelihood ratio at the true parameter value and Equation (7.20) provide approximations of the variable $l_n(\widehat{\xi}_n)$ but the limiting distribution under the hypothesis is not standard. Under the sequence of alternatives K_n, $l_n(\widehat{\xi}_n)$ tends to infinity. A bootstrap test based on the estimated log-likelihood ratio is therefore consistent.

7.6   Tests of comparison of k samples

The tests of homogeneity of a population of Sections 7.3 and 7.4 are extended to k sub-populations, following the same approach. The hypothesis is the equality of the hazard functions of k independent sub-populations

$$H_0 : \lambda_j(t) = \lambda_k(t), \; t \in [0, T], \; j = 1, \ldots, k-1$$

and we first consider a sequence of local alternatives where the neighborhoods of $\lambda_{jn}\lambda_{kn}^{-1}$ are defined by constants, for all $j < k$

$$K_n : \begin{cases} \lambda_{jn}(t) = \lambda_{kn}(t)\{1 + n^{-\frac{1}{2}}\rho_{jn}\}, \; j = 1, \ldots, k-1, \\ \lambda_{kn}(t) = \lambda_k(t)\{1 + n^{-\frac{1}{2}}\gamma_n(t)\}, \; t \in [0, T], \end{cases}$$

where $(\rho_{jn})_{j=1,\ldots,k,n\geq 1}$ is a real sequence converging to a strictly positive limit $(\rho_j)_{j=1,\ldots,k}$ and $(\gamma_n)_{n\geq 1}$ is a sequence of functions converging uniformly in $[0, T]$ to a strictly positive function γ, as n tends to infinity. Let $J_{in}(t) = 1_{\{Y_{in}(t)>0\}}$ in $[0, T]$, for $j = 1, \ldots, k$. The statistic (7.9) is extended to k sub-populations as $S_n(T) = V_n^{-\frac{1}{2}}(T)W_n(T)$, with

$$W_n(T) = \sum_{i=1}^{k-1} \int_0^T \frac{1}{\sum_{j=1}^k Y_{jn}} (Y_{kn}J_{in}\,dN_{in} - Y_{in}J_{kn}\,dN_{kn})$$

$$= \sum_{i=1}^{k-1} \int_0^T \frac{Y_{kn}Y_{in}}{\sum_{j=1}^k Y_{jn}} \left(\frac{J_{in}\,dN_{in}}{Y_{in}} - \frac{J_{kn}\,dN_{kn}}{Y_{kn}} \right),$$

$$V_n(T) = \sum_{i=1}^{k-1} \int_0^T \frac{Y_{kn}Y_{in}}{(\sum_{j=1}^k Y_{jn})^2}(dN_{in} + dN_{kn})$$

$$+ 2\sum_{i=1}^{k-1} \sum_{j\neq i, j=1}^{k-1} \int_0^T \frac{Y_{jn}Y_{in}}{(\sum_{j=1}^k Y_{jn})^2} J_{kn}\,dN_{kn}.$$

Let $Y_n = \sum_{j=1}^k Y_{jn}$. The double sum in the expression of V_n is written $2\sum_{i=1}^{k-1} \sum_{j\neq i, j=1}^{k-1} Y_{jn}Y_{in} = (Y_n - Y_{kn})^2 - \sum_{i=1}^{k-1} Y_{in}^2$. Moreover $\sum_{i=1}^{k-1} Y_{kn}Y_{in} = Y_nY_{kn} - Y_{kn}^2$ hence the integrand of $Y_n^{-2}J_{kn}\,dN_{kn}$ is $Y_n^2 - Y_nY_{kn} - \sum_{i=1}^{k-1} Y_{in}^2$. The mean of $\sum_{i=1}^{k-1} \int_0^T Y_{kn}Y_{in}Y_n^{-2}\,dN_{in}$ equals the mean of $\sum_{i=1}^{k-1} \int_0^T Y_{kn}Y_{in}^2 Y_n^{-2}\,d\Lambda_0$. The expectation of $V_n(T)$ under H_0 reduces therefore to

$$\sigma_{nT}^2 = EV_n(T) = E\int_0^T Y_{kn}(Y_n^2 - Y_nY_{kn})Y_n^{-2}\,d\Lambda_0.$$

An unbiased estimator of the variance of $W_n(T)$ under H_0 is finally

$$V_{0n}(T) = \int_0^T \left(1 - \frac{Y_{kn}}{Y_n}\right) dN_{kn}.$$

Condition 7.6.

(1) The variance $\sigma_T^2 = \int_0^T \{1 - y_k(\sum_{j=1}^k y_j)^{-1}\}y_k\,d\Lambda_0$ is finite.
(2) The variance $\sigma_{k,\rho,\gamma}^2 = \sum_{i=1}^{k-1} \int_0^T (\rho_i + \gamma)^2 y_i\lambda_0\,ds + \int_0^T \gamma^2 y_k\lambda_0\,ds$ is finite.

Theorem 7.9. *Under Condition 7.6, the statistic $S_n(T)$ converges weakly under H_0 to a normal variable S. Under K_n, it converges weakly to $S + \sigma_T^{-1} \sum_{i=1}^{k-1} \int_0^T \rho_i y_k y_i (\sum_{j=1}^k y_j)^{-1} \lambda_0$.*

The asymptotic expansion of the log-likelihood ratio statistic is expressed with the variance $\sigma_{k,\rho,\gamma}^2$.

Theorem 7.10. *Under Conditions 7.2 and 7.6, the log-likelihood ratio statistic has an expansion $l_n(T) = Y_n - \frac{1}{2} Z_n + o_p(1)$ where Y_n is a centered process under H_0. Under H_0 and K_n, the variance of Y_n is asymptotically equivalent to Z_n, as n tends to infinity. The statistic $n^{-\frac{1}{2}} l_n(T)$ converges weakly under H_0 to a Gaussian variable Y with mean $-\frac{1}{2}\sigma_{\rho,\gamma}^2$ and variance $\sigma_{\rho,\gamma}^2$. Under the local alternatives K_n, it converges weakly to $Y + \sigma_{\rho,\gamma}^2$.*

Proof. The likelihood ratio statistic for H_0 against K_n is proportional to $\exp(l_n)$ such that

$$l_n = \sum_{i=1}^{k-1} \left[\int_0^T \log\{(1 + n^{-\frac{1}{2}}\rho_{in})(1 + n^{-\frac{1}{2}}\gamma_n)\} dN_{in} - \int_0^T (\lambda_{in} - \lambda_i) Y_{in} \, ds \right]$$
$$+ \int_0^T \log(1 + n^{-\frac{1}{2}}\gamma_n) dN_{kn} - n^{-\frac{1}{2}} \int_0^T \gamma_n \lambda_k Y_{kn} \, ds.$$

By second order Taylor expansions, it is approximated as

$$l_n = \sum_{i=1}^{k-1} \left[n^{-\frac{1}{2}} \left\{ \int_0^T (\rho_{in} + \gamma_n) dN_{in} + \int_0^T \gamma_n dN_{2n} \right\} \right.$$
$$- (2n)^{-1} \left\{ \int_0^T (\rho_{in}^2 + \gamma_n^2) dN_{in} + \int_0^T \gamma_n^2 dN_{kn} \right\}$$
$$\left. - \int_0^T (\lambda_{in} - \lambda_i) Y_{in} \, ds - n^{-\frac{1}{2}} \int_0^T \gamma_n \lambda_k Y_{kn} \, ds \right] + R_n$$
$$= \sum_{i=1}^{k-1} \left[n^{-\frac{1}{2}} \left\{ \int_0^T (\rho_{in} + \gamma_n) dM_{in} + \int_0^T \gamma_n dM_{kn} \right\} \right.$$
$$\left. - (2n)^{-1} \left\{ \int_0^T (\rho_{in} + \gamma_n)^2 Y_{in} \lambda_0 \, ds + \int_0^T \gamma_n^2 Y_{kn} \lambda_0 \, ds \right\} \right] + R_n,$$

where $M_{in} = N_{in} - \int_0^{\cdot} Y_{in} \lambda_0 \, ds$ is a local martingale under H_0 and R_n converges in probability to zero. It follows that l_n converges weakly under H_0 to a Gaussian variable with mean $\mu_k = -\frac{1}{2}\sigma^2(k,\rho,\gamma)$ and variance $\sigma^2(k,\rho,\gamma)$. The probability distributions of l_n under H_0 and K_n are contiguous therefore R_n converges in probability to zero under the alternative. Under K_n, the first order term of the approximation of l_n is

$\sum_{i=1}^{k-1} [n^{-\frac{1}{2}} \{\int_0^T (\rho_{in} + \gamma_n) \, dM_{in} + \int_0^T \gamma_n \, dM_{kn}\} + n^{-\frac{1}{2}} \{\int_0^T (\rho_{in} + \gamma_n)(\lambda_{in} - \lambda_0) Y_{in} \, ds + \int_0^T \gamma_n (\lambda_{kn} - \lambda_0) Y_{kn} \, ds\}]$ and its asymptotic mean is $\sigma_{k,\rho,\gamma}^2$. □

Theorem 7.11. *The log-likelihood ratio test is asymptotically equivalent under H_0 and K_n to the test defined by the statistic $S_n(T)$.*

Proof. The proof uses the same notations as in Theorems 7.11 and 7.4. The log-likelihood ratio l_n is approximated under H_0 in a partition $(I_{n,l})_{l=1,\ldots,m}$ by $l_{n,m} = \theta_{n,m}^t \Delta_{n,m} - \frac{1}{2}\sigma_{n,m}^2$, as m tends to infinity, where

$$\theta_{n,m} = (\rho_{1,n}, \ldots, \rho_{k-1,n}, \gamma_{n,m,1}, \ldots, \gamma_{n,m,m})^t,$$

$$\Delta_{n,m} = n^{-\frac{1}{2}} (M_{1,n,0}(T), \ldots, M_{k-1,n,0}(T), \sum_{j=1}^k M_{jn,0}(t_{m,1}),$$

$$\sum_{j=1}^k \int_{t_{m,1}}^{t_{m,2}} dM_{jn,0}, \ldots, \sum_{j=1}^k \int_{t_{m,m-1}}^{t_{m,m}} dM_{jn,0})$$

belong to $\mathbb{R}^{m+k}$ and $\sigma_{n,m}^2 = \theta_{n,m}^t \Gamma_m \theta_{n,m}$, where $\Gamma_{n,m}$ is a symmetric matrix of $\mathbb{R}^{m+k-1} \times \mathbb{R}^{m+k-1}$ defined by diagonal sub-matrices

$$\Gamma_{m,j,j} = \sum_{l=1}^m C_{j,m,l}, \ j = 1, \ldots, k-1$$

$$\Gamma_{m,k+i,k+i} = \sum_{j=1}^k C_{j,m,i}, \ i = 1, \ldots, m,$$

with $C_{j,m,l} = \int_{t_{m,l-1}}^{t_{m,l}} y_j \lambda_0 \, ds$, and by the rectangular sub-matrices with terms

$$\Gamma_{m,j,k+i} = \Gamma_{m,k+i,j}^t = C_{j,m,i}, \ j = 1, \ldots, k-1, \ i = 1, \ldots, m.$$

Let $e_1 = (1, 0, \ldots, 0)$, $e_2 = (0, 1, \ldots, 0)$, ..., $e_{k-1} = (0, \ldots, 0, 1, 0, \ldots, 0)$ be first $k-1$ unit vectors of $\mathbb{R}^{m+k-1}$ with one component 1 among the first m components and the other components 0. The Γ_m-orthogonal projections of the vectors e_j in the space Θ_m are

$$u_{j,m} = \left(0, \ldots, 0, 1, 0, \ldots, 0, \frac{-C_{j,m,1}}{\sum_{i=1}^k C_{i,m,1}}, \ldots, \frac{-C_{j,m,m}}{\sum_{i=1}^k C_{i,m,m}}\right)^t,$$

for $j = 1, \ldots, k-1$, with only one nonzero component among the first $k-1$ components. Then $\Gamma_m u_{j,m}$ is proportional to the unit vector e_j of $\mathbb{R}^{m+k-1}$, with the jth component

$$(\Gamma_m u_{j,m})_j = \sum_{l=1}^m C_{j,m,l} \frac{\sum_{i_1 \neq j, i_1 = 1}^k C_{i_1,m,l}}{\sum_{i=1}^k C_{i,m,l}},$$

therefore $u_{j,m}^t \Gamma_m u_{j,m} = e_j^t \Gamma_m u_{j,m}$ for every $j = 1, \ldots, k-1$.

Under H_0, $\theta_{n,m}$ are the sum of its projections on the vectors $u_{j,m}$, which must be zero, and its projection on the sub-space orthogonal to the space generated by the vectors $\{u_{j,m}\}_{j=1,\ldots,k-1}$, corresponding to nuisance parameters, hence H_0 reduces to $\theta_{n,m}^t \Gamma_m u_j = 0$, for $j = 1, \ldots, k-1$. The log-likelihood ratio test statistic is therefore asymptotically equivalent to a test statistic based on the statistic

$$
W_{n,m} = n^{-\frac{1}{2}} \sum_{j=1}^{k-1} u_j^t \Delta_{n,m,j}
$$

$$
= n^{-\frac{1}{2}} \sum_{j=1}^{k-1} \left(M_{jn,0}(T) - \sum_{l=1}^{m} \frac{C_{j,m,l}}{\sum_{i=1}^{k} C_{i,m,l}} \sum_{\nu=1}^{k} \int_{t_{j,m,l-1}}^{t_{m,l}} dM_{\nu n,0} \right).
$$

The limit as m and n tend to infinity of $n^{-\frac{1}{2}} W_{n,m}$ is

$$
W_n = n^{-\frac{1}{2}} \sum_{j=1}^{k-1} \left(M_{jn,0}(T) - \sum_{\nu=1}^{k} \int_{0}^{T} \frac{y_j}{\sum_{i=1}^{k} y_i} dM_{\nu n0} \right)
$$

$$
= n^{-\frac{1}{2}} \sum_{j=1}^{k-1} \int_{0}^{T} \frac{y_j y_k}{\sum_{i=1}^{k} y_i} \left(\frac{dM_{jn,0}}{y_j} - \frac{dM_{kn,0}}{y_k} \right).
$$

By the uniform convergence of the processes $n^{-1} Y_{jn}$ to y_{jn}, for $j = 1, \ldots, k$, W_n asymptotically equivalent to the Student statistic under H_0. Under K_n, l_n has a similar expansion $l_{n,m} = \theta_{n,m}^t \Delta_{n,m} + \frac{1}{2} \sigma_{n,m}^2$ and the asymptotic equivalence is still satisfied.   □

7.7   Goodness of fit tests for k samples

The notations of Section 7.6 are extended to $k-1$ hazard ratios in $[0,T]$. The hypothesis is $H_0 : \lambda_j = \xi_j \lambda_k$ in $[0,T]$, for $j = 1, \ldots, k-1$, with positive functions ξ_j. A sequence of local alternatives K_n to the hypothesis is

$$
K_n : \begin{cases} \lambda_{jn}(t) = \xi_j(t)\lambda_{kn}(t)(1 + n^{-\frac{1}{2}} \rho_{jn}), \ j = 1, \ldots, k-1, \\ \lambda_{kn}(t) = \lambda_k(t)\{1 + n^{-\frac{1}{2}} \gamma_n(t)\}, \ t \in [0,T], \end{cases}
$$

and the baseline hazard function λ_k is denoted λ_0 under H_0. Its primitive Λ_0 is unbiasedly estimated, with the notation $\xi_k = 1$, by

$$
\widehat{\Lambda}_{n\xi}(t) = \int_{0}^{t} 1_{\{\sum_{j=1}^{k} Y_{jn} > 0\}} \frac{\sum_{j=1}^{k} dN_{jn}}{\sum_{j=1}^{k} \xi_j Y_{jn}}, \tag{7.21}
$$

and the variance $V_{\Lambda,\xi,n}(t) = \int_0^t 1_{\{\sum_{j=1}^k Y_{jn}>0\}} \{\sum_{j=1}^k \xi_j Y_{jn}\}^{-1} d\Lambda_0$ of the asymptotically Gaussian variable $n^{\frac{1}{2}}(\widehat{\Lambda}_{n\xi} - \Lambda_0)(t)$ is unbiasedly estimated by

$$V_{\Lambda,\xi,n}(t) = \int_0^t 1_{\{\sum_{j=1}^k Y_{jn}>0\}} \frac{\sum_{j=1}^k dN_{jn}}{(\sum_{j=1}^k \xi_j Y_{jn})^2}.$$

Its asymptotic behavior under H_0 and K_n is similar to those of the test for two sub-populations (Proposition 7.2). A test statistic

$$S_{n\xi}(T) = V_{n\xi}^{-\frac{1}{2}}(T) W_{n\xi}(T) \tag{7.22}$$

for H_0 against K_n is defined by a variable $W_{n\xi}(T)$ and its estimated variance $V_{n\xi}(T)$ depending on the functions $\xi_1, \ldots, \xi_{k-1}$ of the hypothesis H_0

$$\begin{aligned}
W_n(T) &= \sum_{i=1}^{k-1} \int_0^T \frac{1}{\sum_{j=1}^k \xi_j Y_{jn}} \left(Y_{kn} \frac{J_{in}}{\xi_i} dN_{in} - Y_{in} J_{kn} dN_{kn} \right) \\
&= \sum_{i=1}^{k-1} \int_0^T \frac{Y_{kn} Y_{in}}{\sum_{j=1}^k \xi_j Y_{jn}} \left(\frac{J_{in} dN_{in}}{\xi_i Y_{in}} - \frac{J_{kn} dN_{kn}}{Y_{kn}} \right).
\end{aligned}$$

The functions y_i, ξ_i and η_i, $i = 1, \ldots, k-1$, are supposed to satisfy the following conditions.

Condition 7.7.

(1) The variances $\sigma_{i,\xi,T}^2 = \int_0^T y_i y_k^2 \xi_j^{-1} (\sum_{j=1}^k \xi_j y_j)^{-2} d\Lambda_0$, $i = 1, \ldots, k-1$, and $\int_0^T (y - y_k)^2 y_k (\sum_{j=1}^k \xi_j y_j)^{-2} d\Lambda_0$ are finite.

(2) The integrals $\int_0^T \xi_i (\eta_i + \gamma)^2 y_k d\Lambda_0$ and $\int_0^T \gamma^2 y_k \lambda_0 ds$ are finite.

Under Condition 7.7 the variance $\sigma_{n\xi,T}^2$ of $W_{n\xi}(T)$ under H_0 is finite

$$\begin{aligned}
\sigma_{n\xi,T}^2 &= E \int_0^T \left\{ \left(\frac{Y_{kn}}{\sum_{j=1}^k \xi_j Y_{jn}} \right)^2 \sum_{i=1}^{k-1} \frac{Y_{in}}{\xi_i} + \left(\frac{Y_n - Y_{kn}}{\sum_{j=1}^k \xi_j Y_{jn}} \right)^2 Y_{kn} \right\} d\Lambda_0 \\
&= \int_0^T \left\{ \frac{y_k^2}{(\xi_i y_i + y_k)^2} \sum_{i=1}^{k-1} \frac{y_i}{\xi_i} + \frac{(y - y_k)^2}{(\sum_{j=1}^k \xi_j y_j)^2} y_k \right\} d\Lambda_0 + o(1)
\end{aligned}$$

and it is estimated from (7.21) by

$$V_{n\xi}(T) = \int_0^T \left\{ \left(\frac{Y_{kn}}{\sum_{j=1}^k \xi_j Y_{jn}} \right)^2 \sum_{i=1}^{k-1} \frac{Y_{in}}{\xi_i} + \left(\frac{Y_n - Y_{kn}}{\sum_{j=1}^k \xi_j Y_{jn}} \right)^2 Y_{kn} \right\} d\widehat{\Lambda}_{n\xi}.$$

Under Conditions 7.2 and 7.7, the variable $V_{n\xi}(T)$ converges in probability to $\sigma_{\xi,T}^2$ under H_0 and K_n. The weak convergence of the statistic $S_{n\xi}(T)$ is proved like Theorem 7.1 and its asymptotic power for the test of H_0 against K_n is deduced.

Theorem 7.12. *Under Conditions 7.2 and 7.7, the statistic $S_{n\xi}(T)$ converges weakly under the hypothesis H_0 to a normal variable S. Under the local alternatives K_n, $S_{n\xi}(T)$ converges weakly to $S + \sigma_{\xi,T}^{-1} \sum_{i=1}^{k-1} \int_0^T \rho_i y_k y_i (\sum_{j=1}^k \xi_j y_j)^{-1} \lambda_0$.*

Let $\sigma_{\xi,\eta,\gamma}^2 = \sum_{i=1}^{k-1} \int_0^T (\eta_i + \gamma)^2 y_1 \xi_i \lambda_0 + \int_0^T \gamma^2 y_k \lambda_0$. The expansion and the limiting distributions of the log-likelihood test statistic $l_{n,\xi}$, with the scaling and location constant $\sigma_{\xi,\eta,\gamma}^2$, is deduced from expansions of $\log \lambda_{in} - \log \lambda_i$ in the expression of $l_{n,\xi}$ and by Rebolledo's weak convergence theorem under H_0 and K_n.

Theorem 7.13. *Under Conditions 7.2 and 7.4, the log-likelihood ratio statistic of goodness-of-fit has the asymptotic expansions $l_{n,\xi} = Y_{n,\xi} - \frac{1}{2} Z_{n,\xi}$ under H_0 and $l_{n,\xi} = Y_{n,\xi} + \frac{1}{2} Z_{n,\xi}$ under K_n, where $E_0 Y_{n,\xi} = 0$. Under H_0 and K_n, the limit $\sigma_{\xi,\eta,\gamma}^2$ of $Var Y_{n,\xi}$ is the limit in probability of $Z_{n,\xi}$, as n tends to infinity.*

Proof. Under H_0, the log-likelihood ratio statistic for H_0 against K_n is developed as

$$l_{n,\xi} = \sum_{i=1}^{k-1} \left[\int_0^T \log\{(1 + n^{-\frac{1}{2}}\eta_{in})(1 + n^{-\frac{1}{2}}\gamma_n)\} \, dN_{in} \right.$$

$$\left. - \int_0^T (\lambda_{in} - \lambda_i)\xi_i Y_{in} \, ds \right] + \int_0^T \log(1 + n^{-\frac{1}{2}}\gamma_n) \, dN_{kn}$$

$$- n^{-\frac{1}{2}} \int_0^T \gamma_n \lambda_k Y_{kn} \, ds$$

$$= \sum_{i=1}^{k-1} \left[n^{-\frac{1}{2}} \left\{ \int_0^T (\eta_{in} + \gamma_n) \, dN_{in} + \int_0^T \gamma_n \, dN_{2n} \right\} \right.$$

$$- (2n)^{-1} \left\{ \int_0^T (\eta_{in}^2 + \gamma_n^2) \, dN_{in} + \int_0^T \gamma_n^2 \, dN_{kn} \right\}$$

$$\left. - \int_0^T (\lambda_{in} - \lambda_i)\xi_i Y_{in} \, ds - n^{-\frac{1}{2}} \int_0^T \gamma_n \lambda_k Y_{kn} \, ds \right] + R_n$$

$$= \sum_{i=1}^{k-1} \left[n^{-\frac{1}{2}} \left\{ \int_0^T (\eta_{in} + \gamma_n) \, dM_{in} + \int_0^T \gamma_n \, dM_{kn} \right\} \right.$$

$$\left. - (2n)^{-1} \left\{ \int_0^T (\eta_{in} + \gamma_n)^2 \xi_i Y_{in} \, \lambda_0 \, ds + \int_0^T \gamma_n^2 Y_{kn} \, \lambda_0 \, ds \right\} \right] + R_{0n},$$

under H_0, and the remainder term R_{0n} converges in probability to zero. It follows that $l_{n,\xi}$ converges weakly under H_0 to a Gaussian variable with mean $\mu_k = -\frac{1}{2}\sigma^2(\xi, \eta, \gamma)$ and variance $\sigma^2(\xi, \eta, \gamma)$, hence the probability distributions of $l_{n,\xi}$ under H_0 and K_n are contiguous. This implies that R_n converges in probability to zero under the alternative. Under K_n, the sign in the last term of the expansion of $l_{n,\xi}$ is reversed and its asymptotic mean follows. □

Theorem 7.14. *Under Conditions 7.2 and 7.4, the tests defined by the log-likelihood ratio statistic and by the statistic $S_{n,\xi}(T)$ are asymptotically equivalent under H_0 and K_n.*

Proof. The proof of Theorem 7.3 is modified by the multiplicative hazard function $\xi\lambda_0$ for λ_1 in the expression of the integrals $C_{j,m,k}$ and by splitting all functions η_{in} in the partition. The parameter $\Delta_{n,m}$ is a vector of $\mathbb{R}^{km}$ corresponding to m constants for the approximation of the functions η_{in} and γ_n. The $k-1$ vectors $u_{j,m}$ have its $(j-1)m$th to jmth components equal to 1 and its other $(k-1)m$ components are zero, its last m components are equal to $-C_{j,m,l}\{\sum_{s=1}^m C_{j,m,s}\}^{-1}$, for $l = 1,\ldots,m$ and $j = 1,\ldots,k-1$. The expression of $W_{n\xi}(T)$ follows from the limit of $u_{i,m}^t\Delta_{n,m}$ as m tends to infinity. □

7.8 Tests of independence of two censored variables

For a right-censored variable on $\mathbb{R}_+^2$, the observations are $T = (T_1, T_2)$ and the bivariate indicator $\delta = (\delta_1, \delta_2)$, where $T_k = X_k \wedge C_k$, $\delta_k = 1_{\{X_k \leq C_k\}}$, for $k = 1, 2$. Let F, F_1 and F_2 be the joint and marginal distribution functions of X, S, S_1 and S_2 the associated survival functions defined by

$$S(t) = P(X \in [t, \infty[) = 1 - F_1^-(t_1) - F_2^-(t_2) + F^-(t), \; t = (t_1, t_2) \in \mathbb{R}_+^2,$$

and $S_k(t) = P(X_k \geq t_k)$. Let G be the distribution function of the censoring time C such that $P(X < C) > 0$. The independence of the variables T_1 and T_2 is equivalent to the factorization of the joint survival function as $S(t) = S_1(t_1)S_2(t_2)$. A joint cumulative hazard function of the variable X in $\mathbb{R}_+^2$ is

$$\Lambda(t) = \int_{[0,t]} S^{-1}\, dF,$$

defined inside the support of X, it is finite in every compact subset strictly included in its support. Reciprocally, the bivariate distribution function of

X, $F(t) = P(X \in [0,t])$, is the unique solution of equation

$$F(t) = F_1(t_1) + F_2(t_2) + \int_{[0,t]} S \, d\Lambda \tag{7.23}$$

implicitly defined by the marginal functions F_1, F_2 and by Λ, or equivalently by the set of the three hazard functions $(\Lambda_1, \Lambda_2, \Lambda)$. The independence of the components is therefore equivalent to the factorization $\Lambda(t) = \Lambda_1(t_1)\Lambda_2(t_2)$.

For a n-sample $(T_i, \delta_i)_{i=1,\ldots,n}$ of (T, δ), the observed point processes are

$$N_n^{(j,k)}(t) = \sum_{i=1}^n 1_{\{\delta_{i,1}=j\}} 1_{\{\delta_{i,2}=k\}} 1_{\{T_i \le t\}}, \ (j,k) \in \{0,1\}^2,$$

$$Y_n(t) = \sum_{i=1}^n 1_{\{T_i \ge t\}}, \ t = (t_1, t_2) \in \mathbb{R}_+^2,$$

the process $N_n^{(1,1)}$ is simply denoted N_n and the related processes for the marginal observations are denoted N_{1n} and Y_{1n}, N_{2n} and Y_{2n}. The distribution of $N_n^{(j,k)}$, (j,k) in $\{0,1\}^2$, are determined by the four sub-distribution functions

$$F^{(j,k)}(t) = P(\delta_1 = j, \delta_2 = k, T \le t), t \in \mathbb{R}_+^2, (j,k) \in \{0,1\}^2.$$

For a bivariate process or function H, the stochastic integral $\int H \, dN_n^{(j,k)}$, $(j,k) \ne (0,0)$, is a sum of marginal jumps $\Delta_1 H$ or $\Delta_2 H$ or of rectangular variations ΔH with values

$$\Delta_1 H(T_i) = H(T_i) - H(T_{i,1}^-, T_{i,2}) = H(ds_1, T_{i,2}), \text{ if } (j,k) = (1,0),$$

$$\Delta_2 H(T_i) = H(T_i) - H(T_{i,1}, T_{i,2}^-) = H(T_{i,1}, ds_2), \text{ if } (j,k) = (0,1),$$

$$\Delta H(T_i) = \Delta_1 \Delta_2 H(T_i) = H(]T_i^-, T_i]), \text{ if } (j,k) = (1,1).$$

The log-likelihood of a sample with the distribution function F, observed in the interval $[0,T]$, is

$$l_n(F) = \sum_{i=1}^n \sum_{(j,k)\ne(0,0)} \log \int_{[0,T]} S \, dN_n^{(j,k)} + \sum_{i=1}^n (1-\delta_{1i})(1-\delta_{2i}) \log S(T_i).$$

Under the hypothesis of independent of the components of the variable X, the log-likelihood becomes

$$l_n(F_1 F_2) = \sum_{i=1}^n [\delta_{1i}\delta_{2i}\{\log f_1(T_{1i}) + \log f_2(T_{12})\} + \delta_{1i}(1 - \delta_{2i})\{\log f_1(T_{1i})$$

$$+ \log S_2(T_{12})\} + (1 - \delta_{1i})\delta_{2i}\{\log S_1(T_{1i}) + \log f_2(T_{12})\}$$

$$+ \sum_{i=1}^n (1 - \delta_{1i})(1 - \delta_{2i})\{\log S_1(T_{1i}) + \log S_2(T_{12})\}.$$

A nonparametric log-likelihood ratio test statistic $l_n(\widehat{F}_n) - l_n(\widehat{F}_{1n}\widehat{F}_{2n})$ is not easily calculated and a test of independence is also defined from the hazard functions. An estimator of Λ has the same form as the estimator $\widehat{\Lambda}_{kn}$ of the marginal hazard functions

$$\widehat{\Lambda}_n(t) = \int_{[0,t]} \{Y_n > 0\} Y_n^{-1} \, dN_n \qquad (7.24)$$

and it behaves like the one-dimensional estimator. Let $\widetilde{H} = F\bar{G}$.

Proposition 7.3 (Pons, 1986). *If $P(X < C) > 0$, on every compact sub-interval $[0, \tau]$ of the support of X, $n^{1/2}(\widehat{\Lambda}_n - \Lambda)$ converges weakly to a centred Gaussian process $G_{\bar{H}}$ with covariances $Cov(G_{\bar{H}}(s), G_{\bar{H}}(t))$ given by*

$$\int_{[0, s \wedge t]} \bar{H}^{-2} \, d\bar{H} + \int_{u \in [0,s]} \int_{v \in [0,t]} \bar{H}(u \wedge v) \bar{H}^{-2}(u) \, d\widetilde{H}(u) \, \bar{H}^{-2}(v) \, d\widetilde{H}(v)$$

$$-2 \int_{u \in [0,s]} \int_{v \in [u,t]} \bar{H}^{-2}(u) \, d\widetilde{H}(u) \, \bar{H}^{-2}(v) \, d\widetilde{H}(v).$$

A test of independence of the components of the variable X under right-censoring relies on the factorization of the joint hazard function Λ, through the statistic

$$S_n(\tau) = n^{\frac{1}{2}} \{\widehat{\Lambda}_n(\tau) - \widehat{\Lambda}_{1n}(\tau_1)\widehat{\Lambda}_{2n}(\tau_2)\},$$

defined at the end of the observation set strictly included in the support of the variable X. Another statistic is defined as

$$S_n = n^{\frac{1}{2}} \sup_{t \in [0,\tau]} |\widehat{\Lambda}_n(t) - \widehat{\Lambda}_{1n}(t_1)\widehat{\Lambda}_{2n}(t_2)|.$$

By the differentiability property of S_n as a functional of the distribution function F, the process $S_n(\tau)$ converges weakly under the hypothesis H_0 to a centered Gaussian process and a bootstrap test can be performed. The asymptotic distribution of the bootstrap statistic conditionally on the data sample is the empirical distribution of the statistic S_n.

7.9 Comparison of two bivariate distributions

Let $X_A = (X_{A1}, X_{A2})$ and $X_B = (X_{B1}, X_{B2})$ be two independent bivariate random variables under right-censoring and let $T_A = (T_{A1}, T_{A2})$ and $T_B = (T_{B1}, T_{B2})$ be the observed censored times, with bivariate censoring indicators $\delta_A = (\delta_{A1}, \delta_{A2})$ and $\delta_B = (\delta_{B1}, \delta_{B2})$. The hypothesis H_0 is now the equality of the distributions of X_A and X_B. The definition of the joint distribution function by its marginals and the joint hazard function leads to compare two bivariate distribution functions through the comparison of the sets of functions (F_1, F_2, Λ) defining them by (7.3). After a test of equality of the marginal distributions F_{A1} and F_{B1}, F_{A2} and F_{B2}, we consider a comparison of their joint hazard functions Λ_A and Λ_B. The hazard functions Λ_A and Λ_B of X_A and X_B are estimated from two independent samples of respective sizes n_A and n_B by (7.24) and the estimators are denoted $\widehat{\Lambda}_{An}$ and $\widehat{\Lambda}_{Bn}$. The total sample size is $n = n_A + n_B$ such that $n^{-1}n_k$ converges to a real $\rho_k > 0$ for $k = A, B$, as n tend to infinity. The processes $W_{kn}(t) = n_k^{\frac{1}{2}} (\int_{[0,\cdot]} Y_{kn_k}^{-1} 1_{\{Y_{An}>0\}}\, dN_{kn_k} - \Lambda_0)$ converge weakly to a centered Gaussian process on $\mathbb{R}_+^2$, with variance $\sigma_k^2 = \int_{[0,\cdot]} Y_{kn_k}^{-1}\, d\Lambda_0$, for $k = A, B$. It follows that their difference

$$n^{\frac{1}{2}}\{\widehat{\Lambda}_{An} - \widehat{\Lambda}_{Bn})(t) = n^{\frac{1}{2}} \int_{[0,t]} \left(\frac{dN_{An_A}}{Y_{An_A}} - \frac{dN_{Bn_B}}{Y_{Bn_B}} \right)$$

$$= \int_{[0,t]} \left\{ \left(\frac{n}{n_A} \right)^{\frac{1}{2}} n_A^{\frac{1}{2}} \frac{dN_{An_A}}{Y_{An_A}} - \left(\frac{n}{n_B} \right)^{\frac{1}{2}} n_B^{\frac{1}{2}} \frac{dN_{Bn_B}}{Y_{Bn_B}} \right\}$$

converges weakly under H_0 to a centered Gaussian process on $\mathbb{R}_+^2$ and it tends to infinity under alternatives. Test statistics are defined by weighting this process like W_n defined for testing the homogeneity of a right-censored sample in $\mathbb{R}$ (Section 7.3). The variable $W_n(T)$ is replaced by

$$W_n(T) = n^{-\frac{1}{2}} \int_0^T \frac{Y_{An}Y_{Bn}}{Y_{An} + Y_{Bn}} (d\widehat{\Lambda}_{An} - d\widehat{\Lambda}_{Bn})$$

$$= n^{-\frac{1}{2}} \int_0^T \frac{Y_{An}Y_{Bn}}{Y_{An} + Y_{Bn}} (d\widehat{\Lambda}_{An} - d\widehat{\Lambda}_{Bn})$$

$$= n^{-\frac{1}{2}} \left\{ \int_0^T \frac{Y_{Bn}}{Y_{An} + Y_{Bn}} d(N_{An} - Y_{An}\Lambda_0) \right.$$

$$\left. - \int_0^T \frac{Y_{An}}{Y_{An} + Y_{Bn}} d(N_{Bn} - Y_{Bn}\Lambda_0) \right\},$$

its variance under the hypothesis is

$$\sigma_{ABn}^2 = n^{-1} E_0 \int_0^T \frac{Y_{Bn}Y_{An}}{Y_{An} + Y_{Bn}} d\Lambda_0$$

which is asymptotically equivalent to

$$\sigma_{AB}^2 = \rho_A \rho_B \int_0^T \frac{y_B y_A}{y_A + y_B} d\Lambda_0$$

and it is estimated by $V_n(T)$

$$V_n(T) = n^{-1} \int_0^T \frac{Y_{An}Y_{Bn}}{(Y_{An} + Y_{Bn})^2} 1_{\{Y_{An}+Y_{Bn}>0\}} (dN_{An} + dN_{Bn}).$$

The weak convergence of $V_n(T)^{-\frac{1}{2}} W_n(T)$ is proved as previously.

7.10 Tests for left-censored samples

Let X be a random variable left-censored by an independent variable $U > 0$ defined on a probability space $(\Omega, \mathcal{A}, P)$, the observations are a censored variable $T = \max\{U, X\}$ and the censoring indicator $\eta = 1_{\{U \leq X\}}$. The indicator has the value 1 if x is observed and 0 if it is censored. A counting process N_n for the observed variables is

$$N_n(t) = \sum_{i=1}^n \eta_i 1_{\{T_i \leq t\}} = \sum_{i=1}^n 1_{\{U_i \leq X_i \leq t\}}. \tag{7.25}$$

The number of left-censored variables until t is

$$N_n^c(t) = \sum_{1 \leq i \leq n} (1 - \eta_i) 1_{\{T_i \leq t\}} = \sum_{1 \leq i \leq n} 1_{\{X_i \leq U_i \leq t\}}$$

and their sum is $N_n + N_n^c = n - Y_n^+$. Let F and G be the distribution functions of T and, respectively, U, then

$$P\{T \leq t, \eta = 1\} = \int_{\tau_F}^t G \, dF,$$

$$P\{T \leq t, \eta = 0\} = \int_{\tau_F}^t F^- \, dG$$

and $P\{T \leq t\} = FG$. Let $t_F = \inf\{s; F(s) > 0\}$, a retro-hazard function is defined for every $t > t_F$ by

$$d\bar{\Lambda}(t) = P\{t - dt < T < t, \eta = 1 \mid T < t\} = -\frac{dF}{F^-}(t), \quad \text{on the set } \{T < t\}, \tag{7.26}$$

with the convention $\frac{0}{0} = 0$, the (retro-) cumulative hazard function is

$$\bar{\Lambda}(t) = \int_{\tau_F \vee t}^\infty d\bar{\Lambda}(s) = \int_{\tau_F \vee t}^\infty \frac{dF}{F^-}(t). \tag{7.27}$$

The definition (7.27) of $\bar{\Lambda}$ by F is equivalent to the definition of F by $\bar{\Lambda}$ as

$$F(t) = \exp\{-\bar{\Lambda}^c(t)\} \prod_{s > t} \{1 + \Delta\bar{\Lambda}(s^-)\}, \ t > t_F, \tag{7.28}$$

where $\bar{\Lambda}^c(t)$ is the continuous part of $\bar{\Lambda}(t)$ and $\prod_{s>t}\{1+\Delta\bar{\Lambda}(s^-)\}$ its right-continuous discrete part. A right-continuous empirical estimator of the cumulative hazard function (7.27), with left-hand limits, is defined on the interval $]T_{n:1}, T_{n:n}]$ by

$$\widehat{\bar{\Lambda}}_n(t) = \int_t^\infty 1_{\{Y_n<n\}} \frac{dN_n}{n-Y_n}.$$

A product-limit estimator of the function F is defined on $]T_{n:1}, T_{n:n}]$ from the expression of $\widehat{\bar{\Lambda}}_n$ by

$$\widehat{F}_n(t) = \prod_{T_{n:k}\geq t} \left\{1+d\widehat{\bar{\Lambda}}_n(T_{n:k})\right\}^{\eta_{n:k}} = \prod_{T_{n:k}\geq t} \left\{1 - \frac{1}{n-Y_n^-(T_{n:k})}\right\}^{\eta_{n:k}}.$$

Let $\tau_1 = \inf\{s; \bar{F}(s)\bar{G}(s) < 1\}$ and $\tau_2 = \inf\{s; \bar{F}(s)\bar{G}(s) = 0\}$. On every interval $[a, \tau_2]$ such that $a > \tau_1$, the process $\bar{A}_n = n^{1/2}(\widehat{\bar{\Lambda}}_n - \bar{\Lambda})$ converges weakly to a centered Gaussian process $\bar{B}$ with independent increments and its covariance at s and t larger than τ_1 is $\bar{C}(s \vee t)$ such that

$$\bar{C}(t) = \int_t^{\tau_2} H^{-1}(1+\Delta\bar{\Lambda})\, d\bar{\Lambda}.$$

For every $t > 0$

$$\widehat{F}_n(t \vee T_{n:1}) = \int_{t\vee T_{n:1}}^{\tau_2} \widehat{F}_n^-(s)\, d\widehat{\bar{\Lambda}}_n(s).$$

Let X and U be random variables with continuous distribution functions such that $\bar{C}(\tau_1)$ and $\bar{C}(\tau_2)$ are finite, for every $t > \tau_1$

$$\frac{F - \widehat{F}_n}{F}(t \vee T_{n:1}) = \int_{t\vee T_{n:1}}^{\tau_2} \frac{\widehat{F}_n^-}{F}(s)\, \{d\widehat{\bar{\Lambda}}_n(s) - d\bar{\Lambda}(s)\},$$

and $n^{1/2}(F-\widehat{F}_n)F^{-1}$ converges weakly to a centered Gaussian process with covariance $\bar{C}(s \vee t)$.

By the definition of F from $\bar{\Lambda}$, a test for the comparison of the distributions of two left-censored samples observed on an interval $[a, \tau]$ relies on the comparision of their retro-hazard functions $\bar{\Lambda}_1$ and $\bar{\Lambda}_1$. A family of test statistics is defined by a weighting predictable process $W_n = W(n - Y_{1n}, n - Y_{2n})$, in the form

$$U_n = n^{\frac{1}{2}} \int_a^\tau W_n(s) 1_{\{Y_{1n}(s)<n\}} 1_{\{Y_{2n}(s)<n\}} (d\widehat{\bar{\Lambda}}_{2n} - d\widehat{\bar{\Lambda}}_{1n})(s).$$

Under the conditions of Proposition 7.1 for the convergence of the processes W_n and Y_{kn}, its asymptotic variance

$$\sigma_T^2 = \int_a^\tau w^2(s) 1_{\{y_1(s)>0\}} 1_{\{y_2(s)<1\}} \frac{1-y}{(1-y_1)(1-y_2)}(s)\, d\bar{\Lambda}_0(s)$$

is estimated using the estimator $\widehat{\bar{\Lambda}}_n$ defined for the whole sample, by

$$V_n = n \int_a^\tau W_n^2(s) \left\{ 1_{\{Y_{1n}^-(s)<n\}} \frac{1}{n-Y_{1n}(s)} + 1_{\{Y_{2n}(s)<n\}} \frac{1}{n-Y_{2n}^-(s)} \right\} d\widehat{\bar{\Lambda}}_n(s).$$

Under conditions similar to Condition 7.2 for the process $n-Y_n$, the statistic $S_n(T) = V_n^{-\frac{1}{2}}(T)W_n(T)$ converges weakly under the hypothesis H_0 to a normal variable S. Under the local alternatives K_n, it converges weakly to $S + \rho\sigma_T$.

The log-likelihood ratio statistic for the hypothesis H_0 of equal hazard functions in $[a, \tau]$ against local alternatives K_n of Section 7.3, is written as

$$l_n = \sum_{j=1}^2 \int_a^\tau \log \frac{\lambda_j}{\lambda_{jn}} \, dN_{jn} - \sum_{j=1}^2 \int_a^\tau (\bar{\lambda}_j - \bar{\lambda}_{jn})\{n - Y_{jn}(s)\} \, ds.$$

Arguing as in Section 7.3, the log-likelihood ratio statistic has expansions like in Theorem 7.2 with the asymptotic variance

$$\sigma_{\rho,\gamma}^2 = \int_a^\tau (\rho + \gamma)^2 (1 - y_1)\bar{\lambda}_0 + \int_a^\tau \gamma^2 (1 - y_2)\bar{\lambda}_0$$

and its limiting distributions under H_0 and the alternatives are deduced. The log-likelihood ratio test is still asymptotically equivalent to the test defined by the statistic $S_n(T)$. All results proved above for the right-censored variables are extended to left-censored variables with these notations.

7.11 Tests for the mean residual life and excess life

The residual lifetime distribution of a variable X, with distribution function F, is defined in $\mathbb{R}_+$ as

$$R_x(t) = P(X - x \le t | X > x) = \bar{F}^{-1}(x) \int_0^t F(ds+x) = 1 - \bar{F}^{-1}(x)\bar{F}(t+x),$$

with values in $[0, 1]$, for every positive x. The mean residual lifetime at x is the real function

$$m_x = E(X - x | X > x) = \int_0^\infty t \, dR_x(t) = \bar{F}^{-1}(x) \int_0^\infty s \, F(ds + x).$$

The probability of excess is

$$P_t(t + x) = \Pr(T > t + x \mid T > t) = 1 - \frac{F(t + x) - F(t)}{\bar{F}(t)}$$

$$= \exp\left\{ - \int_t^{t+x} \lambda(s) \, ds \right\},$$

where λ is the hazard function of X, it satisfies $P_t(t + x) = 1 - R_t(x)$. The probability of excess is estimated by the product-limit estimator on the interval $]T_{1:n}, T_{n:n}[$

$$\widehat{P}_{n,t}(t + x) = \prod_{1 \leq i \leq n} \left\{ 1 - \frac{1_{\{t < T_i \leq t+x\}} J_n(T_i)}{Y_n(T_i)} \right\}, \qquad (7.29)$$

with $Y_n(t) = \sum_{i=1}^{n} 1_{\{T_i \geq t\}}$ and $J_n = 1_{\{Y_n > 0\}}$. This estimator can be used under independent right or left censorship, using only the uncensored times T_i in the product (7.29) and $Y_n(t)$ is the number of observable times larger than t. By the product definition of the estimator of the probability $P_t(t + x)$, for $t < \tau_F$ such that $\int_0^{\tau_F} \bar{F}^{-1} d\Lambda < \infty$, it satisfies

$$\widehat{P}_{n,t}(t + x) = \int_{t \vee T_{1:n}}^{(t+x) \wedge T_{n:n}} \widehat{P}_{n,t}(t + s^-) \, d\widehat{\Lambda}_n(s),$$

$$\frac{P_t - \widehat{P}_{n,t}}{P_t}(t + x) = \int_{t \vee T_{1:n}}^{(t+x) \wedge T_{n:n}} \frac{1 - \widehat{F}_n(s^-)}{1 - F(s)} \, d(\widehat{\Lambda}_n - \Lambda)(s).$$

The process defined for t and $t + x$ in $[T_{1:n}, T_{n:n}]$ by $B_{P,n}(t) = n^{1/2}\{(P_t - \widehat{P}_{n,t})P_t^{-1}\}(t + x)$ converges weakly on $[T_{1:n}, T_{n:n}]$ to a centered Gaussian process B_P, with independent increments (Pons, 2011).

A test of a constant mean residual lifetime is a test of the hypothesis $E(X 1_{\{X > x\}}) = (x + a)\bar{F}(x)$, for a constant $a > 0$ and for every $x > 0$, this is a goodness of fit test to a parametric family for a nonparametric regression function m_x. A test of $H_0 : R_x(t) = R(t)$ for all $x > 0$ and $t > 0$ is equivalent to $R_x(t) = F(t)$ and to $P_t(t + x) = P_0(x) \equiv \bar{F}(x)$ for all $x > 0$ and $t > 0$. For a variable X having a probability density function, the derivative of $R_x(t)$ with respect to x is zero if and only if the intensity funtion of X satisfies $\lambda(t + x) \equiv \lambda(t)$, i.e. X is an exponential variable under H_0.

For a variable X having a probability density function, a test of an increasing mean residual lifetime is a test of the hypothesis $H_0 : m_x^{(1)} > 0$ for every $x > 0$ and it is equivalent to $m_x \lambda_x > \bar{F}(x)$ for every $x > 0$. Let $\widehat{\lambda}_{x,n}$ be a kernel estimator of the intensity funtion of F, a one-sided test of level α for H_0 has the rejection domain $D_{n,\alpha} = \{\inf_{x>0}\{\widehat{m}_{x,n}\widehat{\lambda}_n(x) - \widehat{\bar{F}}_n(x)\} > c_\alpha\}$ such that $P_0(D_{n,\alpha}) = \alpha$. By the weak convergence of $\widehat{m}_{x,n}$ and $\widehat{\lambda}_{x,n}$ to m_x and respectively λ_x, if $h_n = o(n^{-\frac{1}{5}})$, $\inf_{x>0}(nh_n)^{\frac{1}{2}}(\widehat{m}_{x,n}^{(1)} - m_x^{(1)})$ converges to the minimum of a Gaussian process with mean zero a finite variance. A bootstrap test can be used.

7.12 Tests for right or left-truncated samples

Under right or left truncation, the observations are similar to the censored observations but nothing is known about the variables of interest before left-truncation or after the right-truncation, so the number of truncated variables is unknown and the observations are restricted to what happens in the truncation interval. Let U and X be independent random variables with respective distribution functions F_U and F_X, X is left-truncated by U if X is observed conditionally on $\{U < X\}$ and U is right-truncated by X if U is observed conditionally on $\{U < X\}$ which happens with probability $\alpha = \int \bar{F}_X(s)\,dF_U(s)$. The sub-distribution probability

$$p(u) = P(U \leq u < X | U < X) = \alpha^{-1}\bar{F}_X(u)F_U(u)$$

has a support (τ_1^*, τ_2^*). Considering n possibly truncated individuals, we define the counting processes

$$N_{X,n}(t) = \sum_{i=1}^n 1_{\{U_i < X_i \leq t\}}, \quad Y_{X,n}(t) = \sum_{i=1}^n 1_{\{U_i < t \leq X_i\}},$$

$$\bar{N}_{U,n}(t) = \sum_{i=1}^n 1_{\{t \leq U_i < X_i\}}, \quad \bar{Y}_{U,n}(t) = \sum_{i=1}^n 1_{\{U_i \leq t < X_i\}}.$$

The cumulative hazard and retro-hazard functions for truncation are

$$\Lambda_X(t) = \int_0^t \frac{dF_X}{\bar{F}_X} = \int_0^t \frac{dF_X^*}{p}, t < \tau_2^*,$$

$$\bar{\Lambda}_U(t) = \int_t^\infty \frac{dF_U}{F_U^-} = \int_t^\infty \frac{dF_U^*}{p}, \tau_1^* < t,$$

their empirical estimators are defined as

$$\widehat{\Lambda}_{X,n}(t) = \int_0^t 1_{\{Y_{X,n}>0\}}\frac{dN_{X,n}}{Y_{X,n}}, \, t < \tau_2^*,$$

$$\widehat{\bar{\Lambda}}_{U,n}(t) = \int_t^\infty 1_{\{\bar{Y}_{U,n}>0\}}\frac{d\bar{N}_{U,n}}{\bar{Y}_{U,n}}, \, \tau_1^* < t$$

(Pons, 2008). On every finite interval $[0, a]$ such that $F_U(a)\bar{F}_X(a) > 0$, the processes $A_{X,n} = n^{1/2}(\widehat{\Lambda}_{X,n} - \Lambda_X)$ and $\bar{A}_{U,n} = n^{1/2}(\widehat{\bar{\Lambda}}_{U,n} - \bar{\Lambda}_U)$ converge weakly to continuous centered Gaussian processes B_X and $\bar{B}_U$ respectively, with independent increments and respective variances at t

$$C_X(t) = \int_0^t p^{-1}\,d\Lambda_X, \quad \bar{C}_U(t) = \int_t^\infty p^{-1}\,d\bar{\Lambda}_U.$$

Tests of comparison and goodness of fit for the (retro)-hazard functions are defined as in the previous sections and all asymptotic results apply under truncation.

7.13 Comparison of censored or truncated regression curves

Tests for regression functions are modified under censoring or truncation of the regression variables. On a probability space $(\Omega, \mathcal{F}, P)$, let (X, Y) be a random variable with values in a separable and complete metric space $(\mathcal{X} \times \mathcal{Y}, \mathcal{B})$ and let C be an independent right-censoring time of X with values in $\mathcal{X}_1$. Let F be the joint distribution function of (X, Y), let f_X be the marginal density of X and let G be the distribution function of C. For every x in $\mathcal{X}_1$, the conditional mean function of Y given $X = x$ is $m(x) = E(Y|X = x)$, under the censorship of X it is modified by the censoring indicator as $m(x) = E(\delta Y | \delta = 1, X = x)$. Its empirical estimator under the right-censoring is defined from a sequence of independent random variables distributed like (X, Y, C)

$$\widehat{m}_{nh_n}(x) = \frac{\sum_{i=1}^n \delta_i Y_i K_{h_n}(x - X_i)}{\sum_{i=1}^n \delta_i K_{h_n}(x - X_i)}, \tag{7.30}$$

for every x in $\mathcal{X}_h = \{s \in \mathcal{X}; [s - h, s = h] \subset \mathcal{X}\}$. The denominator $\widehat{d}_{nh_n}$ of $\widehat{m}_{nh_n}$ is a kernel estimator of $d = \bar{G} f_X$ and the numerator $\widehat{\mu}_n(x)$ has the expectation

$$\mu(x) = E(\delta Y 1_{\{X=x\}}) = \bar{G}(x) \int_{\mathcal{Y}} y \dot{F}_x(x, dy) = \bar{G}(x) m(x) f_X(x).$$

The variance of Y is supposed to be finite and its conditional variance under censoring is $\sigma^2(x) = E(\delta Y^2 | \delta = 1, X = x) - m^2(x) = f_X^{-1}(x) w_2(x) - m^2(x)$ where $w_2(x) = \int y^2 f_{XY}(x, y) \, dy$.

Under Conditions 1.1 and 3.1, the bias of the estimator of m has the asymptotic expansion $b_{m,n,h}(x) = E\widehat{m}_{n,h}(x) - m(x) = h^s b_m(x) + o(h^s)$, its variance has the expansion $v_{m,n,h}(x) = (nh)^{-1}\{\sigma_m^2(x) + o(1)\}$, with functions b_m depending on the bias functions b_μ of $\widehat{\mu}_{n,h}(x)$ and b_d of $\widehat{d}_{n,h}(x)$, they depend on the second derivatives of the functions $\bar{G} E(Y 1_{X=x})$ and $\bar{G} f_X$, the variance function $\sigma_m^2(x)$ depends on the variances of $\widehat{\mu}_{n,h}(x)$ and $\widehat{d}_{n,h}(x)$ and on their covariance

$$b_m(x) = \frac{m_{sK}}{s!} \bar{G}^{-1}(x) f_X^{-1}(x) \{b_\mu^{(s)}(x) - m(x) b_d^{(s)}(x)\},$$

$$\sigma_m^2(x) = \kappa_2 \bar{G}^{-1}(x) f_X^{-1}(x) \sigma^2(x),$$

$$(nh)^{\frac{1}{2}}(\widehat{m}_{nh} - m) = \bar{G}^{-1} f_X^{-1}(nh)^{\frac{1}{2}}\{(E\widehat{\mu}_{n,h} - \mu) - m(E\widehat{d}_{n,h} - d)\} + r_n.$$

With the L^2-optimal bandwidth of the estimator $\widehat{m}_{n,h}$, $h_n = O(n^{-\frac{1}{2s+1}})$, the process $U_{n,h} = n^{\frac{s}{2s+1}}\{\widehat{m}_{n,h} - m\} \mathcal{I}_{\{\mathcal{I}_{X,h}\}}$ converges in distribution to

$\sigma_m W_1 + \gamma_s b_m$ where W_1 is a centered Gaussian process on $\mathcal{I}_X$ with variance function 1 and covariance function 0, and γ_s is defined in Condition 1.1. The tests defined in Sections 2.9 to 4.9 without censoring are extended to censored regression variables from this limiting distribution. All results extend to left-censoring by replacing the indicator δ by η as in Section 7.10. The notations of the estimator, its bias and variance are straightforwardly modified with this indicator and the asymptotic behavior of tests about the function $E(Y|X = x) \equiv E(\eta Y|\eta = 1, X = x)$ are deduced.

For a n-sample of (X, Y, C) under independent censorship, the joint distribution function of $(X \wedge C, Y)$ is $\int_0^x \int_0^y \bar{G}(u) \, F(du, dv)$ and the conditional distribution function of Y given $X \wedge C$ is

$$F_{Y|X}(y; x) = \frac{\int_0^x \int_0^y \bar{G}(u) \, F(du, dv)}{\int_0^x \bar{G}(u) \, dF_X(u)},$$

it reduces to $F_Y(y)$ under the independence of X and Y. Tests of independence of these variables rely on the empirical estimators of $F_{Y|X}$ and $F_Y(y)$ under independent right-censoring of X. If (X, Y) have a probability density function, the logarithm of the conditional likelihood ratio of Y given $\delta = 1$ and X, under F with respect to F_0 is

$$l_n(x) = \sum_{i=1}^n \delta_i \log \frac{f_{Y|X}}{f_{0,Y|X}}(Y_i; X_i) + (1 - \delta_i) \log \frac{\bar{F}_{Y|X}}{\bar{F}_{0,Y|X}}(Y_i; C_i)$$

$$= \sum_{i=1}^n \delta_i \log \frac{\lambda_{Y|X}}{\lambda_{0,Y|X}}(Y_i; X_i) - (\Lambda_{Y|X} - \Lambda_{0,Y|X})(Y_i; X_i \wedge C_i),$$

with a conditional hazard function $\Lambda_{Y|X}(y; x) = \int_0^x \lambda_{Y|X}(y; s) \, ds$

$$\lambda_{Y|X}(y; x) = \frac{f_{Y|X}(y; x)}{\int_x^\infty f_{Y|X}(y; s) \, ds} = P(Y = y, X = x | Y = y, X \geq x)$$

and $\bar{F}_{Y|X}(y; s) = \exp\{-\int_0^x \lambda_{Y|X}(y; s) \, ds\}$. The cumulative hazard function $\Lambda_{Y|X}$ has a consistent kernel estimator

$$\widehat{\Lambda}_{nh}(y; x) = \int_0^x \frac{\sum_{i=1}^n \delta_i K_h(y - Y_i) K_h(t - X_i)}{\sum_{i=1}^n K_h(y - Y_i) 1_{\{X_i \geq t\}}} \, dt$$

$$= \int_0^x \frac{\int_0^t K_h(y - v) K_h(t - u) \widehat{F}_n^c(du, dv)}{\int_0^t K_h(y - v) 1_{\{u \geq t\}} \widehat{F}_n(du, dv)} \, dt,$$

where $\widehat{F}_n^c(du, dv) = n^{-1} \sum_{i=1}^n \delta_i 1_{\{X_i \wedge C_i \leq t\}}$ is the empirical estimator of the joint distribution function of (X, Y) under right-censoring of X. Tests of comparison and of goodness of fit of conditional hazard function are obtained from the asymptotic properties of their estimators.

7.14   Observation in random intervals

Let (X, Y) be the variable set of a regression model $Y = m(X) + \varepsilon$, with an error variable such that $E\varepsilon = 0$ and $\sigma^2 = E\varepsilon^2$ is finite. The variables X and ε are supposed to be independent with distribution functions F_X and F_ε. The distribution function of Y conditionally on X is defined by

$$F_{Y|X}(y; x) = P(Y \leq y | X = x) = F_\varepsilon(y - m(x)), \qquad (7.31)$$

with a continuous function m. The joint and marginal distribution functions of X and Y are denoted $F_{X,Y}$, with support $I_{Y,X}$, F_X, with bounded support I_X and, respectively F_Y, such that $F_Y(y) = \int F_\varepsilon(y - m(s)) \, dF_X(s)$ and $F_{X,Y}(x, y) = \int 1_{\{s \leq x\}} F_\varepsilon(y - m(s)) \, dF_X(s)$.

Consider model (7.31) with an independent censoring variable C for Y. For observations by intervals, only C and the indicators that Y belongs to the interval $]-\infty, C]$ or $]C, \infty[$ are observed. The function $F_{Y|X}$ cannot be directly estimated and efficient estimators for m and $F_{Y|X}$ are maximum likelihood estimators. Let $\delta = I_{\{Y \leq C\}}$ and assume that F_ε is C^2. Conditionally on C and $X = x$, the log-likelihood of (δ, C) is

$$l(\delta, C) = \delta \log F_\varepsilon(C - m(x)) + (1 - \delta) \log \bar{F}_\varepsilon(C - m(x))$$

and its derivatives with respect to $m(x)$ and F_ε are

$$\dot{l}_{m(x)}(\delta, C) = -\delta \frac{f_\varepsilon}{F_\varepsilon}(C - m(x)) + (1 - \delta) \frac{f_\varepsilon}{\bar{F}_\varepsilon}(C - m(x)),$$

$$\dot{l}_\varepsilon a(\delta, C) = \delta \frac{\int_{-\infty}^{C - m(x)} a \, dF_\varepsilon}{F_\varepsilon(C - m(x))} + (1 - \delta) \frac{\int_{C - m(x)}^{\infty} a \, dF_\varepsilon}{\bar{F}_\varepsilon(C - m(x))}$$

for every function a such that $\int a \, dF_\varepsilon = 0$ and $\int a^2 \, dF_\varepsilon < \infty$.

With the function $a_F = -f'_\varepsilon f_\varepsilon^{-1}$, $\dot{l}_\varepsilon a_F = \dot{l}_{m(x)}$ then the estimator of $m(x)$ must be determined from the estimator of F_ε, this is possible through the conditional probability function of the observations

$$B(t; x) = P(Y \leq C \leq t | X = x) = \int_{-\infty}^{t} F_\varepsilon(s - m(x)) \, dF_C(s).$$

Let $\widehat{F}_{C,n}$ the empirical estimator of F_C, the function $B(t; x)$ has the kernel estimator

$$\widehat{B}_{n,h}(t; x) = \frac{\sum_{i=1}^{n} K_h(x - X_i) I_{\{Y_i \leq C_i \leq t\}}}{\sum_{i=1}^{n} K_h(x - X_i)}.$$

It satisfies the same asymptotic properties as the kernel estimator of a regression function if the distribution functions F_C and F_ε and the regression function m have densities in $C^2(\mathbb{R})$ and under Conditions 1.1 and 3.1. An estimator $\widehat{H}_n(t;x)$ of the conditional distribution function

$$H(t;x) = F_{\varepsilon,n}(t - m(x)) = P(Y \leq t | X = x)$$

is deduced by deconvolution. Since ε is centered, an estimator of the regression function m is

$$\widehat{m}_n(x) = \int_0^\infty t\, d\widehat{H}_n(t;x).$$

A comparison of two variables Y_1 and Y_2 from observations of occurrence in the random intervals $]-\infty, C]$ or $]C, \infty[$ defined by the same variable C relies on the comparison of the conditional distributions B_k of the variables $C_k \geq Y_k$, given $X_k = x$, for $k = 1,2$. Let $(X_{1i}, Y_{1i}, C_{1i})_{i \leq n_1}$ and $(X_{2i}, Y_{2i}, C_{2i})_{i \leq n_2}$ be the underlying variables of two samples with respective sizes n_1 and n_2 and let $n = n_1 + n_2$. With the optimal bandwidth, the estimator of B for a n-sample has the bias $b_{B,n,h}(t;x) = h^s b_B(t;x) + o(h^s)$ and the variance $v_{B,n,h}(t;x) = (nh)^{-1}\{\sigma_B^2(t;x) + o(1)\}$, where

$$b_B(t;x) = \frac{m_s K}{s!} f_X^{-1}(x)[\{f_X(x)B(t;x)\}^{(s)}(x) - B(t;x)f_X^{(s)}(x)],$$
$$v_B(t;x) = \kappa_2 f_X^{-1}(x)\{B(t;x) - B^2(t;x)\}.$$

A test statistic of the hypothesis H_0 of equality of the functions B_1 and B_2 has a bias correction defined by estimating the function $b_B(t;x)$ under H_0

$$T_n = \sup_{(t,x)\in\mathbb{R}^2} \left| \left(\frac{n_2}{n}\right)^{\frac{s}{2s+1}} \widehat{B}_{1,n_1,h_1}(t;x) - \left(\frac{n_1}{n}\right)^{\frac{s}{2s+1}} \widehat{B}_{2,n_2,h_2}(t;x) \right.$$

$$\left. -\left\{\gamma_1\left(\frac{n_2}{n}\right)^{\frac{s}{2s+1}} - \gamma_2\left(\frac{n_1}{n}\right)^{\frac{s}{2s+1}}\right\}\widehat{b}_{B,n,h}(x) \right| \qquad (7.32)$$

According to Proposition 4.11 and under the same conditions, the statistic T_n converges weakly under H_0 to the supremum of the centered Gaussian process $\sup_{(t,x)\in\mathbb{R}^2} |(1-p)^{\frac{s}{2s+1}} B_1(t;x) - p^{\frac{s}{2s+1}} B_2(t;x)|$. Under the alternative, it tends to infinity. Under a sequence of local alternatives, it has the limit given in Proposition 4.12.

7.15 Exercises

The following exercises are proved by Taylor expansions of the means and variances of the statistics under alternatives K_n in neighborhoods of the hypotheses.

7.15.1. Compare the asymptotic local power of the Anderson-Darling type test and the test defined without censoring by the statistic (7.9) for tests of homogeneity in parametric models for the distribution functions.

7.15.2. Compare the asymptotic local power of the Cramer-von Mises type test and the test defined without censoring by the statistic (7.9) in nonparametric models of the distribution functions.

7.15.3. Write an expansion and determine the limiting distribution of the log-likelihood ratio statistic for homogeneity against a change of location of the hazard functions, under right-censoring.
Hints.   The model of a change of location in the hazard functions is $\Lambda_2(t) = \Lambda_1(t - \theta)$ and $\theta = 0$ under the null hypothesis H_0. The parameter is estimated by maximum of likelihood and its asymptotic behavior is deduced from a first order expansion of $\dot{l}_\theta$ under the change of location. An expansion of the likelihood ratio test statistic is deduced and its limiting distribution under H_0 and the alternative follow.

7.15.4. Determine the first order approximations of the bias and the variance of the estimator of $\Lambda_{Y|X}$ in Section 7.13.
Hints. The expansions follow the same lines as in the nonparametric regression, in (7.5).

Chapter 8

Sequential tests

8.1 Introduction

In sequential tests, a series of tests is performed each according to the result of the preceeding test. The total number of observations is not predetermined and the stopping rule of the sampling and of the consecutive tests is only a statistical rule: they go on until the hypothesis is rejected or until the test statistic exits of the region of uncertainty, if the hypothesis or the alternative are accepted with controlled probabilities. The tests are performed on the basis of a sequence of samples drawn from a large number of trials. The theory of sequential tests has been developed from Wald's approach (1945, 1947) using properties of random stopping times for sums of independent variables and for discrete martingales (Wald and Wolfowitz 1948, Blackwell and Girshick 1947, Stein and Wald 1947, Albert 1947).

In the observation of processes, the variables appear continuously and the sample size increases until a stopping time of the process. In sequential tests, the stopping time is decided from the past observations of the process by a statistical test. With independent observation of a variable, the test statistic converges to its mean as the sample size increases. A series of tests can be performed by increasing the size of one sample or by a series of samples. The principle of sequential testing improves the asymptotic tests. Several kinds of tests can be considered.

(1) A series of tests which is stopped by the rejection of the hypothesis.
(2) Tests where the rejection domain of the hypothesis and the domain of its acceptance are separated by a region where no decision is possible. The series of tests is stopped as the sample size is sufficiently large to decide for the hypothesis or the alternative.

In a sequential control of the quality, products must satisfy measurement constraints which define the hypothesis. The sampling is stopped after a random number of satisfactory controls for an improvement of the production when a degradation appears. In the second case, the sequential test stops after a random number of tests on sub-samples with predefinite sizes. The decision of the sequential test may be earlier than the decision of a single test with a predefinite sample size equal to the sum of all sub-sample sizes. Recursive algorithms have been developed for the sequential calculus of the test statistics in continuous sampling. Let $N_k = n_1 + \cdots + n_k$ be the total size of a sample with k sub-samples for a variable X, the empirical mean of X is calculated recursively as

$$\bar{X}_{N_k} = \frac{N_{k-1}}{N_k}\bar{X}_{N_{k-1}} + \frac{n_k}{N_k}\bar{X}_{n_k}$$

and a recursive formula for the empirical variance of X is

$$\widehat{\sigma}^2_{N_k} = \frac{N_{k-1}}{N_k}\widehat{\sigma}^2_{N_{k-1}} + \frac{n_k}{N_k}\widehat{\sigma}^2_{n_k}.$$

This algorithm fastens the calculus of a sequential Student test with an increasing sample size. This method is generally applied to the calculus of the statistics in sequential tests.

Consider independent sub-samples having deterministic sizes n_j. A sequential test with critical values $c_{n_j,\alpha}$ converging to c_α has, conditionally on the stopping time N_n^* of the test procedure, the power

$$\beta_{n,N_n^*}(\alpha) = \sum_{j=1}^{N_n^*} P(T_{n_j} > c_{n_j,\alpha}).$$

By Chernov's theorem and from Section 1.6, the sequential likelihood ratio test is asymptotically equivalent to

$$\beta_{n,N_n^*}(\alpha) = \sum_{j=1}^{N_n^*} \exp\{-n_j I_1(n_j, c_\alpha)\}.$$

Due to the random stopping time of the series of tests, the power is modified according to Wald's identity.

For a test statistic T_n, let $F_{0,T_{n,c}}$ and $F_{1,T_{n,c}}$ be the distribution function of $T_n 1_{\{T_n > c\}}$ under the hypothesis H_0 and, respectively, the alternative H_1. By concavity,

$$\log P_1(T_n > c) \geq E_0\left\{\log\frac{dF_0}{dF_1}(T_n)1_{\{T_n > c\}}\right\} \geq -H(F_{0,T_{n,c}}, F_{1,T_{n,c}}).$$

In sequential tests

$$\beta_{n,N_n^*}(\alpha) \geq \sum_{j=1}^{N_n^*} \exp\{-H_j(F_{0,T_{n_j},c_{n_j},\alpha}, F_{1,T_{n_j},c_{n_j},\alpha})\}.$$

In sampling of processes, the observations are not independent and the properties of the variable stopping time rely on the asymptotic behavior of the process. Stationary processes satisfying required properties converge and sequential test can be performed in the same way as with an independent sampling of variables. The sequential tests are therefore extended to all tests of the previous chapters.

Other approaches for the processes are the construction of sequential bayesian estimators. In likelihood ratio tests of simple hypothesis and alternative, each density or intensity has a prior probability and the probability distribution of the observations is $\pi P_1 + (1 - \pi)P_0$, with P_0 and P_1 the sampling probabilities under H_0 and, respectively, the alternative H_1 (Peskir and Shiryaev, 2000).

8.2   Definitions and properties

Let $(T_n)_{n \leq 1}$ be a sequence of test statistics for a null hypothesis H_0 concerning the distribution of a sample with the size n. The kth test $\phi_{n,k}$ rejects the hypothesis if its value is 1 and it accepts H_0 if its value is 0

$$\phi_{n,k} = \begin{cases} 1 & \text{if } T_{n,k} > c_n, \\ 0 & \text{if } T_{n,k} \leq c_n. \end{cases} \tag{8.1}$$

The level of the test is the probability to reject the hypothesis H_0

$$\alpha_k = E_0 \phi_{n,k} = P_0(T_{n,k} > c_n),$$

$P_0(T_{n,k} \leq c_n) = 1 - \alpha_k$ and the probability to accept the hypothesis under the alternative is

$$1 - \beta_k = E_1(1 - \phi_{n,k}) = P_1(T_{n,k} \leq c_n).$$

If $\phi_{n_k,k}$ differs from 1, the $(k+1)$th test is performed according to the same rule and the sequential tests stop as soon as a test rejects the hypothesis. The number of consecutive tests is

$$K_n = \inf\{k; \phi_{n,k} = 1\}$$

and the levels of the kth test are evaluated from the consecutive results. If $K_n = k$ and k independent tests of respective levels $\alpha_1, \ldots, \alpha_k$ are consecutively performed, the level of the sequential test is

$$\alpha = \prod_{j=1}^{k-1} (1 - \alpha_j)\alpha_k.$$

In order to have an expression which does not depend on k, the level of the kth test should be increasing and satisfy $\alpha_k = \alpha\{1 - (k-1)\alpha\}^{-1}$, then α_k increases with k, the maximum of k is the largest integer such that $1 - (k-1)\alpha > 0$.

Assuming that each test has the same asymptotic level x, with a constant x in $]0, 1[$ the asymptotic level of the kth sequential test is $a_k = (1 - a_{k-1})x$, hence

$$a_k = \sum_{j=1}^{k} (-1)^{j+1} x^j = \frac{x(1 + (-1)^{k+1}x^k)}{1 + x} + \alpha(-x)^k$$

and the level of the sequential test converges to $a = (1 + x)^{-1}x$ as k tends to infinity, so that $x = (1 - a)^{-1}a$ is the asymptotic level of each partial kth test, as k increases. Moreover

$$E_0 K_n = \sum_{k \geq 1} k P_0(K_n = k) = \sum_{k \geq 1} k a_k$$

therefore $E_0 K_n$ may be infinite if all tests have the same level and the stopping time tends to infinity. It is necessary that the sequence of levels α_k decreases to zero and $\sum_{k \geq 1} k a_k$ is finite to ensure that $E_0 K_n$ is finite.

For tests with two critical values, b_n and c_n, there exists a constant κ between 0 and 1 such that for a value of the statistic between b_n and c_n the test has the value κ

$$\phi_{n,k} = \begin{cases} 1 & \text{if } T_{n,k} > c_n, \\ \kappa & \text{if } b_n < T_{n,k} \leq c_n, \\ 0 & \text{if } T_{n,k} \leq b_n, \end{cases} \tag{8.2}$$

The errors of the test are the probability to reject the hypothesis under H_0

$$\alpha_k = E_0 \phi_{n,k} = P_0(T_{n,k} > c_n) + \kappa P_0(b_n < T_{n,k} \leq c_n)$$

and the probability to accept the hypothesis under the alternative

$$1 - \beta_k = E_1(1 - \phi_{n,k}) = P_1(T_{n,k} \leq b_n) + (1 - \kappa)P_1(b_n < T_{n,k} \leq c_n).$$

The probability to accept the hypothesis under H_0 is

$$1 - \alpha_k = E_0(1 - \phi_{n,k}) = P_0(T_{n,k} \leq b_n) + (1 - \kappa)P_0(b_n < T_{n,k} \leq c_n)$$

and the probability to reject the hypothesis H_0 under the alternative is

$$\beta_k = E_1\phi_{n,k} = P_1(T_{n,k} > c_n) + \kappa P_1(b_n < T_{n,k} \leq c_n).$$

A sequence of tests $(\phi_{n,k})_{k \geq 1}$ is performed until the stopping variable

$$K_n = \min\{k \geq 1, T_{n,k} \notin]b_n, c_n]\}.$$

The variable K_n is infinite if l_n belongs to $]b_n, c_n]$ for every n. Then $P_0(T_{n,K_n} \in]b_n, c_n]) = P_1(T_{n,K_n} \in]b_n, c_n]) = 0$, for every n, and the levels of the test ϕ_{n,K_n} are

$$\alpha_n = P_0(T_{n,K_n} > c_n) \leq E_0 \alpha_{K_n},$$
$$1 - \beta_n = P_1(T_{n,K_n} \leq b_n) \leq E_1(1 - \beta_{K_n}).$$

Proposition 8.1. *Let $(T_{n,k})_{n \geq 1}$ be a sequence of statistics with mean zero and variance 1 under H_0, and with means $\mu_{n,k}$ and variances $\sigma_{n,k}^2$ under the alternative. Under H_0, the statistic T_{n,K_n} has the mean 0 and the variance 1, under the alternative its mean is $\mu_n = E_1\mu_{n,K_n}$ and its variance is $\sigma_n^2 = E_1\sigma_{n,K_n}^2 + \sum_{k \geq 1} \mu_{n,k}^2 Var 1_{\{K_N=k\}}.$*

Proof. Since $E_0 T_{n,K_n} = \sum_{k \geq 1} E_0(T_{n,k})P_0(K_N = k)$, the statistic T_{n,K_n} is centered under H_0 and its variance satisfies

$$Var T_{n,K_n} = \sum_{k \geq 1} E_0(T_{n,k}^2)P_0(K_N = k) = 1.$$

Under the alternative, the mean of T_{n,K_n} is

$$\mu_n = \sum_{k \geq 1} E_1 T_{n,k} P_1(K_N = k) = E_1\mu_{n,K_n}$$

and its variance is

$$\sigma_n^2 = E_1(T_{n,K_n} - \mu_n)^2 = \sum_{k \geq 1} E_1 T_{n,k}^2 P_1(K_N = k) - E_1^2 \mu_{n,K_n}$$
$$= \sum_{k \geq 1} (\sigma_{n,k}^2 + \mu_{n,k}^2)P_1(K_N = k) - E_1^2 \mu_{n,K_n}$$
$$= E_1\sigma_{n,K_n}^2 + \sum_{k \geq 1} \mu_{n,k}^2 \{P_1(K_N = k) - P_1^2(K_N = k)\}.$$

$\square$

From the Bienaymé-Chebychev inequality, it follows that the levels of the sequential test are bounded, with $\alpha_n \leq c_n^{-2}$ and $\beta_n \geq 1 - \sigma_n^{-2}(b_n - \mu_n)^2$. Assuming that the sample sizes are sufficiently large to use the asymptotic levels α and β of the test and its asymptotic means and variances, the sequences $(b_n)_{n \geq 1}$ and $(c_n)_{n \geq 1}$ converge to constants b and c such that $\alpha \leq c^{-2}$ and $1 - \beta \leq \sigma^{-2}(b - \mu)^2$.

Example. The number of defaults in a system of n independent components failing with a probability p has a Binomial distribution $\mathcal{B}(n, p)$ and the system stops if the failure probability p is larger than p_0. Let $X_i = 1$ if the ith component has a failure and zero if it works, the likelihood ratio is $\prod_{i=1}^{n} (\frac{p}{p_0})^{X_i} (\frac{1-p}{1-p_0})^{1-X_i}$ and the tests statistic for the hypothesis $H_0 : p \geq p_0$ relies on the Binomial variable $N_n = \sum_{i=1}^{n} X_i$, the level of the test with a critical level k_n is

$$\alpha_n = P_0(N_n > k_n) = 1 - \sum_{k=0}^{k_n} P_0(N_n = k)$$

$$= 1 - \frac{(1-p_0)^{k_n}}{2} \frac{1 - \{p_0(1-p_0)^{-1}\}^{k_n+1}}{1 - p_0(1-p_0)^{-1}}$$

and α determines the critical level k_n of the Binomial variable N_n. The mean under H_0 of N_n is np_0 and its variance is $np_0(1-p_0)$, the variable $X_n = (N_n - np_0)\{np_0(1-p_0)\}^{-\frac{1}{2}}$ is asymptotically normal.

In a sequential test, independent sub-samples of $n_1, \ldots, n_\tau$ components are sequentially tested and the stopping time $\tau \leq n$ is the index of the sample that fails to the test and we assume that $\alpha\tau < 1$. The level of the sequential test

$$P_0(N_\tau > k_\tau) = \sum_{j=1}^{n} P_0(N_j > k_j, N_i \leq k_i, i = 1, \ldots, j - 1)$$

$$= \sum_{j=1}^{n} P_0(N_j > k_j) \prod_{i=1}^{j-1} P_0(N_i \leq k_i)$$

$$= \sum_{j=1}^{n} \alpha_j \prod_{i=1}^{j-1} (1 - \alpha_i).$$

The power of the tests follows a similar equality.

Let $X_1, \ldots, X_n, \ldots$ be a sample of a random variable X with a distribution function F, with mean zero and a finite variance σ^2. Wald's lemma asserts that the stopping time ν of exit of $\sum_{i=1}^{n} X_i$ from $[-b, a]$ satisfies

$\lim_{n\to\infty} P(\nu < \infty) = 1$.   Albert (1947) proved that the stopping time of the sum of independent and identically distributed variables having a Laplace transform φ satisfies $E\{e^{(\sum_{i=1}^{\nu} X_i)t}\varphi^{-\nu}(t)\} = 1$.

In a sequence of two-sided tests based on $S_n = n^{-\frac{1}{2}} \sum_{i=1}^n X_i$, the stopping time of testing is $\tau_n = \inf\{n \geq 1; S_n < -a \text{ or } S_n > a\}$.

Lemma 8.1. *The stopping time τ_n defined by the first exit of S_n from $[-a, a]$ satisfies $\lim_{n\to\infty} P_0(\tau_n < \infty) = \alpha$ such that $a = c_{\frac{\alpha}{2}}$, $E_0 S_{\tau_n} = 0$ and $Var_0 S_{\tau_n} = \sigma^2$.*

Proof.   By the central limit theorem, $\lim_{n\to\infty} P_0(n^{-\frac{1}{2}} S_n \notin [-a, a]) = \alpha$ and for every integer N, $p_N = P_0(\tau_n > N) = P_0(-a \leq N^{-\frac{1}{2}} S_N \leq a)$ converges to $1 - \alpha$ as N tends to infinity, therefore $\lim_{n\to\infty} P_0(\tau_n < \infty) = \alpha$. The mean and variance of S_{τ_n} under H_0 are $E_0 S_{\tau_n} = \sum_{k>0} E_0 S_k 1_{\{\tau_n=k\}} = 0$ and $E_0 S_{\tau_n}^2 = \sum_{k>0} E_0 S_k^2 1_{\{\tau_n=k\}} = \sigma^2 \sum_{k>0} P_0(\tau_n = k) = \sigma^2$.   □

8.3   Sequential likelihood ratio test

The results of the previous section apply to the likelihood ratio test and the properties of mean of the statistic under the hypothesis and the alternative provide further results. Let X be a random variable with density f_0 under H_0 and f_1 under the alternative H_1. The likelihood ratio test for the simple hypothesis H_0 against H_1 relies on the statistic

$$l_n = \log \prod_{i=1}^n \frac{f_1}{f_0}(X_i) = \log L_n.$$

Its mean under H_0 is $\mu_{0n} = n \int (\log f_1 - \log f_0) f_0 = n\mu_0 \leq 0$ and under H_1 it is $\mu_{1n} = n \int (\log f_1 - \log f_0) f_1 = n\mu_1 \geq 0$, for every integer n. For j in $\{0, 1\}$, the statistic $n^{-\frac{1}{2}}(l_n - \mu_{jn})$ converges weakly under H_j to a Gaussian variable with mean zero and with variance $\sigma_j^2 = \int (\log f_1 - \log f_0)^2 f_j$. The approximation of $2l_n$ by a χ^2 variables, as n tends to infinity, has been developed for the parametric case and it is not used in this section. An asymptotically normal test of levels α and β is defined for large n by

$$\phi_n = \begin{cases} 1 & \text{if } n^{-\frac{1}{2}} l_n > c_n, \\ \kappa & \text{if } n^{-\frac{1}{2}} l_n \in]b_n, c_n], \\ 0 & \text{if } n^{-\frac{1}{2}} l_n \leq b_n, \end{cases} \tag{8.3}$$

where the critical values b_n and c_n of the test are determined from the $(1 - \alpha)$-quantile of the normal distribution and the means of l_n under H_0

and H_1 as

$$b_n = \sigma_0 c_\alpha + n^{\frac{1}{2}} \mu_0 - \varepsilon_n,$$

$$c_n = \sigma_1 c_\alpha + n^{\frac{1}{2}} \mu_1 + \varepsilon_n. \tag{8.4}$$

The difference of the means $\mu_1 - \mu_0$ is positive and $(\varepsilon_n)_{n\geq 1}$ is a positive sequence of errors for the approximation of $\sigma_{jn}^{-1}(l_n - \mu_{jn})$ by a normal variable as n tends to infinity. The decision of the test is H_0 if $n^{-\frac{1}{2}} l_n < b_n$ and H_1 if $n^{-\frac{1}{2}} l_n \geq c_n$. The interval $]b_n, c_n]$ is a domain where H_0 is accepted with a probability κ and rejected with the probability $1 - \kappa$. Since the interval $]b_n, c_n]$ is not empty, a sequential likelihood ratio test for H_0 against H_1 is performed with the stopping variable

$$N = \min\{n \geq 1, n^{-\frac{1}{2}} l_n \notin]b_n, c_n]\}$$

and N is infinite if $n^{-\frac{1}{2}} l_n$ belongs to $]b_n, c_n]$ for every n. If N is finite, a decision can be taken for H_0 or H_1.

By the weak convergence of the normalized likelihood ratio statistic to normal variables under H_0 and H_1,

$$\alpha_n = P_0(l_n > c_n) = P_0\{n^{-\frac{1}{2}}(l_n - \mu_{0n}) > n^{-\frac{1}{2}}(c_n - \mu_{0n})\}$$

converges to $\alpha = 1 - \Phi(c_\alpha)$ such that $c_\alpha = \lim_{n\to\infty} n^{-\frac{1}{2}}(c_n - \mu_{0n})$ and

$$1 - \beta_n = P_1(l_n \leq c_n) = P_1\{n^{-\frac{1}{2}}(l_n - \mu_{1n}) \leq n^{-\frac{1}{2}}(c_n - \mu_{0n}) + n^{-\frac{1}{2}}(\mu_{0n} - \mu_{1n})\}$$

converges to $1 - \beta = \Phi(c_\alpha + C)$, where $C = \lim_{n\to\infty} n^{-\frac{1}{2}}(\mu_{0n} - \mu_{1n})$ if this sequence converges to a finite limit, otherwise $\beta = 1$.

Let N be a random sample size of the sequential test. By Wald's identity, the mean of the statistic l_N under the hypothesis is

$$E_0 l_N = \mu_0 E_0 N$$

and it is negative, under the alternative, the mean of l_N is $E_1 l_N = \mu_1 E_1 N$ and it is positive. The levels of the sequential test satisfy

$$\alpha = P_0(l_N \geq \sqrt{N} c_N) = \sum_{n\geq 1} P_0(N = n, L_N \geq e^{\sqrt{n} c_n})$$

$$= \sum_{n\geq 1} E_1(L_n^{-1} 1_{\{N=n, l_n \geq \sqrt{n} c_n\}}) \leq E_1(e^{-c_N} 1_{\{l_N \geq \sqrt{N} c_N\}})$$

$$\leq \{(1 - \beta) E_1(e^{-2\sqrt{N} c_N}\}^{\frac{1}{2}}$$

and under H_1

$$\beta = P_1(l_N \leq \sqrt{N} b_N) = \sum_{n\geq 1} P_1(N = n, L_N \leq e^{\sqrt{n} b_n})$$

$$\leq \sum_{n\geq 1} e^{\sqrt{n} b_n} P_0(N = n, l_n \leq \sqrt{n} b_n)$$

$$\leq \{(1 - \alpha) E_0 e^{2\sqrt{N} b_N}\}^{\frac{1}{2}}.$$

A test with a composite hypothesis $\mathcal{F}_0$ and a composite alternative $\mathcal{F}_1$ such that $\mathcal{F}_0 \cap \mathcal{F}_1 = \emptyset$ relies on the statistic

$$l_n = \log \prod_{i=1}^{n} \frac{\sup_{g \in \mathcal{F}_1} g}{\sup_{f \in \mathcal{F}_0} f}(X_i).$$

Denoting f_0 the unknown density of the variable X under H_0, f_1 the unknown density of X under H_1 and $\widehat{f}_{jn}$ the maximum likelihood estimator of f_j in $\mathcal{F}_j$, for $j = 0, 1$, the statistic is written as

$$l_n = \sum_{i=1}^{n} \{\log \widehat{f}_{1n}(X_i) - \log \widehat{f}_{0n}(X_i)\}.$$

The mean of l_n under H_j is $\mu_{jn} = n \int \{\log \sup_{g \in \mathcal{F}_1} g - \log \sup_{f \in \mathcal{F}_0} f\} f_j$, it is negative under H_0 and positive under H_1, for every integer n. A sequential likelihood ratio test is written like (8.3) with critical values depending on the quantile $c_\alpha = \Phi^{-1}(1 - \alpha)$ of the normal distribution, on the estimators of the variance under H_0 and H_1 and on an approximation error ε_n

$$b_n = \widehat{\sigma}_{0n} c_\alpha + n^{\frac{1}{2}} \widehat{\mu}_{0n} - \varepsilon_n,$$
$$c_n = \widehat{\sigma}_{1n} c_\alpha + n^{\frac{1}{2}} \widehat{\mu}_{1n} + \varepsilon_n,$$

with estimators of the means and the variances for densities in $\mathcal{F}_0$ and $\mathcal{F}_1$ respectively. The levels and the power of the sequential test satisfies inequalities as above.

8.4   Sequential algorithms for test statistics

Let $\mathcal{F}$ be the class of the distribution functions of a real random variable X and let θ be a parameter expressed in a closed form $\theta = \varphi(F)$ from a $C^1(\mathcal{F})$ functional and the empirical distribution function F of X. Let $(\mathbb{X}_{n,k})_{k \geq 1}$ be a sequence of independent samples of X with respective sizes n_k defined as $\mathbb{X}_{n,k} = (X_{N_{k-1}}, \ldots, X_{N_k})$ with $N_k = \sum_{j=1,\ldots,k} n_j$ and such that all n_k have the same order. Let $\widehat{F}_{n_k,k}$ be the empirical distribution function of X for the kth sub-sample of size n_k and let $\widehat{F}_{N_k}$ be the empirical distribution function of X for the k first sub-samples, then

$$\widehat{F}_{N_k} = \frac{N_{k-1}}{N_k} \widehat{F}_{N_{k-1}} + \frac{n_k}{N_k} \widehat{F}_{n_k,k} = \widehat{F}_{N_{k-1}} - \frac{n_k}{N_k}(\widehat{F}_{N_{k-1}} - \widehat{F}_{n_k,k}). \quad (8.5)$$

As N_k increases, $\widehat{F}_{N_k}$ converges to F in probability and

$$\varphi(\widehat{F}_{N_k}) = \varphi(\widehat{F}_{N_{k-1}}) - \frac{n_k}{N_k}(\widehat{F}_{N_{k-1}} - \widehat{F}_{n_k,k})\varphi'(\widehat{F}_{N_{k-1}}) + o_p(1)$$

converges to $\varphi(F)$ in probability.

Let $\widehat{\theta}_n = \varphi(\widehat{F}_n)$ be the empirical estimator of a parameter $\theta = \varphi(F)$ calculated from the empirical distribution function of a n-sample of X. A sequential estimation of the parameter is obtained from $\widehat{\theta}_{n,1} = \varphi(\widehat{F}_{n,1})$ as

$$\widehat{\theta}_{n,k} = \varphi(\widehat{F}_{N_k}).$$

As N_k is sufficiently large to ensure that $N_k^{-1} n_k < \varepsilon$ is small, it is calculated by a recursive algorithm from the $(k-1)$th estimator and $\mathbb{X}_{n,k}$ as

$$\widehat{\theta}_{n,k} = \varphi\Big(\frac{N_{k-1}}{N_k}\widehat{F}_{N_{k-1}} + \frac{n_k}{N_k}\widehat{F}_{n_k,k}\Big)$$

$$= \widehat{\theta}_{n,k-1} + \frac{n_k}{N_k}(\widehat{F}_{n_k,k} - \widehat{F}_{N_{k-1}})\varphi'(\widehat{F}_{N_{k-1}}) + o_p(1).$$

As k tends to infinity, $\widehat{F}_{N_{k-1}}$ converges to F and $\widehat{\theta}_{n,k} - \widehat{\theta}_{n,k-1}$ converges to zero, in probability.

Let T_n be a statistic calculated by a smooth functional of the distribution function of X, such as the Cramer-von Mises or the Anderson-Darling statistics. A recursive algorithm for a statistic T_n from k sub-samples of respective sizes n_j of X is defined by a sequence of functionals (φ_n) of $C^1(\mathcal{F})$ converging uniformly in $\mathcal{F}$ to a functional φ of $C^1(\mathcal{F})$, it follows that $\lim_{n\to\infty}\|\varphi_n - \varphi_{n-1}\|_\mathcal{F} = 0$. An approximate recursive algorithm for the calculus of T_{N_k} from $T_{N_{k-1}}$ and (φ_n), starting from T_{n_1} is defined by

$$T_{N_k} = \varphi_{N_k}(\widehat{F}_{N_k}) = \varphi_{N_k}(\widehat{F}_{N_{k-1}}) - \frac{n_k}{N_k}(\widehat{F}_{N_{k-1}} - \widehat{F}_{n_k,k})$$

$$= T_{N_{k-1}} + (\varphi_{N_k} - \varphi_{N_{k-1}})(\widehat{F}_{N_{k-1}})$$

$$- \frac{n_k}{N_k}(\widehat{F}_{N_{k-1}} - \widehat{F}_{n_k,k})\varphi'_{N_k}(\widehat{F}_{N_{k-1}})$$

$$+ o_p(\|\varphi_{N_k} - \varphi_{N_{k-1}}\|_\mathcal{F}) + o_p(N_k^{-1}n_k).$$

The difference of T_{N_k} and $T_{N_{k-1}}$ tends to zero as k tends to infinity and the first order term of the expansion of their difference provides an algorithm to update the sequence of statistics.

The Cramer-von Mises goodness of fit statistic for a simple hypothesis $H_0 : X$ has the distribution function F_0 is defined as $T_n = \varphi_n(\nu_n)$, with C^1 functionals on the space $\mathcal{G}$ of the empirical processes of the variable X, $\varphi_n(u) = \int_\mathbb{R} u^2(x)\,d\widehat{F}_n(x)$. Its first derivative is $\varphi'_n(u) = 2\int_\mathbb{R} u(x)\,d\widehat{F}_n(x)$. For the Cramer-von Mises statistic

$$\varphi_{N_k} - \varphi_{N_{k-1}} = N_k^{-1}n_k(\varphi_{n_k} - \varphi_{N_{k-1}})$$

and the empirical process ν_{N_k} has the expansion

$$\nu_{N_k} = N_k^{-\frac{1}{2}}(N_{k-1}^{\frac{1}{2}}\nu_{N_{k-1}} + n_k^{\frac{1}{2}}\nu_{n_k,k}) = \nu_{N_{k-1}} + u_{n_k,N_k}$$

where

$$u_{n_k,N_k} = \left\{\left(\frac{N_{k-1}}{N_k}\right)^{\frac{1}{2}} - 1\right\}\nu_{N_{k-1}} + \left(\frac{n_k}{N_k}\right)^{\frac{1}{2}}\nu_{n_k,k} = o_p(1).$$

It follows that

$$
\begin{aligned}
T_{N_k} &= \varphi_{N_k}(\nu_{N_k}) = \varphi_{N_k}(\nu_{N_{k-1}}) + u_{n_k,N_k}\varphi'_{N_k}(\nu_{N_{k-1}}) + o_p(u_{n_k,N_k}) \\
&= T_{N_{k-1}} + N_k^{-1}n_k(\varphi_{n_k,k} - \varphi_{N_{k-1}})(\nu_{N_{k-1}}) \\
&\quad + \left(\frac{n_k}{N_k}\right)^{\frac{1}{2}}\nu_{N_{k-1}}\varphi'_{N_k}(\nu_{N_{k-1}}) + o_p(u_{n_k,N_k}).
\end{aligned}
\tag{8.6}
$$

The Anderson-Darling statistic is defined in the same form with the C^1 functional

$$\varphi_n(u) = \int_{\mathbb{R}} u^2 \frac{dH_n}{H_n(1 - H_n)},$$

with a deterministic function $H_n = F_0$ in the goodness for fit test and H_n is the empirical estimator of the distribution function of the whole sample in the homogeneity tests. The difference $\varphi_{N_k} - \varphi_{N_{k-1}}$ equals

$$\int_{\mathbb{R}} u^2 \left\{ \frac{d(H_{N_k} - H_{N_{k-1}})}{H_{N_{k-1}}(1 - H_{N_{k-1}})} - \frac{(H_{N_{k-1}} - H_{N_k})(1 - H_{N_{k-1}} - H_{N_k})\,dH_{N_k}}{H_{N_{k-1}}(1 - H_{N_{k-1}})H_{N_k}(1 - H_{N_k})} \right\},$$

using (8.5) for the variations of the empirical distribution functions, the statistic has the expansion (8.6) with this expression.

Similar expansions hold for multi-samples statistics, replacing $\widehat{F}_n$ by $(n_k + n_j)^{-1}(n_k\widehat{F}_{jn_j} - n_j\widehat{F}_{kn_k})$ in the k-sample test with unequal sub-sample sizes n_j, for all $j = 1, \ldots, k$ and $k \geq 2$.

The estimation of a dependence parameter and the stochastic EM algorithms are other example of this kind of recursive estimators where an initial parameter estimator obtained from a N_{k-1}-sample is provided for the calculus of an estimator from a N_k-sample. The estimation algorithms stop as the variation of the estimators falls below a threshold. The recursive estimators can be used to test the parameter values in goodness of fit tests. The sequence of levels of the tests has the same relationships as in Section 8.2.

8.5   Properties of the record variables

In a n-sample $(X_i)_{i \geq 1}$ of a real random variable X, the record variables are the consecutive maximal variables of the sample $X_n^* = \max_{i=1,\ldots,n} X_i$ and $X_{N_k^*}^* = X_{(k:n)}$ is defined by the integer stopping variables

$$N_k^* = \inf\{k \geq N_{k-1}^*; X_k^* > X_j, j = N_{k-1}^*, \ldots, X_{N_k^*-1}^*\}$$

where the value $X_{(k:n)}$ is reached, for $k \geq 2$, starting from $N_1^* = 1$.

For every n, there exists $k \leq n$ such that $N_{k-1}^* \leq n < N_k^*$, then

$$n^{-1} X_n^* \leq N_{k-1}^{*-1} X_{N_{k-1}^*}^*.$$

Lemma 8.2. *The sequence $n^{-\frac{1}{2}} X_n^*$ is relatively compact. If X is centered and has a finite variance σ_X^2, then $n^{-\frac{1}{2}} X_n^*$ converges weakly to a centered variable with variance σ_X^2 and higher moments zero.*

Proof.   By the Bienaymé-Chebychev inequality, the distribution of X_n^* satisfies

$$P(n^{-\frac{1}{2}} X_n^* > t) = \sum_{i=1}^n P(X_i > n^{\frac{1}{2}} t) = n\{1 - F(n^{\frac{1}{2}} t)\}$$

$$\leq t^{-2} EX^2,$$

and it converges to zero as t tends to infinity, which proves the relative compactness of $n^{-\frac{1}{2}} X_n^*$. The distribution function of a variable $Y_n^* = n^{-\frac{1}{2}} X_n^*$ is $F_n^*(x) = 1 - n\bar{F}(n^{\frac{1}{2}} x)$, with $\bar{F} = 1 - F$, and its moments of Y_n^* are

$$EY_n^* = n \int x \, dF(n^{\frac{1}{2}} x) = n^{\frac{1}{2}} \int y \, dF(y) = n^{\frac{1}{2}} EX = 0,$$

$$EY_n^{*2} = n \int x^2 \, dF(n^{\frac{1}{2}} x) = E(X^2),$$

$$EY_n^{*k} = n^{\frac{2-k}{2}} E(X^k), \quad k \geq 1,$$

they converge to zero as n tends to infinity, for every $k > 2$ and the result is deduced from these moments.   □

It follows that $n^{-1} X_n^*$ converges in probability to zero under the conditions of Lemma 8.2.

The Laplace transform of the variable Y_n^* is $L_{Y_n^*}(t) = n^{\frac{1}{2}} L_X(n^{-\frac{1}{2}} t)$, where L_X is the Laplace transform of X, hence $n^{-\frac{1}{2}} L_{Y_n^*}(n^{\frac{1}{2}} t) = L_X(t)$.

Example 1. Let X be an exponential variable with distribution $\mathcal{E}_\theta$, the probability $P(Y_n^* > t) = n\bar{F}(n^{\frac{1}{2}} t) = n \exp\{-\theta n^{\frac{1}{2}} t\}$ tends to zero as n or t tend to infinity.

Example 2. For a normal variable, the probability $P(Y_n^* > t)$ is asymptotically equivalent to $n^{\frac{1}{2}} \exp\{-\frac{n}{2}t^2\}$ as n tend to infinity, by the Chernov theorem (Appendix A.1). It tends to zero as n or t tend to infinity.

By the independence of the observations, for all integers $m < n$

$$P(X_n^* > t) = P(X_m^* > t) + P(\max_{m < i \leq n} X_i > t)$$
$$\equiv P(X_m^* > t) + P(X_{n-m}^* > t),$$

so the distribution function of the records X_m^* and X_n^* satisfy $F_{X_n^*} = F_{X_m^*} + F_{X_{n-m}^*}$, as the sample size increases. For all integers k and $m \geq 1$

$$F_{X_{km}^*} = k F_{X_m^*}, \tag{8.7}$$

that is also the distribution of the maximum of a sample observed in k independent blocks of size m.

Lemma 8.3. *If F is continuous, the stopping times N_k^* satisfy*

$$P(N_k^* = j) = P(R_j = k) = n^{-1},$$

for all j and k in $\{1, \ldots, n\}$.

Proposition 8.2. *If F is continuous, for every integer $k \leq n$*

$$P(X_{N_k^*}^* > t | N_{k-1}^*) = \frac{(n - N_k^*)(n - N_k^* + 1)}{2n} \bar{F}(t).$$

Proof. The probability of $\{X_{N_k^*}^* > t\}$ conditionally on $N_{k-1}^* = j$ is

$$P_j = P(\max_{j < i \leq N_k^*} X_i > t | N_{k-1}^* = j)$$

$$= \sum_{l=j+1}^{n} P(N_k^* = l)(l - j)\bar{F}(t) = n^{-1} \sum_{l=1}^{n-j} l\bar{F}(t)$$

and the result follows. $\qquad\qquad\qquad\qquad\qquad\qquad\qquad\qquad\square$

The records of sums $S_1 = X_1, \ldots, S_n = \sum_{i=1}^{n} X_i$ of independent and identically distributed observations of a random variable X with mean zero satisfy the above results with the nth convolution of the distribution of X and they are recursively written as

$$S_{n+1}^* = S_n^* \vee S_{n+1} = S_n^* 1_{\{X_{n+1} \leq 0\}} + S_{n+1} 1_{\{X_{n+1} > 0\}}. \tag{8.8}$$

Sequential likelihood ratio tests are performed at the records of the likelihood ratio variables calculated in sub-samples, with the recursive expression

(8.8). Their level and the critical value of the consecutive tests are calculated from Proposition 8.2. In reliability control by blocks of observations, items are measured along time and tests of the hypothesis of no change in a measurement variable X are performed as a sequence of likelihood ratio tests. The global level is calculated by (8.7) as the $\alpha = k\alpha_0$, when the consecutive tests are performed with the same level α_0 in each block, their critical values are deduced from this level as $c_{1-\alpha_0}$.

8.6   Sequential tests for point processes

Failures in a series of components do not necessarily stop a device and the number of failures or defaults in systems are indicators of the risk of their global failure. The frailty of groups of components can be compared through tests of equality of the intensities of the counting processes related to several groups of components or under several stress conditions. Let $N_1, \ldots, N_k$ be Poisson processes with parametric intensity functions λ_0 as a reference model and $\lambda_{\theta_1}, \ldots, \lambda_{\theta_k}$, for the observations of k sub-samples, with a parameter values varying in a discrete or continuous space. They are estimated by $\widehat{\theta}_{jT} = \widehat{\theta}_{jn_j}$ that maximize the log-likelihood ratio from the observation of all failure times T_{ij}, $i = 1, \ldots, n_j$. Let $M_{0T}(s) = \sum_{i=1}^{N_{jT}} 1_{\{T_i \leq s\}} - \Lambda_0(s)$, then

$$l_T(\theta_j) - l_T(\theta_0) = \sum_{i=1}^{N_{jT}} \left\{ \log \frac{\lambda_{\theta_j}}{\lambda_0}(T_{ij}) - \Lambda_{\theta_j}(T) \right\} + \Lambda_0(T)$$

$$= \int_0^T \log \frac{\lambda_{\theta_j}}{\lambda_0} \, dM_{0T} + \int_0^T \left\{ \log \frac{\lambda_{\theta_j}}{\lambda_0} - \frac{\lambda_{\theta_j}}{\lambda_0} + 1 \right\} d\Lambda_0.$$

The intensities of independent Poisson processes N_j observed at T are estimated at T from the log-likelihoods $l_{jT}(\lambda_{jT}) = N_{jT} \log \lambda_{jT} - \lambda_{jT} - \log(N_{jT}!)$, hence $\widehat{\lambda}_{jT} = T^{-1} N_{jT}$, for $j = 1, \ldots, k$. The log-likelihood of $N_j(T)$ is then estimated by $\widehat{l}_{jT} = l_{jT}(\widehat{\lambda}_{jT})$. The comparison of k samples of point processes is performed as in Section 7.6 for continuously observed processes or through the differences of the k estimators $\widehat{l}_{jT}$, $j = 1, \ldots, k$.

When m consecutive observation times are scheduled at fixed or predictable random times $\tau_{j,r}$ for $r = 1, \ldots, m$ and $j = 1, \ldots, k$, the cumulated observations of the k Poisson processes at times $\tau_{j,r}$ are $N_j(\tau_{j,r}) = n_{j,r}$ and their log-likelihoods are $l_{j,\tau_{j,r}} = \sum_{i=1}^{n_{j,r}} \log \lambda_{j\tau_{j,r}} - \lambda_{j\tau_{j,r}} - \log(n_{j,r}!)$. A number of failures $n_{j,r} - n_{j,r-1}$ occur in the sub-interval $]\tau_{j,r-1}, \tau_{j,r}]$ and an estimator of the Poisson parameter λ_j is defined as

$$\widehat{\lambda}_{j,r} = (\tau_{j,r} - \tau_{j,r-1})^{-1}(N_j(\tau_{j,r}) - N_j(\tau_{j,r-1})),$$

for all $j = 1, \ldots, k$ and $r = 1, \ldots, m$. The sequence of estimators of the parameter λ_j must remain in the same range along the observation times. The log-likelihoods are cumulated from the beginning of the observations over m sub-intervals of total length $T = \sum_{r=1,\ldots,m} \tau_{j,r}$ and the global estimators of the intensities are $\widehat{\lambda}_{jT} = T^{-1} \sum_{r=1,\ldots,m} n_{j,r} = T^{-1} N_{jT}$, for $j = 1, \ldots, k$. The estimated log-likelihood statistics provide independent sequences of tests along the observation time, due to the independence of the increments of the Poisson processes, their levels are therefore cumulated as in (8.7).

Asymptotic confidence intervals $\widehat{I}_{j,r}$ for the true parameter value λ_j are defined for each sub-interval by the normal quantiles and the estimated variances $\widehat{\lambda}_{j,r}$, their intersection is not empty if the parameter λ_j is constant along the observation times. By the independence of the increments of the process, the probability of the confidence intervals is

$$P(\lambda_j \in \cap_{r=1,\ldots,m} \widehat{I}_{j,r}) = 1 - \sum_{r=1,\ldots,m} P(\widehat{\lambda}_{j,r}^{-\frac{1}{2}} |\widehat{\lambda}_{j,r} - \lambda_j| > c_{\alpha_0})$$

$$= 1 - m\alpha_0 + o(1) = 1 - \alpha + o(1),$$

as in (8.7) for the sequence of independent goodness of fit tests. Since the actual parameter value λ_j are unknown, they are estimated by the estimator over cumulated intervals until the previous observation time $\tau_{j,r-1}$ if the hypothesis H_0 was accepted at that time, the sequential test stops at the first time $\tau_{j,r}$ of a rejection of H_{0j}. An aymptotically normal test statistic relies on the difference

$$\widehat{\lambda}_{j,r}^{-\frac{1}{2}} |\widehat{\lambda}_{j,r} - \widehat{\overline{\lambda}}_{j,r-1}|,$$

where $\widehat{\overline{\lambda}}_{j,r} = \tau_{j,r}^{-1} \sum_{l=1,\ldots,r} n_{jl}$. Comparison of the independent Poisson processes are performed in the same way and the tests have the same form in parametric models of Poisson processes.

With continuous observations, functional intensities λ_j of point processes with independent increments can be estimated. The whole cumulative intensity function is

$$\Lambda_j(t) = \sum_{l=1}^{r-1} \int_{\tau_{j,l-1}}^{\tau_{j,l}} d\Lambda_{j,l} + \int_{\tau_{j,r-1}}^{t} d\Lambda_{j,r}, \; t \in]\tau_{j,r-1}, \tau_{j,r}].$$

It is estimated from the counting process restricted on a sub-interval $]\tau_{j,r-1}, \tau_{j,r}]$ by

$$\widehat{\Lambda}_{j,r}(t) = \widehat{\Lambda}_{j,r}(\tau_{j,r-1}) + \int_{\tau_{j,r-1}}^{t} dN_j(s), \; t \in]\tau_{j,r-1}, \tau_{j,r}].$$

The tests of Section 6.3 apply to the functions λ_j. Tests of homogeneity and goodness of fit tests are performed sequentially from these estimators, by intervals as above. Similar sequential tests by intervals apply to the processes with multiplicative intensities.

8.7   Sequential tests for hazard functions

The tests of Chapter 7 apply to the components of systems. If the components are independent, the counting process N_n of the failure times has a predictable compensator $\tilde{N}_n = \int_0^{\cdot} Y_n^{-1} d\Lambda$ where Y_n is the predictable counting process of the components still working. Tests are sequentially performed at times τ_j until $\tau = \tau_K$ such that the hypothesis is rejected at the Kth test.

Tests of the hypothesis H_0 of equal hazard functions $\lambda_2 = \lambda_1$, for two independent sub-samples, relies on the log-likelihood ratio statistic. The likelihood ratio of the counting processes N_{jn} with hazard functions λ_j, for $j = 1, 2$, is written as their sum and λ_0 denotes the unknown hazard function under the hypothesis H_0

$$l_{nt} = \sum_{j=1,2} \frac{dP_{N_{jn}}}{dP_{0N_{jn}}}(t) = \sum_{j=1,2} \left\{ \int_0^t \log \frac{\lambda_j}{\lambda_0} \, dN_{jn} - \int_0^t (\lambda_j - \lambda_0) Y_{jn} \, ds \right\}.$$

Let $\mathbb{F}_{jn}$ be the filtration generated by the processes N_{jn} and Y_{jn}, and let $\mathbb{F}_n$ be the filtration generated by all observations. The local martingales $M_{jn,0}(t) = N_{jn}(t) - \int_0^t Y_{jn}\lambda_0 \, ds$ and $M_{jn}(t) = N_{jn}(t) - \int_0^t Y_{jn}\lambda_j \, ds$, $t > 0$, defined with respect to $\mathbb{F}_{jn}$, are related to the counting processes N_{jn}, $j = 1, 2$, under H_0 and, respectively, alternatives with hazard functions $\lambda_1 \neq \lambda_2$.

Under H_0, the local martingales $Z_{jn}^{(0)}(t) = n^{-\frac{1}{2}} \int_0^t \log(\lambda_j \lambda_0^{-1}) \, dM_{jn,0}$ are centered and their respective variances are $v_{jt}^{(0)} = \int_0^t \{\log(\lambda_j \lambda_0^{-1})\}^2 \bar{H}_0 \, d\lambda_0$, for every t. The processes Z_{jn} converge in probability to centered Gaussian processes Z_j with independent increments and with variances v_j, under H_0, as n tends to infinity. Under fixed alternatives, the processes $Z_{jn}^{(1)} = n^{-\frac{1}{2}} \int_0^t \log(\lambda_j \lambda_0^{-1}) \, dM_{jn}$ are centered and their respective variances are $v_{jt}^{(1)} = \int_0^t \{\log(\lambda_j \lambda_0^{-1})\}^2 \bar{H}_j \, d\lambda_j$, for every t, and for $j = 1, 2$.

Under H_0, $S_{nt} = n^{-\frac{1}{2}} l_{nt}$ is written as

$$S_{nt} = \sum_{j=1,2} \left\{ Z_{jn}^{(0)}(t) + n^{-\frac{1}{2}} \int_0^t \left(\lambda_0 \log \frac{\lambda_j}{\lambda_0} - \lambda_j + \lambda_0 \right) Y_{jn}\, ds \right\}$$

$$= \sum_{j=1,2} \left\{ Z_{jn}^{(0)}(t) + n^{-\frac{1}{2}} \int_0^t \phi_j(x) \lambda_0(x) Y_{jn}\, ds \right\},$$

with the notation

$$\phi_j(x) = \log \frac{\lambda_j}{\lambda_0}(x) - \frac{\lambda_j}{\lambda_0}(x) + 1 \leq 0,$$

and $\phi_j(x) = 0$ only under H_0. Under alternatives

$$S_{nt} = \sum_{j=1,2} \left\{ Z_{jn}^{(1)}(t) + n^{-\frac{1}{2}} \int_0^t \left(\lambda_j \log \frac{\lambda_j}{\lambda_0} - \lambda_j + \lambda_0 \right) Y_{jn}\, ds \right\}$$

$$= \sum_{j=1,2} \left\{ Z_{jn}^{(1)}(t) - n^{-\frac{1}{2}} \int_0^t \psi_j(x) \lambda_j(x) Y_{jn}\, ds \right\},$$

with the notation

$$\psi_j(x) = \log \frac{\lambda_0}{\lambda_j}(x) - \frac{\lambda_0}{\lambda_j}(x) + 1 < 0.$$

Under alternatives, the mean of S_{nt} is therefore positive for every fixed $t > 0$ and it diverges. Log-likelihood ratio tests of hypotheses concerning the hazard functions are one-sided tests with rejection of the hypothesis if the value of S_{nt} is above a critical value.

Let $n^{-1} E_0 Y_{jn}(x) = \bar{H}_0(x)$ and $n^{-1} E_K Y_{jn}(x) = \bar{H}_j(x)$, $j = 1, 2$. For all T_{ji} and $s < x$, $P(T_{ji} < x | \mathcal{F}_{js}) = 1_{\{T_{ji} < s\}} + \{1 - \bar{H}_j^{-1}(s) \bar{H}_j(x)\} 1_{\{T_{ji} \geq s\}}$, hence $E(Y_{jn}(x) | \mathcal{F}_{js}) = Y_{jn}(s) \bar{H}_j^{-1}(s) \bar{H}_j(x)\}$. From the martingale property, for all $s < t$

$$E_0(S_{nt} | \mathcal{F}_s) = S_{ns} + n^{-\frac{1}{2}} \sum_{j=1,2} \int_s^t \phi_j(x) E_0(Y_{jn}(x) | \mathcal{F}_{js}) \lambda_0(x)\, dx$$

$$= S_{ns} + n^{-\frac{1}{2}} \sum_{j=1,2} Y_{jn}(s) \bar{H}_0^{-1}(s) \int_s^t \phi_j \bar{H}_j\, d\Lambda_0$$

and under alternatives

$$E(S_{nt} | \mathcal{F}_s) = S_{ns} - n^{-\frac{1}{2}} \sum_{j=1,2} Y_{jn}(s) \int_s^t \psi_j(x) E(Y_{jn}(x) | \mathcal{F}_{js}) \lambda_j(x)\, dx$$

$$= S_{ns} - n^{-\frac{1}{2}} \sum_{j=1,2} Y_{jn}(s) \bar{H}_j^{-1}(s) \int_s^t \psi_j \bar{H}_j\, d\Lambda_j.$$

A sequential test of H_0 is performed by tests at a sequence of fixed or random times τ_l, $l = 1, \ldots, k$, until the rejection of the hypothesis at a random integer K. When the statistic $n^{-\frac{1}{2}} l_{n\tau_{l-1}}$ is smaller than the critical value $c_{n,l-1}$ of the test at the level α_{l-1}, the behavior of the variation $A_{n,l} = n^{-\frac{1}{2}}(l_{n\tau_l} - l_{n\tau_{l-1}})$ determines whether the test performed with the observations in the time interval $]\tau_{l-1}, \tau_l]$ lead to accept or reject to hypothesis the lth test with the level α_l. As for l_{nt}, the conditional increments $n^{-\frac{1}{2}}\{E(l_{n\tau_l}|\mathcal{F}_{\tau_{l-1}}) - l_{n\tau_{l-1}}\}$ diverge as n tends to infinity and the lth test rejects the hypothesis if $\{A_{n,l} > c_{n,l}\}$, it accepts the hypothesis if $\{A_{n,l} \leq c_{n,l}\}$, if the $N_{jn}(\tau_l) - N_{jn}(\tau\tau_{l-1})$ is sufficiently large for the normal approximations.

The limits of the local likelihood ratio test statistic under H_0 and K_n are given in Chapter 6.11, they apply to the sequential tests of the hypothesis H_0. All tests of Chapter 6.11 can be performed as sequential tests.

8.8   Sequential tests for regressions and diffusion processes

The sequential comparison of regression functions $m(x) = E(Y|X = x)$ or drift functions of diffusions follow the same principle. They are based on kernel estimators and the conditions for their convergence are the same when the drift function is estimated from an approximation by discretization of the sample path of the diffusion process X, in the form

$$Y_i = \delta\alpha(X_i) + \beta(X_i)\varepsilon_i,$$

where $Y_i = X_{t_{i+1}} - X_{t_i}$ and $\varepsilon_i = W_{t_{i+1}} - W_{t_i}$, as $\delta = t_{i+1} - t_i$ is sufficiently small. The error variable ε_i is independent of X_i, centered, have identical variances and they are mutually independent. The kernel estimators $\widehat{m}_{n,h}(x)$ of a regression function $m(x)$ and $\widehat{\alpha}_{n,h}(x)$ of a drift function $\alpha(x)$ are both the ratio of the kernel estimators of a mean function $\mu(x) = E(Y_t 1_{\{X_t=x\}})$ and of the density f of the variable X_t. The increment $\Delta\widehat{m}_{n,h}(\xi_l) = \widehat{m}_{n,h}(\xi_l) - \widehat{m}_{n,h}(\xi_{l-1})$ between two testing points ξ_{l-1} and ξ_l of sequential tests is

$$\Delta\widehat{m}_{n,h}(\xi_l) = \frac{\widehat{\mu}_{n,h}(\xi_l) - \widehat{\mu}_{n,h}(\xi_{l-1})}{\widehat{f}_{n,h}(\xi_l)}$$
$$- \frac{\widehat{f}_{n,h}(\xi_l) - \widehat{f}_{n,h}(\xi_{l-1})}{\widehat{f}_{n,h}(\xi_l)}\widehat{m}_{n,h}(\xi_{l-1}). \tag{8.9}$$

This expression provides a sequential algorithm for the calculus of the estimator $\widehat{m}_{n,h}(\xi_l)$ from $\widehat{\mu}_{n,h}(\xi_{l-1})$ and $\widehat{f}_{n,h}(\xi_{l-1})$ and from the increments of

$\widehat{f}_{n,h}$ and $\widehat{m}_{n,h}$ between ξ_{l-1} and ξ_l. Other estimators are defined from the sequence of sub-samples with n_l observations in the interval $I_l =]\xi_{l-1}, \xi_l]$. The lth estimators are defined in $I_{l,h} =]\xi_{l-1} + h, \xi_l - h]$ by

$$\widehat{f}_{n,h,l}(x) = n_l^{-1} \sum_{i=1}^{n_l} K_h(X_i - x) 1_{\{]\xi_{l-1}, \xi_l]\}}(X_i),$$

$$\widehat{\mu}_{n,h,l}(x) = n_l^{-1} \sum_{i=1}^{n_l} Y_i K_h(X_i - x) 1_{\{]\xi_{l-1}, \xi_l]\}}(X_i) n$$

they are $n_l^{\frac{s}{2s+1}}$-consistent in $I_{l,h}$. The estimators defined in $I_{l,h}$ are indepen-dent from the past estimators up to ξ_{l-1}. Sequential estimators of f and μ are defined from (8.9), replacing the variation of the global estimators $\widehat{f}_{n,h}$ and $\widehat{m}_{n,h}$ in I_l by the variations of $\widehat{f}_{n,h,l}$ and $\widehat{\mu}_{n,h,l}$ in $I_{l,h}$, due to the bias on the edge of the intervals. The sequential estimators of f and μ are sums of independent estimators calculated from sub-samples.

The Kolmogorov-Smirnov test statistic (4.14) for the comparison of two regression curves splits into a sequence of statistics restricted to the inter-vals $I_{l,h}$

$$T_{n,l} = \max\left[T_{n,l-1}, \sup_{x \in I_{l,h}} \left|\left(\frac{n_{1,l} n_{2,l}}{n_l}\right)^{\frac{s}{2s+1}} \{\widehat{m}_{1,n_{1,l},h_1}(x) - \widehat{m}_{2,n_{2,l},h_2}(x)\}\right.\right.$$

$$\left.\left. - \left\{\gamma_1\left(\frac{n_{2,l}}{n_l}\right)^{\frac{s}{2s+1}} - \gamma_2\left(\frac{n_{1,l}}{n_l}\right)^{\frac{s}{2s+1}}\right\} \widehat{b}_{m,n_l,h}(x)\right|\right]$$

and the actualization of the estimators from I_{l-1} to I_l is calculated recur-sively by (8.9).

In the comparison of two drift functions of diffusion processes, the L^2 statistic $d_T = \int_{\mathcal{I}_X} \{\widehat{\alpha}_{1t}(x) - \widehat{\alpha}_{2t}(x)\}^2 \, dx$ splits into a sequence of statistics at the points ξ_l

$$d_{\xi_l}(\widehat{\alpha}_{1,h}, \widehat{\alpha}_{2,h}) = d_{\xi_{l-1}}(\widehat{\alpha}_{1,h}, \widehat{\alpha}_{2,h}) + \int_{I_l} \{\widehat{\alpha}_{1,h}(x) - \widehat{\alpha}_{2,h}(x)\}^2 \, dx$$

with a recursive estimation of the function α by (8.9) from the discretization of the sample path of the diffusion. The drift function α_t is estimated from the continuous process by (6.23) in an interval $[0, T]$ by $\widehat{\alpha}_{T,h} = \widehat{\mu}_{T,h} \widehat{f}_{T,h}^{-1}$ with $\widehat{\mu}_{T,h}(x) = (\delta T)^{-1} \int_0^T K_h(x - X_s) \, dX_s$ and $\widehat{f}_{T,h}(x) = T^{-1} \int_0^T K_h(x - X_s) \, ds$. The increment of $\widehat{\alpha}_{T,h}$ in the interval I_l between two tests is written like in (8.9) for the regression

$$\Delta\widehat{\alpha}_{T,h}(\xi_l) = \frac{\widehat{\mu}_{T,h}(\xi_l) - \widehat{\mu}_{T,h}(\xi_{l-1})}{\widehat{f}_{T,h}(\xi_l)} - \frac{\widehat{f}_{T,h}(\xi_l) - \widehat{f}_{T,h}(\xi_{l-1})}{\widehat{f}_{T,h}(\xi_l)} \widehat{\alpha}_{T,h}(\xi_{l-1}).$$

Replacing the global estimators by a sequence of estimators $\widehat{f}_{n,h,l}$ and $\widehat{\mu}_{n,h,l}$, $l = 1, \ldots, k$, restricted to the consecutive sub-samples yields sequential estimators of f and μ as sums of independent estimators calculated from sub-samples, like in the sequential estimation for the regression functions.

A Kolmogorov-Smirnov type test can be defined in I_l as

$$T_{T,l,l-1} = \sup_{x \in I_{l,h}} |\widehat{\alpha}_{1,T,h}(x) - \widehat{\alpha}_{2,T,h}(x) - \widehat{\alpha}_{1,T,h}(\tau_{l-1}) + \widehat{\alpha}_{2,T,h}(\tau_{l-1})|$$

or using bias corrected estimators. If the durations between the times of tests are sufficiently large, new estimators of the functions m and α caculated from the observations restricted to the sub-intervals I_l are independent and a sequence of independent tests can be performed. Using all observations until ξ_l improves the tests by increasing the sample size of each test and the lth test statistic is $T_{n,l} = \max\{T_{n,l-1}, T_{T,l,l-1}\}$, $l \geq 1$, starting with $T_{n,1} = T_{n-1}$. They are record variables and the global level of the test is obtained from Section 8.5.

A heteroscedastic functional regression is defined by $Y = m(X) + \sigma(X)\varepsilon$ such that the observations ε_i are mutually independent, centered and with variance 1, conditionally on X_i. Its variance function $\sigma^2(x)$ is estimated from the discretized observation of the diffusion process by the mean squared estimated errors

$$\widehat{\sigma}_n^2(x) = n^{-1} \sum_{i=1}^{n} K_h(x - X_i)\{Y_i - \widehat{m}_{n,h}(X_i)\}^2.$$

It is similar to the estimator (6.25) built for the variance of a continuously observed diffusion process from the estimation error Z_t defined by (6.24). A sequential L^2 test statistic is defined for the comparison of the variances of diffusions as for their drift function and its asymptotic behavior is deduced from Section 8.5, the observations on the sub-intervals being independent with the sequence of estimators restricted to the sub-samples.

Appendix A

Functional estimation and probability

A.1 Chernov's theorem

Let $(\Omega, \mathcal{F}, P)$ be a probability space and let X be a real variable defined on $(\Omega, \mathcal{F}, P)$, with distribution function F. The Laplace transform of a centered variable X is $L_X(t) = \int_{\mathbb{R}} e^{tx} \, dF(x)$.

Theorem A.1 (Chernov's theorem). *On a probability space $(\Omega, \mathcal{F}, P)$, let $(X_i)_{i=1,\ldots,n}$ be a sequence of independent and identically distributed real random variables with mean zero, having a finite Laplace transform, and let $S_n = \sum_{i=1}^{n} X_i$. For every $a > 0$ and $n > 0$*

$$\log P(S_n > a) = \inf_{t>0}\{n \log L_X(t) - at\}.$$

A Gaussian variable X with mean zero and variance σ^2 has the Laplace transform $L_X(t) = e^{\frac{1}{2}t^2\sigma^2}$ and a variable with mean m and variance σ^2 has the Laplace transform $L_X(t) = e^{\frac{1}{2}t^2\sigma^2} e^{mt}$. The logarithm of the probability in Chernov's theorem for the sum S_n of independent and Gaussian variables is $-I_n(a) = -\frac{(a-m)^2}{2n\sigma^2}$ and

$$\log P(S_n > a) = \exp\{-I_n(a)\} = \exp\Big\{-\frac{(a-m)^2}{2n\sigma^2}\Big\}.$$

A.2 Martingales in $\mathbb{R}_+$

In a probability space $(\Omega, \mathcal{F}, P)$, a filtration $(\mathcal{F}_t)_{t\geq 0}$ is defined as a right-continuous and increasing sequence of σ-algebras: for every $0 < s < t$

$$F_s \subset \mathcal{F}_t, \ \mathcal{F}_t = \cap_{s>t>0}\mathcal{F}_s, \ \mathcal{F} = \mathcal{F}_\infty.$$

A random variable τ is a stopping time with respect to the filtration $(\mathcal{F}_t)_{t\geq 0}$ if for every $t > 0$, the set $\{\tau \leq t\}$ belongs to $\mathcal{F}_t$.

A time-continuous martingale $X = (X_t)_{t \geq 0}$ on the filtered probability space $(\Omega, \mathcal{F}, (\mathcal{F}_t)_{t \geq 0}, P)$ is a stochastic process such that

$$E(X_t | \mathcal{F}_s) = X_s \text{ for every } s < t \geq 0. \tag{A.1}$$

The Brownian motion $(B_t)_{t \geq 0}$ on a filtered probability space $(\Omega, \mathcal{F}, (\mathcal{F}_t)_{t \geq 0}, P)$ is a martingale with independent increments, defined by Gaussian marginals, a mean zero and the variance $B_t^2 = t$. Moreover, $(B_t^2 - t)_{t \geq 0}$ is a martingale with respect to the filtration $(\mathcal{F}_t)_{t \geq 0}$.

On a space $(\Omega, \mathcal{F}, P, \mathbb{F})$, let M be a square integrable martingale. There exists an unique increasing and predictable process $<M>$ such that

$$M^2 - <M>$$

is a martingale. The process $<M>$ is the process of the predictable quadratic variations of M. It satisfies

$$E\{(M_t - M_s)^2 | \mathcal{F}_s\} = E(M_t^2 | \mathcal{F}_s) - E M_s^2 = E(<M>_t | \mathcal{F}_s) - <M>_s,$$

for every $0 < s < t$. It defines a scalar product for square integrable martingales M_1 and M_2 with mean zero

$$<M_1, M_2> = \frac{1}{2}(<M_1 + M_2, M_1 + M_2> \\ - <M_1, M_1> - <M_2, M_2>), \tag{A.2}$$

then $E<M_1, M_2>_t = E M_{1t} M_{2t}$ for every $t > 0$. Two square integrable martingales M_1 and M_2 are orthogonal if and only if $<M_1, M_2> = 0$ or, equivalently, if $M_1 M_2$ is a martingale. Let $\mathcal{M}_0^2$ be the space of the right-continuous square integrable martingales with mean zero, provided with the norm $\|M\|_2 = \sup_t (E M_t^2)^{\frac{1}{2}} = \sup_t (E<M>_t)^{\frac{1}{2}}$.

A process $(M_t)_{t \geq 0}$ is a local martingale if there exists an increasing sequence of stopping times $(S_n)_n$ such that $(M(t \wedge S_n))_t$ belongs to $\mathcal{M}^2$ and S_n tends to infinity. The above properties of the martingales are then extended to local martingales. A point process N in $\mathbb{R}_+$ with nonpredictable jump times has a predictable compensator $\widetilde{N}$ such that $M = N - \widetilde{N}$ is a local martingale in $\mathbb{R}_+$ that satisfies $<M> = \widetilde{N}$.

The stochastic integral of a predictable process B on $(\Omega, \mathcal{A}, P, \mathbb{F})$ with respect to an increasing process A on $(\Omega, \mathcal{A}, P, \mathbb{F})$, such that $E \int_0^\infty |B_t| \, dA(t) < \infty$, is defined as the limit of a stepwise process defined from a partition with an increasing number of sub-intervals of $\mathbb{R}_+$. For all $t > 0$ and $\omega \in \Omega$

$$\int_0^t B(\omega, s) \, dA(\omega)(s) = \lim_{n \to \infty} \sum_{1 \leq i \leq n} B(\omega, t_i)\{A(\omega, t_{i+1}) - A(\omega, t_i)\}.$$

The integral of a predictable process B on $(\Omega, \mathcal{A}, P, \mathbb{F})$ with respect to a point process $N = (T_n)_{n \geq 1}$ is $\int_0^t B(s)\,dN(s) = \sum_{T_n \leq t} B(T_n)$, with the convention $\int_0^T B(s)\,dM(s) = \int_{T_{1:N(T)}}^T B(s)\,dM(s)$. Its mean is $E \int_{T_{1:N(T)}}^t B(s)\,d\widetilde{N}(s)$ and its variance is $E \int_{T_{1:N(T)}}^t B^2(s)\,d\widetilde{N}(s)$, if B is square integrable with respect to $\widetilde{N}$. The process $M = N - \widetilde{N}$ is then a local martingale of L^2 and the stochastic integral $\int_0^t B(s)\,dN(s)$ can be viewed as the difference of the integrals of B with respect to N and $\widetilde{N}$.

The integral of a predictable process B with respect to a local square integrable martingale is defined as the L_2-limit of the integral of B^2 with respect to the increasing predictable compensator related to M^2.

A.3   Weak convergence

Let $(\Omega, \mathcal{F}, P)$ be a probability space and let $(S, \mathcal{S}, d)$ be a measurable metric space. The distribution of a random variable X defined from $(\Omega, \mathcal{F}, P)$ into a metric space $(S, \mathcal{S})$ is the image probability $P_X = P(X^{-1})$ of X on $(S, \mathcal{S})$

$$P_X(A) = P\{w : X(w) \in A\} = P(X \in A), \ A \in \mathcal{S}.$$

The distribution function of a random variable X defined in a k-dimensional real space with the Borel σ-algebra $(\mathbb{R}^k, \mathcal{B}_k, d_k)$ is $F(x) = P_X(]-\infty, x]$ and for every measurable function h defined from $(\mathbb{R}^k, \mathcal{B}_k, d_k)$ into $\mathbb{R}$, $\int_\Omega h(X)\,dP = \int_S h(x)\,dF(x)$.

A sequence of random variables X_n converges weakly to a random variable X if the sequence of probabilities P_{X_n} converges weakly to the probability P_X. For every continuous and bounded function $f : (S, \mathcal{S}) \to (\mathbb{R}, \mathcal{B})$

$$\lim_{n \to \infty} \int f(X_n)\,dP = \int f(X)\,dP,$$

i.e. $\lim_{n \to \infty} Ef(X_n) = Ef(X)$. This is equivalent to the convergence of the characteristic function of X_n to the characteristic function of X. For every $t \in \mathbb{R}^k$

$$\lim_{n \to \infty} \phi_{X_n}(t) = \lim_{n \to \infty} E e^{it^T X_n} = \phi_X(t) = E e^{it^T X}.$$

The following theorems are proved by Billingsley (1968), with equivalent moment criteria for the tightness of sequences of processes. A probability in a measurable space $(S, \mathcal{S})$ is tight if for every $\varepsilon > 0$, there exists a compact subset K_ε of S such that $P(K_\varepsilon) \geq 1 - \varepsilon$. Equivalently, a sequence of random variables X_n in $(S, \mathcal{S})$ is tight if for every $\varepsilon > 0$, there exists a compact subset K_ε of S such that $P(X_n \in K_\varepsilon) \geq 1 - \varepsilon$ for every n.

Theorem A.2 (Prokhorov). *Let P_n be a sequence of probabilities on (S, S). If P_n is tight, then it is relatively compact. If S is a complete and separable metric space and if P_n is relatively compact, then it is tight.*

Theorem A.3. *A sequence of processes $(X_n(t))_{t \in I, n \geq 1}$ converges weakly to a process X in a complete and separable metric space if the sequence of their image probabilities is tight and if their finite dimensional distributions converge weakly to the finite dimensional distributions of the process X.*

The tightness of a sequence of processes $(X_n(t))_{t \in I, n \geq 1}$ defined in a complete and separable metric space is equivalent to the following convergence of their modulus of continuity. For all $\eta > 0$ and $\varepsilon > 0$, there exists $\delta > 0$ such that

$$\lim_{n \to \infty} \sup P \left\{ \sup_{|t-s| < \delta} |X_n(t) - X_n(s)| > \varepsilon \right\} < \eta.$$

The space C_I of the real continuous functions x defined in $I \subseteq \mathbb{R}$ provided with the uniform norm $\|x\| = \sup_{t \in I} |x(t)|$ and with the distance

$$d(x, y) = \|x - y\|, \ x, y \in C_I$$

is a complete and separable metric space.

Let D_I be the space of the real right-continuous functions x with left-hand limits defined in $I \subseteq \mathbb{R}$. The Skorohod distance of x and y in D_I is the smallest $\varepsilon > 0$ such that there exists a strictly increasing and continuous map g from I to I, with $g(0) = 0$, $g(1) = 1$ and

$$d_K(x, y) = \sup \left\{ t \in I : |g(t) - t| \leq \varepsilon, \quad \sup_{t \in I} |x(t) - y(g(t))| \leq \varepsilon \right\}.$$

The space (D_I, d_K) is a complete and separable metric space.

Gaussian martingales are characterized by their variance.

Theorem A.4. *Let k increasing functions $v_1, v_2, \cdots, v_k$ be defined in $(\mathbb{R}, \mathcal{B}(\mathbb{R}))$, with values zero at 0. Then there exist k independent centered Gaussian processes in $C(\mathbb{R})$, $Z_1, \cdots, Z_k$, with independent increments and respective variances v_i, $i = 1, \cdots, k$. Reciprocally, let $Z_1, \cdots, Z_k$ be centered martingales in $(\mathbb{R}, \mathcal{B}(\mathbb{R}))$ such that $<Z_i, Z_j>(t) = v_i(t)\delta_{i,j}$ for all $i, j = 1, \cdots, k$, then they are independent centered Gaussian processes with independent increments.*

Let $\varepsilon > 0$, a martingale M is written as the sum $M = M^\varepsilon + M_\varepsilon$ such that the jumps of M^ε have a size larger than ε and the jumps of M_ε are smaller than ε.

Theorem A.5 (Rebolledo). *Let $Z = (Z_1, \cdots, Z_k)$ be centered Gaussian martingale in $C^k(\mathbb{R})$ a variance matrix V and let $M_n = (M_{1n}, \ldots, M_{kn})$ be a sequence of martingales satisfying*

(1) for every $\varepsilon > 0$, $<M_{in}^{\varepsilon}>(t)$ converges in probability to zero as n tends to infinity, for $i = 1, \cdots, k$,

(2) $<M_{in}, M_{jn}>$ converges in probability to v_{ij}, uniformly in $\mathbb{R}$, as n tends to infinity, for all $i, j = 1, \cdots, k$.

Then M_n converges weakly to Z in $(D(\mathbb{R}), d_K)^k$.

A.4 Algebra

The inversion of a symmetric matrix has several equivalent expressions. Let $I = \begin{pmatrix} I_{11} & I_{12} \\ I_{21} & I_{22} \end{pmatrix}$ be a block decomposition of a symmetric matrix such that the square sub-matrices I_{11} and I_{22} have an inverse. With the notations

$$A = I_{11} - I_{12}I_{22}^{-1}I_{21},$$
$$B = I_{21}I_{11}^{-1},$$
$$C = I_{22} - I_{21}I_{11}^{-1}I_{12}.$$

A.5 Estimation of densities and curves

Let f be the real density in $C^1(\mathcal{I})$ of a random variable X with distribution function F and let $X_1, \ldots, X_n$ be a sample of independent observations of X. The histogram of f, with a bandwidth $\delta = \delta_n$ tending to zero as n tends to infinity, is

$$\widetilde{f}_{n\delta}(x) = \frac{1}{n\delta} \sum_{i=1}^{n} 1_{\{X_i \in \Delta_{jx}\}},$$

where $(\Delta_j)_{j \in J}$ is a partition of the support of X in sub-intervals of length δ and Δ_{jx} is the interval containing x. Writing $\Delta_j =]a_j - \frac{\delta}{2}, a_j + \frac{\delta}{2}]$, the bias $\widetilde{b}_{n\delta} = E\widetilde{f}_{n\delta}(x) - f(x)$ of the histogram is $\widetilde{b}_{n\delta} = \delta^{-1}\{F(a_j + \frac{\delta}{2}) - F(a_j - \frac{\delta}{2})\} - f(x)$, hence $\widetilde{b}_{n\delta} = \delta f^{(1)}(x) + o(\delta)$ and it tends to zero as n tends to infinity.

The empirical process $\nu_n = n^{\frac{1}{2}}(\widehat{F}_n - F)$ has the covariance function $\mu_2(x, y) = F(x \wedge y) - F(x)F(y)$. Then the empirical process of the histogram, $\tilde{\nu}_{n\delta} = (n\delta)^{\frac{1}{2}}(\tilde{f}_{n\delta} - f)$, has the variance

$$\tilde{v}_{n\delta}(x) = \delta^{-1}\left[F\left(a_j + \frac{\delta}{2}\right) - F^2\left(a_j + \frac{\delta}{2}\right) - F\left(a_j - \frac{\delta}{2}\right) + F^2\left(a_j - \frac{\delta}{2}\right)\right]$$
$$= f(x) + o(1).$$

If x and y belong to the same sub-interval of the partition, its covariance at x and y is $\tilde{v}_{n\delta}(x) = \tilde{v}_{n\delta}(y)$. If they belong to disjoint sub-intervals Δ_{jx} and Δ_{ky}, the covariance of $\tilde{\nu}_{n\delta}(x)$ and $\tilde{\nu}_{n\delta}(y)$ is zero since the X_i are independent. The mean squared error of the histogram is asymptotically equivalent to $\int\{\delta_n^2 f^{(1)2}(x) + (n\delta_n)^{-1}f(x)\}\,dx$, it is minimal for a bandwidth $\delta_n = O(n^{-\frac{1}{3}})$ where it converges with the rate $(n^{-\frac{1}{3}})$, so it is smaller than the optimal convergence bandwidth of the kernel estimator for every class C^s, $s > 1$.

The upper centered moments of $\tilde{\nu}_{n\delta}$ are calculated by expansions of the sums $(\tilde{f}_{n\delta}(x) - E\tilde{f}_{n\delta}(x))^k$, the following moments have a first order term equal to zero

$$E\tilde{\nu}_n^3(x) = (n\delta)^{\frac{3}{2}}\left((n\delta)^{-2}\{f(a_j) + o(1)\} + (n\delta)^{-1}\{f^2(a_j) + o(1)\}\right.$$
$$+ \{f^3(a_j) + o(1)\} - 3f(x)[(n\delta)^{-1}\{f(a_j) + o(1)\}$$
$$+ (1 - n^{-1})\{f^3(a_j) + o(1)\}] + 3f^2(x)\{f(a_j) + o(1)\} + f^3(x))$$
$$= -2(n\delta)^{\frac{1}{2}}\{f^2(x) + o(1)\},$$

this entails $\mu_3(x) = E\tilde{\nu}_{n\delta}^3(x) = -(n\delta)^{\frac{1}{2}}\{f^2(x) + o(1)\}$. By the same expansions $\mu_4(x) = E\tilde{\nu}_{n\delta}^4(x) = 3n\delta\{f^3(x) + o(1)\}$ and for x and y in disjoint sub-intervals of the partition, $\mu_{2,2}(x, y) = E\{\tilde{\nu}_{n\delta}^2(x)\tilde{\nu}_{n\delta}^2(y)\} = f(x)f(y) + o(1)$, it equals $\mu_4(x)$ if they belong to the same sub-interval. Therefore the process $\tilde{\nu}_n$ diverges as $n\delta$ tends to infinity.

Integrating $\tilde{\nu}_{n\delta}^2$ with respect to a density $w(x)$

$$n\delta_n E \int \{\tilde{f}_{n\delta_n}(x) - f(x)\}^2 w(x)\,dx$$

$$= \int f(x)w(x)\,dx + n\delta_n^3 \int f^{(1)2}(x)w(x)\,dx + o(1)$$

and it converges to $Ew(X) + a\int f^{(1)2}(x)w(x)\,dx$ if $\lim_{n\to\infty} n\delta_n^3 = a$, so an optimal bandwidth for the L^2 error of the estimation of f by a histogram has the rate $n^{-\frac{1}{3}}$.

A density f of $C^2(\mathbb{R}^2)$ has a bivariate kernel estimator

$$\widehat{f}_{n,h}(x) = \int K_{h_1,h_2}(x_1 - s, x_2 - t)\, d\widehat{F}_n(s,t)$$

with a bivariate kernel $K(u,v) = K_1(u)K_2(v)$ such that the kernels K_j belong to $C^2(\mathbb{R})$ and satisfy Condition 1.1 with $m_{2K_j} \neq 0$, for $j = 1, 2$. For all bivariate $x = (x_1, x_2)$ and $h = (h_1, h_2)$, let hu be the vector with components $h_k u_k$, the mean and variance of $\widehat{f}_{n,h}(x)$ have the asymptotic expansions

$$E\widehat{f}_{n,h}(x) = \int K_{h_1,h_2}(x_1 - s, x_2 - t)f(s,t)\, ds\, dt$$

$$= \int K(u,v)f(x_1 - h_1 u_1, x_2 - h_2 u_2)\, du_1\, du_2$$

$$= f(x) + \frac{m_{2K}}{2} h^t f^{(2)}(x) h + o(\|h\|^2),$$

$$Var\widehat{f}_{n,h}(x) = \frac{\kappa_2^2}{nh_1 h_2} f(x) + o((nh_1 h_2)^{-1}),$$

where $m_{2K} = \int u^2 K(u)\, du$ and $\kappa_2 = \int K^2(u)\, du$ for the real kernel. The first partial derivatives $\widehat{f}^{(1)}_{n,h,k}(x)$, $k = 1, 2$, of the kernel density estimator satisfy

$$\widehat{f}^{(1)}_{n,h_1}(x) = \int K^{(1)}_{h_1}(x_1 - s)K_{h_2}(x_2 - t)\, d\widehat{F}_n(s,t),$$

$$\widehat{f}^{(1)}_{n,h_2}(x) = \int K_{h_1}(x_1 - s)K^{(1)}_{h_2}(x_2 - t)\, d\widehat{F}_n(s,t).$$

From the properties of the integrals of the kernel derivatives and its moments (Lemma 2.1 in Pons, 2011), the mean and variance of $\widehat{f}_{n,h}$ have the expansions

$$E\widehat{f}^{(1)}_{n,h,1}(x) = \int K^{(1)}_{h_1}(x_1 - s)K_{h_2}(x_2 - t)f(s,t)\, ds\, dt$$

$$= \frac{1}{h_1} \int K^{(1)}(u)K(v)f(x_1 - h_1 u, x_2 - h_2 v)\, du\, dv$$

$$= \int K^{(1)}(u_1)K(u_2)\{f(x) - (hu)^t f^{(1)}(x)$$

$$+ \frac{1}{2}(hu)^t f^{(2)}(x)hu + o(\|h\|^2)\}\, du_1\, du_2$$

$$= f^{(1)}(x) + \frac{m_{2K}}{2} h_1^2 f_1^{(3)}(x) + o(h_1^2), \tag{A.3}$$

$$Var\widehat{f}^{(1)}_{n,h,1}(x) = \frac{\kappa_2 \int K^{'2}}{nh_1^3 h_2} f(x) + o((nh_1^3 h_2)^{-1}) \tag{A.4}$$

and the expressions are similar for $\widehat{f}^{(1)}_{n,h,2}$.

Lemma A.1. *The estimator $\widehat{f}_{n,h}^{(1)}$ of the first derivative of a monotone density f of $C^3(\mathbb{R}^2)$ converges to $f^{(1)}$ with the optimal rate $n^{-\frac{1}{4}}$, as h_1 and h_2 are $O(n^{-\frac{1}{8}})$.*

The matrix $\widehat{f}_{n,h}^{(2)}$ of its second derivatives has the components

$$\widehat{f}_{n,h,11}^{(2)}(x) = (nh_1^2h_2^2)^{-1} \sum_{i=1}^{n} K^{(1)}\left(\frac{x_1 - X_{1i}}{h_1}\right) K^{(1)}\left(\frac{x_2 - X_{2i}}{h_2}\right),$$

$$\widehat{f}_{n,h,20}^{(2)}(x) = (nh_1^3)^{-1} \sum_{i=1}^{n} K^{(2)}\left(\frac{x_1 - X_{1i}}{h_1}\right) K_{h_2}(x_2 - X_{2i}),$$

$$\widehat{f}_{n,h,02}^{(2)}(x) = (nh_2^3)^{-1} \sum_{i=1}^{n} K_{h_1}(x_1 - X_{1i}) K^{(2)}\left(\frac{x_2 - X_{2i}}{h_2}\right)$$

has the asymptotic mean and variance

$$E\widehat{f}_{n,h}^{(2)}(x) = f^{(2)}(x) + \frac{m_{2K}}{2} h^t f^{(4)}(x)h + o(\|h\|^2),$$

$$Var\widehat{f}_{n,h,20}^{(2)}(x) = \frac{cste}{nh_1^5h_2} f(x) + o((nh_1^5h_2)^{-1}),$$

$$Var\widehat{f}_{n,h,11}^{(2)}(x) = \frac{cste}{nh_1^3h_2^3} f(x) + o((nh_1^3h_2^3)^{-1}).$$

Lemma A.2. *The estimator $\widehat{f}_{n,h}^{(2)}$ of the second derivative of a monotone density f of $C^4(\mathbb{R}^2)$ converges to $f^{(2)}$ with the optimal rate $n^{-\frac{1}{5}}$, as h_1 and h_2 are $O(n^{-\frac{1}{10}})$.*

Bibliography

Aalen, O. (1978). Non-parametric inference for a family of counting processes, *Ann. Statist.* **6**, pp. 701–726.

Abramovitch, L. and Singh, K. (1985). Edgeworth corrected pivotal statistics and the bootstrap, *Ann. Statist.* **13**, pp. 116–132.

Albers, W., Bickel, P. J. and van Zwet, W. R. (1976). Asymptotic expansion for the power of distribution free tests in the one-sample problem, *Ann. Statist.* **4**, pp. 108–156.

Albert, G. E. (1947). A note on the fundamental identity of sequential analysis, *Ann. Math. Statist.* **18**, pp. 593–596.

Andersen, P., Borgan, O., Gill, R. and Keiding, N. (1982). Linear nonparametric tests for comparison of counting processes, *Intern. Statist. Review* **50**, pp. 219–258.

Anderson, T. W. and Darling, D. A. (1952). Asymptotic theory of certain "goodness of fit" criteria based on stochastic processes, *Ann. Statist.* **23**, pp. 193–212.

Begun, J., Hall, W., Huang, W.-M. and Wellner, J. (1983). Information and asymptotic efficiency in parametric-nonparametric models, *Ann. Statist.* **11**, pp. 432–452.

Beran, R. (1982). Robust estimation in models for independent non-identically distributed data, *Ann. Statist.* **10**, pp. 415–428.

Beran, R. (1986). Simulated power functions, *Ann. Statist.* **14**, pp. 151–173.

Beran, R. (1988). Prepivoting test statistics, a bootstrap view of asymptotic refinements, *J. Amer. Statist. Soc.* **83**, pp. 687–673.

Bickel, P. J. (1974). Edgeworth expansions in nonparametric statistics, *Ann. Statist.* **2**, pp. 1–20.

Bickel, P. J. and Freedman, D. A. (1981). Some asymptotic theory for the bootstrap, *Ann. Statist.* **9**, pp. 1196–1217.

Bickel, P. J. and van Zwet, W. R. (1978). Asymptotic expansion for the power of distribution free tests in the two-sample problem, *Ann. Statist.* **5**, pp. 937–1004.

Billingsley, P. (1968). *Convergence of probability measures* (Wiley, New York).

Blackwell, D. and Girshick, M. A. (1947). A lower bound for the variance of some unbiased sequential estimates, *Ann. Math. Statist.* **18**, pp. 277–280.

Breslow, N. and Crowley, J. (1974). A large sample study of the life table and product limit estimates under random censorship, *Ann. Statist.* **2**, pp. 437–453.

Cairoli, R. and Walsh, J. B. (1975). Stochastic integrals in the plane, *Acta. Math.* **134**, pp. 111–183.

Cochran, W. G. (1947). χ^2 test of goodness of fit, *Annals of Math. Statist.* **23**, pp. 315–345.

Donsker, M. D. (1952). Justification and extension of Doob's heuristic approach to the Kolmogorov-Smirnov theorems, *Ann. Statist.* **23**, pp. 277–281.

Doob, M. D. (1952). Heuristic approach to the Kolmogorov-Smirnov theorems, *Ann. Statist.* **20**, pp. 393–403.

Eggermont, P. P. B. and LaRiccia, V. N. (2000). Maximum likelihood estimation of smooth monotone and unimodal densities, *Ann. Statist.* **28**, pp. 922–947.

Fattorini, L., Greco, L. and Naddeo, S. (2002). The use of the chi-square and Kolmogorov-Smirnov statistics in permutation-based pairwise comparison, *Metron* **60**, pp. 11–20.

Feller, W. (1966). *An Introduction to Probability Theory and its Applications* (Vol. 2, Wiley, New York).

Genest, C. and Rivest, L.-P. (1993). Statistical inference procedures for bivariate Archimedian copulas, *J. Amer. Statist. Soc.* **88**, pp. 1034–1043.

Gill, R. D. (1983). Large sample behaviour of the product-limit estimator on the whole line, *Ann. Statist.* **11**, pp. 49–58.

Hájek, J. and Sidák, Z. (1967). *Theory of Rank Tests* (Academic Press, New York).

Hall, P. (1986). On the bootstrap and confidence intervals, *Ann. Statist.* **23**, pp. 1431–1452.

Hall, P. and Huang, L. S. (2001). Nonparametric kernel regression subject to monotonicity constraints, *Ann. Statist.* **29**, pp. 624–647.

Hodges, J. L. and Lehman, E. L. (1961). Comparison of the normal scores and Wilcoxon tests, *Proceedings of the 4th Berkeley Symposium on Mathematical Statistics* **1**, pp. 307–318.

Huber-Carol, C. (1986). *Théorie de la robustesse* (In Lecture Notes in Probability and Statistics, Springer, Berlin Heidelberg).

Huber-Carol, C., Balakrishnan, N., Nikulin, M. S. and Mesbah, M. (2002). *Goodness-of-Fit Tests and Validity of Models* (Birkhauser, New York).

Kaplan, M. and Meier, P. A. (1958). Nonparametric estimator from incomplete observations, *J. Am. Statist. Ass.* **53**, pp. 457–481.

Kendall, M. G. and Stuart, A. (1947). *Advanced Theory of Statistics* (Griffin, Pensylvania).

Laplace, P.-S. (1774). Mémoire sur la probabilité des causes par les évènemens, *Mémoires de mathématiques et de physique présentés à l'Académie royale des sciences par divers savans*, Paris **VI**, pp. 621–656.

LeCam, L. (1956). On the asymptotic theory of estimation and testing hypotheses, *Proceedings of the 2nd Berkeley Symposium on Mathematical Statistics* **1**, pp. 129–156.

Lehmann, E. L. (1959). *Testing Statistical Hypotheses* (Wiley, New York).

Lehmann, E. L. and Romano, J. P. (1999). *Testing Statistical Hypotheses* (Springer, New York).

Mason, D. M. and Shao, Q.-M. (2001). Bootstrapping the Student *t*-statistic, *Ann. Statist.* **29**, pp. 1435–1450.

Millar, P. W. (1981). *The minimax principle in asymptotic statistical theory* (In Ecole d'été St Flour, Springer, Berlin Heidelberg).

Muliere, P. and Nikitin, Y. A. (2002). Scale-invariant test of normality based on Polya's characterization, *Metron* **60**, pp. 21–33.

Nelson, W. (1972). Theory and application of hazard plotting for censored failure data, *Technometrics* **14**, pp. 945–966.

Nikitin, Y. Y. (1984). Local asymptotic Bahadur optimality and characterization problems, *Theor. Probab. Appl.* **29**, pp. 79–92.

Nikitin, Y. Y. (1987). On the Hodges-Lehmann asymptotic efficiency of nonparametric tests of goodness of fit anf homogeneity, *Theor. Probab. Appl.* **32**, pp. 77–85.

Oakes, D. (1989). Bivariate survival models induced by frailties, *J. Amer. Statist. Assoc.* **84**, pp. 487–493.

Paulson, E. (1947). A note on the efficiency of the Wald sequential test, *Ann. Math. Statist.* **18**, pp. 447–450.

Pearson, K. (1920a). The fundamental problem of practical statistics, *Biometrika* **13**, pp. 1–16.

Pearson, K. (1920b). Notes on the history of correlation, *Biometrika* **13**, pp. 25–45.

Peskir, G. and Shiryaev, A. N. (2000). Sequential testing problem for poisson processes, *Ann. Statist.* **28**, pp. 837–859.

Pons, O. (1980). Test non paramétrique sur la loi d'un processus ponctuel, *C.R. Acad. Sci. Paris, Ser. A* **290**, pp. 189–192.

Pons, O. (1981). Tests sur la loi d'un processus ponctuel, *C.R. Acad. Sci. Paris, Ser. A* **292**, pp. 91–94.

Pons, O. (1986a). A test of independence between two censored survival times, *Scand. J. Statist.* **13**, pp. 173–185.

Pons, O. (1986b). Vitesse de convergence des estimateurs à noyau pour l'intensité d'un processus ponctuel, *Statistics* **17**, pp. 577–584.

Pons, O. (2004). Estimation of semi-Markov models with right-censored data, *Handbook of Statistics* **23**, pp. 175–194.

Pons, O. (2007). Bootstrap of means under stratified sampling, *Electron. J. Statist.* **1**, pp. 381–391.

Pons, O. (2008). *Statistique de processus de renouvellement et markoviens* (Hermès Science Lavoisier, London and Paris).

Pons, O. (2009). *Estimation et tests dans les modèles de mélanges de lois et de ruptures* (Hermès-Science Lavoiser, Paris-London).

Pons, O. (2011). *Functional Estimation for Density, Regression Models and Processes* (World Scientific Publishing, Singapore).

Pons, O. (2012). *Inequalities in Analysis and Probability* (World Scientific Publishing, Singapore).

Rao, C. (1967). *Linear Statistical Inference and its Applications* (Wiley, New York).

Rebolledo, R. (1977). Remarques sur la convergence en loi des martingales vers des martingales continues, *C.R. Acad. Sci. Paris, Ser. A* **285**, pp. 517–520.

Rebolledo, R. (1978). Sur les applications de la théorie des martingales à l'étude statistique d'une famille de processus ponctuels, *Journée de Statistique des Processus Stochastiques, Lecture Notes in Mathematics* **636**, pp. 27–70.

Rebolledo, R. (1980). Central limit theorem for local martingales, *Z. Wahrsch. verw. Gebiete* **51**, pp. 269–286.

Redner, R. (1981). Note on the consistency of the maximum likelihood estimate for nonidentifiable distributions, *Ann. Statist.* **9**, pp. 225–228.

Rothman, E. D. and Woodroofe, M. (1972). A Cramer-von Mises type statistic for testing symmetry, *Annals of Math. Statist.* **43**, pp. 2035–2038.

Scheffe, N. (1975). *Linear Statistical Inference* (Wiley, New York).

Scholtz, F. W. and Stephens, M. A. (1987). k-sample Anderson-Darling tests, *Amer. J. Statist.* **82**, pp. 918–924.

Serfling, R. J. (1980). *Approximation Theorems of Mathematical Statistics* (Wiley, New York).

Shapiro, S. and Wilk, M. B. (1965). An analysis of variance test for normality, *Biometrika* **52**, pp. 591–611.

Shorack, G. R. and Wellner, J. A. (1986). *Empirical Processes and Applications to Statistics* (Wiley, New York).

Singh, K. (1981). On the asymptotic accuracy of Efron's bootstrap, *Ann. Statist.* **9**, pp. 1187–1195.

Stein, C. and Wald, A. (1947). Sequential confidence intervals for the mean of a normal distribution with known variance, *Ann. Math. Statist.* **18**, pp. 427–433.

von Mises, R. (1947). Differentiable statistical functions, *Annals of Math. Statist.* **23**, pp. 309–348.

Wald, A. (1945). Sequential tests of statistical hypotheses, *Ann. Math. Statist.* **16**, pp. 118–186.

Wald, A. (1949). Note on the consistency of the maximum likelihood estimate, *Ann. Math. Statist.* **20**, pp. 595–601.

Wald, A. and Wolfowitz, J. (1948). Optimum character of the sequential probability ratio test, *Ann. Math. Statist.* **19**, pp. 326–339.

Wellner, J. (1982). Asymptotic optimality of the produc-limit estimator, *Ann. Statist.* **10**, pp. 595–602.

Wieand, H. S. (1976). A condition under which the Pitman and Bahadur approaches to efficiency coincide, *Ann. Statist.* **4**, pp. 1003–11011.

Index